W9-BJB-978

Intermediate Algebra

Intermediate Algebra

Charles P. McKeague

CUESTA COLLEGE

Academic Press

New York
San Francisco
London
A Subsidiary of Harcourt Brace Jovanovich, Publishers

COVER ART BY DAN LENORE

COPYRIGHT © 1979, BY ACADEMIC PRESS, INC.
ALL RIGHTS RESERVED
NO PART OF THIS PUBLICATION MAY BE
REPRODUCED OR TRANSMITTED IN ANY
FORM OR BY ANY MEANS, ELECTRONIC OR
MECHANICAL, INCLUDING PHOTOCOPY,
RECORDING, OR ANY INFORMATION
STORAGE AND RETRIEVAL SYSTEM,
WITHOUT PERMISSION IN WRITING
FROM THE PUBLISHER

ACADEMIC PRESS, INC.
111 FIFTH AVENUE,
NEW YORK, NEW YORK 10003

UNITED KINGDOM EDITION PUBLISHED BY
ACADEMIC PRESS, INC. (LONDON) LTD.
24/28 OVAL ROAD, LONDON NW1

ISBN: 0-12-484760-9
Library of Congress Catalog Card Number: 78-57143

PRINTED IN THE UNITED STATES OF AMERICA

To The Instructor

This book grew out of my experiences teaching intermediate algebra at Cuesta Community College in San Luis Obispo. I enjoy teaching algebra at this level and have found the students to be unique and challenging. A number of them come directly from high school, having completed a year or two of high school algebra, while many others are returning to school after an absence of five or more years. Since elementary algebra is a prerequisite for this course, most students have some foundation in the basic concepts of algebra. I feel I have been quite successful in finding ways to teach and motivate these students and have written this text with them in mind. The vocabulary necessary to communicate effectively in algebra is defined and explained clearly, and the properties and definitions are highlighted so that reference back to them is simple. When memorization is necessary I say so.

While writing this book, I also kept you, the instructor, in mind. The text is divided into sections, each of which can be presented easily in a 45- or 50-minute lecture or class period. The problem sets are organized so that the students get a good start on new material before they encounter the more difficult problems. Many problem sets end with an application problem, which I have found to be very helpful in stimulating the students' motivation and interest.

Organization of the Text: The book begins with a preface to the student explaining what study habits are necessary to ensure success in an intermediate algebra course.

The rest of the book is divided into chapters. Each chapter is organized as follows:

1. A preface to the student explains in a very general way what he or she can expect to find in the chapter and some of the applications

associated with topics in the chapter. The preface includes a list of previous material that is used to develop the concepts in the chapter.

2. The body of the chapter, which is divided into sections. Each section contains three main parts:

    a. *Explanations:* The explanations are made as simple and as intuitive as possible without sacrificing mathematical completeness. The ideas, properties, and definitions from Chapters One and Two are used continuously throughout the book. The idea is to make as few rules and definitions as possible and then refer back to them when new situations are encountered. This has been a very successful teaching technique for me, and I feel confident about applying it to this text.

    b. *Examples:* The examples are chosen to clarify the explanations and preview the problem sets. They cover all situations encountered in the problem sets and can be referred to easily when trouble arises.

    c. *Problem Sets:* I have incorporated four main ideas into each problem set.

        i.   *Drill:* There are enough problems in each set to ensure student proficiency in the material.

        ii. *Progressive Difficulty:* The problems increase in difficulty as the problem set progresses.

        iii. *Odd–Even Similarities:* Each pair of consecutive problems is similar. The answers to the odd problems are listed in the back of the book. This gives the students a chance to check their work and then try a similar problem.

        iv. *Application Problems:* Whenever possible I have ended the problem sets with an application problem. My experience is that students are always curious about how the algebra they are learning can be applied. They are also much more likely to put some time and effort into trying an application problem if there is only one problem to try.

3. The chapter summary and review, which lists the new properties and definitions found in the chapter. (Often it also contains a list of common mistakes, clearly marked as such, so that the student can learn to recognize and avoid them.)

4. The chapter test, designed to give the student an idea of how well he or she has mastered the material in the chapter. The problems are representative of all problems in the chapter. All answers for these chapter tests are included in the back of the book to let the student

evaluate his or her grasp of the material, and decide if and where further effort is needed.

An Instructor's Manual, available upon request, provides answers to the even-numbered problems and all answers to the chapter tests. Also included are additional chapter tests, with answers.

ACKNOWLEDGMENTS I would like to thank Mary Leach, of the University of Maryland—Baltimore County, Norman Mittman, of Northeastern Illinois University, and Dell Swearingen, of Linn-Benton Community College, for their reviews of this manuscript. Their comments and suggestions were very helpful.

Special thanks to the College Department of Academic Press for their fine suggestions and patience and their careful editing of this manuscript.

Thanks also to Kate Bauer, Cathie Jarosz, Rona Addes, Mary Ann Airth, and Linda Gracia for typing this manuscript; Mike O'Connell, Ellen Schell, Greg Carstairs, Bill Husted, Jim Domenico, and Rodger Larsen for working the problems; and Diane McKeague for her assistance throughout the process of writing this book.

To The Student

I have seen a number of students over the last few years for whom this quote really makes sense. They don't enjoy math and science classes because they are always worried about whether they will understand the material. Most of them just try to get by and hope for the best. (Are you like that?)

Then there are other students who just don't worry about it. They know they can become as proficient at algebra as they want (and get whatever grade they want, too). I have people in my class who just can't help but do well on the tests I give. Most of my other students think these people are smart. They may be, but that's not the reason they do well. These successful students have learned that topics in mathematics are not always understandable the first time around. They don't worry about understanding the material; they are successful because they work lots of problems.

That's the key to success in mathematics: working problems. The more problems you work, the better you become at working problems. It's that simple. Don't worry about being successful or understanding the material, just work problems. Lots of them. The more problems you work, the better you will understand the material.

That's the answer to the big question of how to master this course.

Here are the answers to some other questions that are often asked by students of intermediate algebra.

How much math do I need to know before taking intermediate algebra?

You should have passed a beginning algebra class. If it has been a few years since you took it, you may have to put in some extra time at the beginning of the semester to get back some of the skills you have lost.

What is the best way to study?

The best way to study is consistently. You must work problems every day. The more time you spend on your homework in the beginning, the easier the sections in the rest of the book will seem. The first two chapters in the book contain the basic properties and methods that the rest of the chapters are built on. Do well on these first two chapters and you will have set a successful pace for the rest of the book.

If I understand everything that goes on in class, can I take it easy on my homework?

Not necessarily. There is a big difference between understanding a problem someone else is working and working the same problem yourself. You are watching someone else think through a problem. There is no substitute for thinking it through yourself. The concepts and properties are understandable to you only if you yourself work problems involving them.

I'm worried about not understanding it. I've passed algebra before but I'm not really sure I understood everything that went on.

There will probably be few times when you can understand absolutely everything that goes on in class. This is standard with most math classes. It doesn't mean you will never understand it. As you read through the book and try problems on your own, you will understand more and more of the material. But if you don't try the problems on your own, you are almost guaranteed to be left confused. By the way, reading the book is important, even if it seems difficult. Reading a math book isn't the same as reading a novel. (You'll see what I mean when you start to read this one.) Reading the book through once isn't going to do it. You will have to read some sections a number of times before they really sink in.

If you have decided to be successful in intermediate algebra, here is a list of things you can do that will help you attain that success.

1. Attend class. There is only one way to find out what goes on in class and that is to be there. Missing class and then hoping to get

the information from someone who was there is not the same as being there yourself.

2. Read the book and work problems every day. Remember, the key to success in mathematics is working problems. Be wild and crazy about it. Work all the problems you can get your hands on. If you have assigned problems to do, do them first. If you don't understand the concepts after finishing the assigned problems, then keep working problems until you do.

3. Do it on your own. Don't be misled into thinking someone else's work is your own.

4. Don't expect to understand it the first time. Most people have to go through a new topic at least twice before they begin to see what is going on.

5. Relax. It's probably not as difficult as you think.

Contents

3

EXPONENTS AND POLYNOMIALS

4

RATIONAL EXPRESSIONS

5

RATIONAL EXPONENTS AND ROOTS

9

THE CONIC SECTIONS

10

RELATIONS AND FUNCTIONS

11

LOGARITHMS

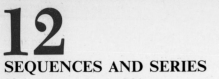

SEQUENCES AND SERIES

APPENDIX A: SYNTHETIC DIVISION

APPENDIX B: QUADRATIC INEQUALITIES IN ONE VARIABLE

APPENDIX C: TABLE OF POWERS, ROOTS, AND PRIME FACTORS

APPENDIX D: TABLE OF COMMON LOGARITHMS

ANSWERS TO ODD-NUMBERED EXERCISES AND CHAPTER TESTS

INDEX

Basic Properties and Definitions

1

To the student:

The material in Chapter 1 is some of the most important material in the book. It is also some of the easiest material to understand. Be sure that you master it. Your success in the following chapters is directly related to how well you understand the material in Chapter 1.

Here is a list of the most essential concepts from Chapter 1 that you will need in order to be successful in the succeeding chapters:

1. You must know how to add, subtract, multiply, and divide positive and negative numbers. You must be consistently accurate in getting correct answers to simple arithmetic problems.

2. You must understand and recognize the commutative, associative, and distributive properties. These are the three most important properties of real numbers. They are used many times throughout the book to justify and explain other rules and properties.

3. You should know the major classifications of numbers. Some rules and properties hold only for specific kinds of numbers. You must therefore know the difference between whole numbers, integers, rational numbers, and real numbers.

Actually, what Chapter 1 is all about is listing the rules of the game. We are playing the game (algebra) with numbers. The properties and definitions in Chapter 1 tell us what we can start with. The rest of the book, then, is simply the playing of the game.

The material in this chapter may seem very familiar to you. You may have a tendency to skip over it lightly because it is familiar. Don't do it. Understanding algebra begins with understanding Chapter 1. Make sure you *understand* Chapter 1, even if you are familiar with it.

1.1
Basic Definitions

This section is, for the most part, simply a list of many of the basic symbols and definitions we will be using throughout the book.

We begin by reviewing the symbols for comparison of numbers and expressions, followed by a review of the operation symbols.

Comparison Symbols

In symbols	*In words*
$a = b$	a is equal to b
$a \neq b$	a is not equal to b
$a < b$	a is less than b
$a \leq b$	a is less than or equal to b
$a \geq b$	a is greater than or equal to b
$a > b$	a is greater than b
$a \ngtr b$	a is not greater than b
$a \nless b$	a is not less than b
$a \Leftrightarrow b$	a is equivalent to b
	(This symbol is usually used when accompanying logical statements.)

Operation Symbols

Operation	*In symbols*	*In words*
Addition	$a + b$	The sum of a and b
Subtraction	$a - b$	The difference of a and b
Multiplication	ab, $a \cdot b$, $a(b)$, $(a)b$, or $(a)(b)$	The product of a and b
Division	$a \div b$, a/b, or $\dfrac{a}{b}$	The quotient of a and b

The key words are *sum, difference, product,* and *quotient.* They are used frequently in mathematics. For instance, we may say the product of 3 and 4 is 12. We mean both the statements $3 \cdot 4$ and 12 are called the product of 3 and 4. The important idea here is that the word *product* implies multiplication, regardless of whether it is written $3 \cdot 4$, 12, 3(4), or (3)4.

The following is an example of translating expressions written in English into expressions written in symbols.

▼   **Example 1**

In English	*In symbols*
The sum of x and 5.	$x + 5$
The product of 3 and x.	$3x$
The quotient of y and 6.	$y/6$
Twice the difference of b and 7.	$2(b - 7)$
The difference of twice b and 7.	$2b - 7$
The quotient of twice x and 3.	$2x/3$ ✗
The product of x and y is less than the sum of x and y.	$xy < x + y$
The quotient of 3 times x and 5 is greater than or equal to the difference of x and 4.	$\dfrac{3x}{5} \geq x - 4$

▲

We now list the basic definitions and operations associated with sets.

DEFINITION   A *set* is a well-defined collection of objects or things. The   *Sets*
objects in the set are called *elements* or *members* of the set.

   The concept of a set can be considered the starting point for all the
branches of mathematics. For instance, most of the important concepts
in statistics are based, in one form or another, on probability theory. The
basic concept in probability theory is the idea of an event, and an event
is nothing more than a set.
   Sets are usually denoted by capital letters and elements of sets by
lower-case letters. To show an element is contained in a set we use the
symbol $\in$. That is,

> $x \in A$ is read "x is an element (member) of set A"
>   or   "x is contained in A."
> The symbol $\notin$ is read "is *not* a member of."

DEFINITION   Set A is a subset of set B, written $A \subset B$, if every element
in A is also an element of B. That is,

> $A \subset B$ if and only if A is contained in B.

We use the braces { } to enclose the elements of a set.
Here are some examples of sets and subsets:

▼   **Example 2**

a. The set of numbers used to count things is $\{1, 2, 3, \ldots\}$. The dots mean the set continues indefinitely in the same manner. This is an example of an infinite set.

b. The set of all numbers represented by the dots on the faces of a regular die is $\{1, 2, 3, 4, 5, 6\}$. This set is a subset of the set in part a. It is an example of a *finite* set, since it has a limited number of elements.

c. The set of all Fords is a subset of the set of all cars, since every Ford is also a car.   ▲

DEFINITION   The set with no members is called the *empty* or *null set*. It is denoted by the symbol $\varnothing$. (*Note:* a mistake is sometimes made by trying to denote the empty set with the notation $\{\varnothing\}$. The set $\{\varnothing\}$ is not the empty set, since it contains one element, the empty set $\varnothing$.)

The empty set is considered a subset of every set.

Operations with Sets

There are two basic operations used to combine sets. The operations are union and intersection.

DEFINITION   The *union* of two sets A and B, written $A \cup B$, is the set of all elements that are either in A or in B, or in both A and B. The key word here is *or*. For an element to be in $A \cup B$ it must be in A or B. In symbols the definition looks like this:

$$x \in A \cup B \quad \text{if and only if} \quad x \in A \text{ or } x \in B$$

DEFINITION   The *intersection* of two sets A and B, written $A \cap B$, is the set of elements in both A and B. The key word in this definition is the word *and*. For an element to be in $A \cap B$ it must be in both A and B, or

$$x \in A \cap B \quad \text{if and only if} \quad x \in A \text{ and } x \in B$$

▼   **Example 3** Let $A = \{1, 3, 5\}, B = \{0, 2, 4\}$, and $C = \{1, 2, 3, \ldots\}$. Then

a. $A \cup B = \{0, 1, 2, 3, 4, 5\}$

b. $A \cap B = \varnothing$ (A and B have no elements in common.)

c. $A \cap C = \{1, 3, 5\} = A$

d. $B \cup C = \{0, 1, 2, 3, \ldots\}$ ▲

We can represent the union and intersection of two sets A and B with pictures. The pictures are called Venn diagrams.

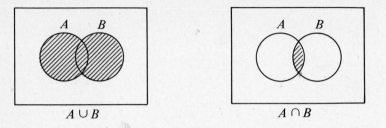

$$A \cup B \qquad\qquad\qquad A \cap B$$

DEFINITION   Two sets with no elements in common are said to be *disjoint* or *mutually exclusive*. Two sets are disjoint if their intersection is the empty set.

A and B are disjoint if and only if
$$A \cap B = \varnothing$$

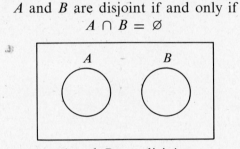

A and B are disjoint.

Along with giving a graphical representation of union and intersection, Venn diagrams can be used to test the validity of statements involving combinations of sets and operations on sets.

▼   **Example 4**   Use Venn diagrams to check the expression

$$A \cap (B \cup C) = (A \cap B) \cup (A \cap C)$$

(Assume no two sets are disjoint.)

We begin by making a Venn diagram of the left side with $B \cup C$ shaded in with vertical lines.

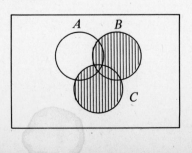

Using the same diagram we now shade in set A with horizontal lines.

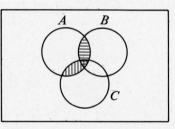

The region containing both vertical and horizontal lines is the intersection of A with $B \cup C$, or $A \cap (B \cup C)$.

We diagram the right side and shade in $A \cap B$ with horizontal lines and $A \cap C$ with vertical lines.

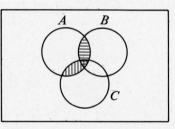

Any region containing vertical or horizontal lines is part of the union of $A \cap B$ and $A \cap C$, or $(A \cap B) \cup (A \cap C)$.

The original statement appears to be true.

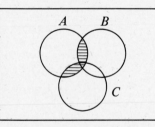 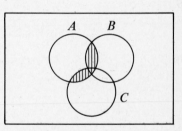

$A \cap (B \cup C)$ $(A \cap B) \cup (A \cap C)$ ▲

Problem Set 1.1

Translate each of the following sentences into symbols:

1. The sum of x and 5.
2. The sum of y and -3.
3. The difference of 6 and x.

4. The difference of x and 6.
5. The product of t and 2 is less than y.
6. The product of $5x$ and y is equal to z.
7. The quotient of $3x$ and $2y$ is greater than 6.
8. The quotient of $2y$ and $3x$ is not less than 7.
9. The sum of x and y is less than the difference of x and y.
10. Twice the sum of a and b is 15.
11. Three times the difference of x and 5 is more than y.
12. The product of x and y is greater than or equal to the quotient of x and y.
13. The difference of s and t is not equal to their sum.
14. The quotient of $2x$ and y is less than or equal to the sum of $2x$ and y.
15. Twice the sum of t and 3 is not greater than the difference of t and 6.
16. Three times the product of x and y is equal to the sum of $2y$ and $3z$.

For problems 17–34, let $A = \{0, 2, 4, 6\}$, $B = \{1, 2, 3, 4, 5\}$, $C = \{1, 3, 5, 7\}$, and $D = \{-2, -1, 0, 1, 2\}$ and find the following:

17. $A \cup B$	18. $A \cup C$
19. $A \cap B$	20. $A \cap C$
21. $C \cup D$	22. $B \cup D$
23. $C \cap D$	24. $B \cap D$
25. $A \cup D$	26. $B \cup C$
27. $A \cap D$	28. $B \cap C$
29. $A \cup (B \cap C)$	30. $C \cup (A \cap B)$
31. $(A \cup B) \cap (A \cup C)$	32. $(C \cup A) \cap (C \cup B)$
33. $A \cap (B \cup D)$	34. $B \cap (C \cup D)$

35. List all the subsets of $\{a, b, c\}$.
36. List all the subsets of $\{0, 1\}$.
37. Give an example, using any sets you want, of two mutually exclusive sets.
38. Give an example of two nonmutually exclusive sets.

Use Venn diagrams to show each of the following regions. Assume sets A, B, and C all intersect one another.

39. $(A \cup B) \cup C$	40. $(A \cap B) \cap C$
41. $A \cap (B \cap C)$	42. $A \cup (B \cup C)$
43. $A \cap (B \cup C)$	44. $(A \cap B) \cup C$
45. $A \cup (B \cap C)$	46. $(A \cup B) \cap C$
47. $(A \cup B) \cap (A \cup C)$	48. $(A \cap B) \cup (A \cap C)$

In this section we will give a definition of real numbers in terms of the real number line. We will then classify all pairs of numbers that add to 0, and do the same for all pairs of numbers whose product is 1. We will end the section with two definitions of absolute value and a list of some special sets.

**1.2
The Real Numbers, Opposites, Reciprocals, and Absolute Value**

The Real Numbers

The real number line is constructed by drawing a straight line and labeling a convenient point with the number 0. Positive numbers are in increasing order to the right of 0, negative numbers are in decreasing order to the left of 0. The point on the line corresponding to 0 is called the origin.

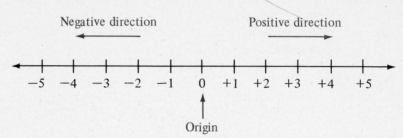

The numbers associated with the points on the line are called *coordinates* of those points. Every point on the line has a number associated with it. The set of all these numbers makes up the set of real numbers.

DEFINITION   A *real number* is any number that is the coordinate of a point on the real number line.

There are many different sets of numbers contained in the real numbers. We will classify some of them later. Among those numbers contained in the real numbers are all positive and negative fractions and decimals.

▼   **Example 1**   Locate the numbers -4.5, $-2\frac{1}{4}$, $-.75$, $\frac{1}{2}$, $\sqrt{2}$, π, and 4.1 on the real number line.

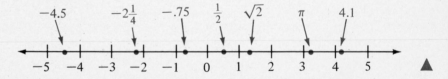

Note   In this book we will refer to real numbers as being on the real number line. Actually, real numbers are *not* on the line; only the points they represent are on the line. We can save some writing, however, if we simply refer to real numbers as being on the number line.

Opposites and Reciprocals

DEFINITION   Any two real numbers, the same distance from 0, but in opposite directions from 0 on the number line, are called *opposites* or *additive inverses*.

▼   **Example 2**   The numbers -3 and 3 are opposites. So are π and $-\pi$, $\frac{3}{4}$ and $-\frac{3}{4}$, and $\sqrt{2}$ and $-\sqrt{2}$.   ▲

The negative sign in front of a number can be read in a number of different ways. It can be read as "negative" or "the opposite of." We say -4 is the opposite of 4 or negative 4. The one we use will depend on the situation. For instance, the expression $-(-3)$ is best read "the opposite of negative 3." Since the opposite of -3 is 3, we have $-(-3) = 3$. In general, if a is any positive real number, then

$$-(-a) = a \qquad \text{(The opposite of a negative is positive.)}$$

One property all pairs of opposites have is that their sum is 0. It is for this reason they are called additive inverses.

DEFINITION   Any two real numbers whose product is 1 are called *reciprocals* or *multiplicative inverses*.

▼   **Example 3**   The following table gives several numbers with the opposite and reciprocal of each:

	Number	Opposite	Reciprocal
a.	5	-5	$\dfrac{1}{5}$
b.	-3	$-(-3) = 3$	$-\dfrac{1}{3}$
c.	$\dfrac{2}{3}$	$-\dfrac{2}{3}$	$\dfrac{1}{\frac{2}{3}} = \dfrac{3}{2}$
d.	$-\dfrac{4}{5}$	$-\left(-\dfrac{4}{5}\right) = \dfrac{4}{5}$	$\dfrac{1}{-\frac{4}{5}} = -\dfrac{5}{4}$
e.	0	$-0 = 0$	undefined
f.	$\sqrt{2}$	$-\sqrt{2}$	$\dfrac{1}{\sqrt{2}}$

Notice 0 does not have a reciprocal. By definition, reciprocals must multiply to 1. From past experience we know multiplying by 0 always results in 0. Hence 0 does not have a reciprocal.   ▲

The Absolute Value of
a Real Number

Sometimes it is convenient to consider only the distance a number is from 0 and not its direction from 0.

DEFINITION   The *absolute value* of a number (also called its *magnitude*) is the distance the number is from 0 on the number line. If x represents a real number, then the absolute value of x is written $|x|$.

Since distances are always represented by positive numbers or 0, the absolute value of a quantity is always positive or 0; the absolute value of a number is never negative.

The above definition of absolute value is geometric in form since it defines absolute value in terms of the number line.

Here is an alternate definition of absolute value that is algebraic since it involves only symbols.

DEFINITION   If x represents a real number, then the *absolute value* (or *magnitude*) of x is written $|x|$, and is given by

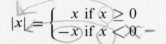

$$|x| = \begin{cases} x & \text{if } x \geq 0 \\ -x & \text{if } x < 0 \end{cases}$$

If the original number is positive or 0, then its absolute value is the number itself. If the number is negative, its absolute value is its opposite (which must be positive). We see that $-x$, as written in this definition, is a positive quantity since x in this case is a negative number.

Note   It is important to recognize that if x is a real number, $-x$ is not necessarily negative. For example, if x is 5, then $-x$ is -5. On the other hand, if x were -5, then $-x$ would be $-(-5)$, which is 5. To assume that $-x$ is a negative number we must know beforehand that x is a positive number. If x is a negative number, then $-x$ is a positive number.

▼   **Example 4**

a. $|5| = 5$ e. $-|-3| = -3$
b. $|-2| = 2$ f. $-|5| = -5$
c. $|-\frac{1}{2}| = \frac{1}{2}$ g. $-|-\sqrt{2}| = -\sqrt{2}$
d. $|\frac{3}{4}| = \frac{3}{4}$ h. $-|\frac{5}{6}| = -\frac{5}{6}$

The last four parts of this example do not contradict the statement that the absolute value of a number is always positive or 0. In part e we are asked to find the *opposite* of the absolute value of -3, $-|-3|$. The absolute value of -3 is 3, the opposite of which is -3. If the absolute value of a number is always positive or 0, then the opposite of its absolute value must always be negative or 0. ▲

Note There is a tendency to confuse the two expressions

$$-(-3) \quad \text{and} \quad -|-3|$$

The first expression is the opposite of -3, which is 3. The second expression is the opposite of the absolute value of -3, which is -3.

$$-(-3) = 3 \quad \text{and} \quad -|-3| = -3$$

Note It is incorrect to write $|x| = x$. This statement is true only if x is a positive number. If x were negative, then, by definition, $|x| = -x$.

We will end this section by listing the major subsets of the set of real numbers.

Subsets of the Real Numbers

$$\textit{Counting numbers} = \{1, 2, 3, \ldots\}$$

These are also called *natural numbers* or the *set of positive integers*. If we include the number 0 in the set of counting numbers, we have the *set of whole numbers:*

$$\textit{Whole numbers} = \{0, 1, 2, 3, \ldots\}$$

The whole numbers along with the opposites of all the counting numbers make up the *set of integers:*

$$\textit{Integers} = \{\ldots -3, -2, -1, 0, 1, 2, 3, \ldots\}$$

The set of all numbers that can be written as the ratio of two integers is called the *set of rational numbers*. (The ratio of two numbers a and b is the same as their quotient a/b.)

$$\textit{Rational numbers} = \{a/b \mid a \text{ and } b \text{ are integers}, b \neq 0\}$$

This is read "the set of all a/b such that a and b are integers, where $b \neq 0$."

▼   **Example 5** The following are all rational numbers. They can all be written as the ratio of two integers.

a. $\dfrac{3}{4}$

b. $-\dfrac{5}{6}$

c. $5\left(\text{because } 5 = \dfrac{5}{1}\right)$

d. $-8\left(\text{because } -8 = \dfrac{-8}{1}\right)$

e. $.75\left(\text{because } .75 = \dfrac{75}{100}\right)$

f. $-.9\left(\text{because } -.9 = -\dfrac{9}{10}\right)$

g. $.333\ldots\left(\text{because } .333\ldots = \dfrac{1}{3}\right)$

h. $-.666\ldots\left(\text{because } -.666\ldots = -\dfrac{2}{3}\right)$ ▲

The real numbers that cannot be written as the ratio of two integers comprise the *set of irrational numbers*. These numbers have decimal representations that are all nonrepeating, nonterminating decimals— that is, decimals with an infinite number of digits past the decimal point in which no repeating pattern of digits can be found.

Irrational numbers = $\{x \mid x \text{ is a nonrepeating, nonterminating decimal}\}$

Since irrational numbers cannot be written as the ratio of integers, and their decimal representations never terminate or repeat, they have to be written in other ways.

▼   **Example 6** The following are irrational numbers:

a. $\sqrt{2}$
b. $-\sqrt{3}$
c. $4 + 2\sqrt{3}$
d. π
e. $\pi + 5\sqrt{6}$ ▲

Another way to define the set of real numbers is as the union of the set of rational numbers and irrational numbers.

Real numbers = $\{x \mid x \text{ is rational or } x \text{ is irrational}\}$

The relationships between the above sets of numbers are shown as follows:

Counting #'s ⊂ Whole #'s ⊂ Integers ⊂ Rational #'s ⊂ Real #'s
Irrational #'s ⊂ Real #'s
Rational #'s ∩ Irrational #'s = ∅
Rational #'s ∪ Irrational #'s = Real #'s

---

1. Locate the numbers -3, -1.75, $-\frac{1}{2}$, 0, $\frac{1}{3}$, 1, 1.3, and 4.5 on the number line.

2. Locate the numbers -3.25, -3, -2.5, -1, $-\frac{1}{3}$, 0, $\frac{3}{4}$, 1.2, and 2 on the number line.

Problem Set 1.2

Copy and complete the following table.

	Number	Opposite	Reciprocal
3.	4		
4.	-3		
5.	$-\frac{1}{2}$		
6.	$\frac{5}{6}$		
7.		-5	
8.		7	
9.		$-\frac{3}{8}$	
10.		$\frac{1}{2}$	
11.			-6
12.			-3
13.			$\frac{1}{3}$
14.			$-\frac{1}{4}$
15.		$-\sqrt{3}$	
16.		$\sqrt{5}$	
17.			$-\sqrt{2}$
18.			$-\frac{3}{5}$
19.	x		
20.	0		

21. Name two numbers that are their own reciprocals.
22. Give the number that has no reciprocal.
23. Name the number that is its own opposite.
24. The reciprocal of a negative number is negative—true or false?

Write each of the following without absolute value symbols:

25. $|-2|$ **26.** $|-7|$
27. $|-\frac{3}{4}|$ **28.** $|\frac{5}{6}|$
29. $|\pi|$ **30.** $|-\sqrt{2}|$

31. $-|4|$　　　　　　　　　　　　**32.** $-|5|$
33. $-|-2|$　　　　　　　　　　　**34.** $-|-10|$
35. $-|-\frac{3}{4}|$　　　　　　　　　　　**36.** $-|\frac{7}{8}|$

Find the value of each of the following expressions:

37. $-(-2)$　　　　　　　　　　**38.** $-(-\frac{3}{4})$
39. $-[-(-\frac{1}{3})]$　　　　　　　　**40.** $-[-(-1)]$
41. $|2| + |3|$　　　　　　　　　　**42.** $|7| + |4|$
43. $|-3| + |5|$　　　　　　　　　**44.** $|-1| + |0|$
45. $|-8| - |-3|$　　　　　　　　**46.** $|-6| - |-1|$
47. $|-10| - |4|$　　　　　　　　**48.** $|-3| - |2|$
49. $|-2| + |-3| - |5|$　　　　　**50.** $|-6| - |-2| + |-4|$

For the set $\{-6, -5.2, -\sqrt{7}, -\pi, 0, 1, 2, 2.3, \frac{9}{2}, \sqrt{17}\}$ list all the elements that are:

51. Counting numbers　　　　　**52.** Whole numbers
53. Rational numbers　　　　　**54.** Integers
55. Irrational numbers　　　　　**56.** Real numbers
57. Nonnegative integers　　　　**58.** Positive integers

Label the following true or false. For each false statement give a counter example (that is, an example that shows the statement is false).

59. Zero has an opposite and a reciprocal.
60. Some irrational numbers are also rational numbers.
61. All whole numbers are integers.
62. Every real number is a rational number.
63. All integers are rational numbers.
64. Some negative numbers are integers.
65. Zero is both rational and irrational.
66. Zero is not considered a real number.
67. The opposite of an integer is also an integer.
68. The reciprocal of an integer is an integer.
69. The counting numbers are a subset of the whole numbers.
70. The rational numbers are a subset of the irrational numbers.

**1.3
Graphing Simple and
Compound
Inequalities**

In this section we will use some of the ideas developed in the first two sections to graph inequalities. The graph of an inequality uses the real number line to give a visual representation of an algebraic expression.

▼　**Example 1**　Graph $\{x \mid x \leq 3\}$.

Solution　We want to graph all the real numbers less than or equal to 3—that is, all the real numbers below 3 and including 3. We label

0 on the number line for reference as well as 3 since the latter is what we call the end point. The graph is as follows.

We use a solid circle at 3 since 3 is included in the graph.   ▲

▼   **Example 2**   Graph $\{x \mid x < 3\}$.

Solution   The graph will be identical to the graph in Example 1 except at the end point 3. In this case we will use an open circle since 3 is not included in the graph.   ▲

   In Section 1.1 we defined the *union* of two sets A and B to be the set of all elements that are in either A or B. The word *or* is the key word in the definition. The *intersection* of two sets A and B is the set of all elements contained in both A and B, the key word here being *and*. We can put the words *and* and *or* together with our methods of graphing inequalities to graph some compound inequalities.

▼   **Example 3**   Graph $\{x \mid x \leq -2 \text{ or } x > 3\}$.

Solution   The two inequalities connected by the word *or* are referred to as a *compound inequality*. We begin by graphing each inequality separately.

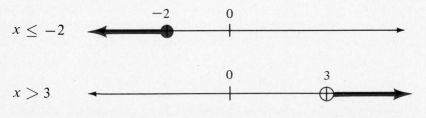

Since the two are connected by the word *or*, we graph their union. That is, we graph all points on either graph.   ▲

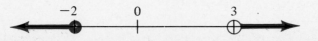

▼   **Example 4**   Graph $\{x \mid x > -1 \text{ and } x < 2\}$.

Solution We first graph each inequality separately.

$x > -1$
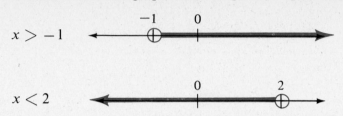

$x < 2$

Since the two inequalities are connected by the word *and,* we graph
their intersection—the part they have in common. ▲

NOTATION Sometimes compound inequalities that use the word *and* as
the connecting word can be written in a shorter form. For example, the
compound inequality $-3 \leq x$ and $x \leq 4$ can be written $-3 \leq x \leq 4$.
The word *and* does not appear when an inequality is written in this
form. It is implied. Inequalities of the form $-3 \leq x \leq 4$ are called
continued inequalities. This new notation is useful because is takes fewer
symbols to write it and because if $-3 \leq x$ and $x \leq 4$, then x must be
between -3 and 4. With the notation $-3 \leq x \leq 4$, it "looks" as though
x is between -3 and 4. The graph of $-3 \leq x \leq 4$ is

▼ **Example 5** Graph $\{x \mid 1 \leq x < 2\}$.

Solution The word *and* is implied in the continued inequality
$1 \leq x < 2$. We graph all the numbers between 1 and 2 on the
number line, including 1 but not including 2. ▲

Problem Set 1.3   Graph the following on the real number line:

1. $\{x \mid x < 1\}$ **2.** $\{x \mid x > -2\}$
3. $\{x \mid x \leq 1\}$ **4.** $\{x \mid x \geq -2\}$
5. $\{x \mid x \geq 4\}$ **6.** $\{x \mid x \leq -3\}$
7. $\{x \mid x > 4\}$ **8.** $\{x \mid x < -3\}$

9. $\{x \mid x > 0\}$ 10. $\{x \mid x < 0\}$
11. $\{x \mid -2 < x\}$ 12. $\{x \mid 3 \geq x\}$
13. $\{x \mid 4 \leq x\}$ 14. $\{x \mid 2 > x\}$

Graph the following compound inequalities:

15. $\{x \mid x < -3 \text{ or } x > 1\}$ 16. $\{x \mid x \leq 1 \text{ or } x \geq 4\}$
17. $\{x \mid x \leq -3 \text{ or } x \geq 1\}$ 18. $\{x \mid x < 1 \text{ or } x > 4\}$
19. $\{x \mid -3 \leq x \text{ and } x \leq 1\}$ 20. $\{x \mid 1 < x \text{ and } x < 4\}$
21. $\{x \mid -3 < x \text{ and } x < 1\}$ 22. $\{x \mid 1 \leq x \text{ and } x \leq 4\}$
23. $\{x \mid -1 \leq x \text{ and } 2 \leq x\}$ 24. $\{x \mid 3 \leq x \text{ and } 4 \leq x\}$
25. $\{x \mid -1 \leq x \text{ or } 2 \leq x\}$ 26. $\{x \mid 3 \leq x \text{ or } 4 \leq x\}$
27. $\{x \mid x < -1 \text{ or } x \geq 3\}$ 28. $\{x \mid x < 0 \text{ or } x \geq 3\}$
29. $\{x \mid x \leq -1 \text{ and } x \geq 3\}$ 30. $\{x \mid x \leq 0 \text{ and } x \geq 3\}$
31. $\{x \mid x > -4 \text{ and } x < 2\}$ 32. $\{x \mid x > -3 \text{ and } x < 0\}$

Graph the following continued inequalities:

33. $\{x \mid -1 \leq x \leq 2\}$ 34. $\{x \mid -2 \leq x \leq 1\}$
35. $\{x \mid -1 < x < 2\}$ 36. $\{x \mid -2 < x < 1\}$
37. $\{x \mid -3 < x < 1\}$ 38. $\{x \mid 1 \leq x \leq 2\}$
39. $\{x \mid -3 \leq x < 0\}$ 40. $\{x \mid 2 \leq x < 4\}$
41. $\{x \mid -4 < x \leq 1\}$ 42. $\{x \mid -1 < x \leq 5\}$

In this section we will list all the things we know to be true of real numbers and the comparison and operation symbols listed in Section 1.1. Mathematics is a game we play with real numbers. The rules of the game are the properties of real numbers listed in this section. We play the game by taking real numbers and their properties and applying them to as many new situations as possible.

The list of properties given in this section is actually just an organized summary of the things we know from past experience to be true about numbers in general. For instance, we know that adding 3 and 7 gives the same answer as adding 7 and 3. The order of two numbers in an addition problem has no effect on the result. Properties like these do not require proof. We know them to be true intuitively. The properties are therefore listed as *axioms*.

We begin our list with the assumptions (axioms) we will use concerning real numbers and the equality and inequality comparison symbols.

In the following two lists, *a*, *b*, and *c* represent real numbers.

**1.4
Properties of Real
Numbers**

Axioms of Equality

1. $a = a$ — Reflexive property
2. If $a = b$, then $b = a$. — Symmetric property
3. If $a = b$ and $b = c$, then $a = c$. — Transitive property
4. If $a = b$, then, wherever a occurs, b may be used, and wherever b occurs, a may be used. — Substitution property

Axioms of Inequality (Order)

1. Exactly one of the following is true.

$$a < b, \ a = b, \ \text{or} \ a > b \qquad \text{Trichotomy property}$$

2. If $a < b$ and $b < c$, then $a < c$. — Transitive property

The substitution property is the most frequently used axiom from the above lists. The other properties will not be referred to very often. They are included to make the list complete.

The more useful properties of real numbers are those used in connection with the operation symbols for addition and multiplication. Subtraction and division are not included in the list since subtraction will be defined in terms of addition, and division in terms of multiplication.

Axioms of Real Numbers under Addition and Multiplication

In the following list, a, b, and c represent real numbers.

Closure Property for Addition

In symbols: $a + b = $ A real number
In words: The sum of two real numbers is always a real number.

Closure Property for Multiplication

In symbols: $a \cdot b = $ A real number
In words: The product of two real numbers is always a real number.

Commutative Property of Addition

In symbols: $a + b = b + a$
In words: The *order* of the numbers in a sum does not affect the result.

Commutative Property of Multiplication

In symbols: $a \cdot b = b \cdot a$
In words: The *order* of the numbers in a product does not affect the result.

Associative Property of Addition

In symbols: $a + (b + c) = (a + b) + c$

In words: The *grouping* of the numbers in a sum does not affect the result.

Associative Property of Multiplication

In symbols: $a(bc) = (ab)c$

In words: The *grouping* of the numbers in a product does not affect the result.

Distributive Property

In symbols: $a(b + c) = ab + ac$

In words: Multiplication *distributes* over addition.

Additive Identity Property

There exists a unique number 0 such that:

In symbols: $a + 0 = a$ and $0 + a = a$

In words: Zero preserves identities under addition. (The identity of the number is unchanged after addition with 0.)

Multiplicative Identity Property

There exists a unique number 1 such that:

In symbols: $a(1) = a$ and $1(a) = a$

In words: The number 1 preserves identities under multiplication. (The identity of the number is unchanged after multiplication by 1.)

Note 0 and 1 are called the *additive identity* and *multiplicative identity,* respectively. Combining 0 with a number, under addition, does not change the identity of the number. Likewise, combining 1 with a number, under multiplication, does not alter the identity of the number. We see that 0 is to addition what 1 is to multiplication.

Additive Inverse Property

For each real number a, there exists a unique number $-a$ such that:

In symbols: $a + (-a) = 0$

In words: Opposites add to 0.

Multiplicative Inverse Property

For every real number a, except 0, there exists a unique real number $1/a$ such that:

In symbols: $a\left(\dfrac{1}{a}\right) = 1$

In words: Reciprocals multiply to 1.

Of all the basic properties listed, the commutative, associative, and distributive properties are the ones we will use most often. They are important because they are used as justifications or reasons for many of the things we will do in the future.

The following example illustrates how we use the properties listed above. Each line contains an algebraic expression that has been changed in some way. The property that justifies the change is written to the right.

▼   **Example 1**

a. $5(x + 3) = 5x + 15$ Distributive property

b. $7(1) = 7$ 1 is the identity element for multiplication

c. $11 + 5 = 5 + 11$ Commutative property of addition

d. $11 \cdot 5 = 5 \cdot 11$ Commutative property of multiplication

e. $4 + (-4) = 0$ Additive inverse property

f. $3 + (x + 1) = (3 + x) + 1$ Associative property of addition

g. $6(\frac{1}{6}) = 1$ Multiplicative inverse property

h. $3 + 9$ is a real number   Closure property for addition

i. $(2 + 5) + x = 5 + (2 + x)$ Commutative and associative properties

j. $(1 + y) + 3 = 3 + (1 + y)$ Commutative property for addition

k. $(5 + 0) + 2 = 5 + 2$ 0 is the identity element for addition   ▲

Notice in part i of Example 1 that both the order and grouping have changed from the left to the right expression, so both the commutative and associative properties were used.

In part j the only change is in the order of the numbers in the expression, not in the grouping, so only the commutative property was used.

Certainly there are other properties of numbers not contained in the list. For example, we know that multiplying by 0 always gives us 0. This property of 0 is not listed with the other basic properties because it is implied by the properties listed. That is, we can prove it is true using the axioms already listed. Remember, we want to keep our list of assumptions as short as possible. We will call any additional properties *theorems*, and then prove them using the properties developed previously.

THEOREM 1.1 For any real number a,

$$a(0) = 0$$

Proof

$a(1) = a$	1 is the identity element for multiplication
$a(1 + 0) = a$	$1 = 1 + 0$ (0 is the additive identity)
$a(1) + a(0) = a$	Distributive property
$a + a(0) = a$	$a(1) = a$ (1 is the multiplicative identity)
$a + a(0) = a + 0$	0 is the identity element for addition

Since 0 was assumed to be unique in the additive identity axiom, we must conclude from the last line that $a(0)$ must be the same as 0.

As a final note on the properties of real numbers we should mention that although some of the properties are stated for only two or three real numbers, they hold for as many numbers as needed. For example, the distributive property holds for expressions like $3(x + y + z + 5 + 2)$. That is,

$$3(x + y + z + 5 + 2) = 3x + 3y + 3z + 15 + 6$$

It is not important how many numbers are contained in the sum, only that it is a sum. Multiplication, you see, distributes over addition, whether there are two numbers in the sum or two hundred.

Name the axiom of equality or inequality that justifies each of the following: ***Problem Set 1.4***

1. $5 = 5$
2. If $4 - 1 = 3$ and $3 = 2 + 1$, then $4 - 1 = 2 + 1$.
3. If $2 < 5$ and $5 < 7$, then $2 < 7$.
4. If $4(a + b) = x$ and $a + b = 2$, then $4(2) = x$.
5. $\pi = \pi$
6. If x is a real number, then $x < 2$, $x = 2$, or $x > 2$.
7. If $a = 7$, then $7 = a$.
8. If $r - s = 7$ and $7 = a + b$, then $r - s = a + b$.

9.   If $x = 6$ and $7 - y = x$, then $7 - y = 6$.
10. If $2(x + y) = 3$, then $3 = 2(x + y)$.

Identify the property of real numbers that justifies each of the following:

11.	$3 + 2 = 2 + 3$	12.	$3(a \cdot b) = (3a)b$
13.	$5(x \cdot y) = (5x)y$	14.	$5x = x \cdot 5$
15.	$\sqrt{2} + 0 = \sqrt{2}$	16.	$1(6) = 6$
17.	$6(0) = 0$	18.	$0(\pi) = 0$
19.	$4 + (-4) = 0$	20.	$y + (-y) = 0$
21.	$x + (y + 2) = (y + 2) + x$	22.	$(a + 3) + 4 = a + (3 + 4)$
23.	$5(x - 2) = 5x - 10$	24.	$(r + 7) + s = (r + s) + 7$
25.	$\sqrt{2} + 3$ is a real number.	26.	$11(x + 3) = 11x + 33$
27.	$8 + (-8) = 0$	28.	$5 + 0 = 5$
29.	$4(5 \cdot 7) = 5 \cdot (4 \cdot 7)$	30.	$6 \cdot (x \cdot y) = (x \cdot y) \cdot 6$
31.	$4 + (x + y) = (4 + y) + x$	32.	$10(\frac{1}{10}) = 1$

Use the given property to complete each of the following:

33. $5 \cdot y = ?$ Commutative property
34. $4 + 0 = ?$ Additive identity property
35. $3 + a = ?$ Commutative property
36. $5(x + y) = ?$ Distributive property
37. $7(1) = ?$ Multiplicative identity property
38. $6(\frac{1}{6}) = ?$ Multiplicative inverse property
39. $2 + (x + 6) = ?$ Associative property
40. $5(x \cdot 3) = ?$ Associative property
41. $4x + 4y = ?$ Distributive property
42. $(a + 4) + 7 = ?$ Associative property
43. $-6 + 6 = ?$ Additive inverse property
44. $8 + 0 = ?$ Additive identity property
45. $7a + 7b + 7c = ?$ Distributive property
46. $(x + 2) \cdot y ?$ Distributive property
47. $7(r \cdot s) = ?$ Associative property
48. $5 \cdot y = ?$ Commutative property
49. $11 \cdot 1 = ?$ Multiplicative identity property
50. $(a + 1) + b = ?$ Associative property

1.5
Addition and
Subtraction of Real
Numbers

The purpose of this section is to review the rules for addition and subtraction of real numbers and the justification for those rules. The goal here is the ability to add and subtract positive and negative real numbers quickly and accurately, the latter being the more important.

Addition of Real
Numbers

We can justify the rules for addition of real numbers geometrically by use of the real number line. Since real numbers can be thought of as having both a distance from 0 (absolute value) and a direction from 0 (positive or negative), we can visualize addition of two numbers as follows.

Consider the sum of -5 and 3.

$$-5 + 3$$

We can interpret this expression as meaning "start at the origin and move 5 units in the negative direction and then 3 units in the positive direction." With the aid of a number line we can visualize the process:

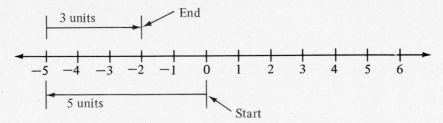

Since the process ends at -2, we say the sum of -5 and 3 is -2.

$$-5 + 3 = -2$$

We can use the real number line in this way to add any combination of positive and negative numbers.

The sum of -4 and -2, $-4 + (-2)$, can be interpreted as starting at the origin, moving 4 units in the negative direction, and then 2 more units in the negative direction:

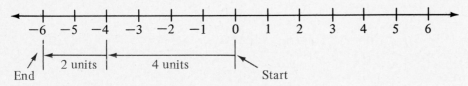

Since the process ends at -6, we say the sum of -4 and -2 is -6.

$$-4 + (-2) = -6$$

We can eliminate actually drawing a number line by simply visualizing it mentally. The following example gives the results of all possible sums of positive and negative 5 and 7.

▼   **Example 1**

$$5 + 7 = 12$$
$$-5 + 7 = 2$$
$$5 + (-7) = -2$$
$$-5 + (-7) = -12$$

▲

Looking closely at the relationships in Example 1 (and trying other similar examples if necessary), we can arrive at the following rule for adding two real numbers.

RULE To add two real numbers with

  a. the *same* sign: simply add absolute values and use the common sign. If both numbers are positive, the answer is positive. If both numbers are negative, the answer is negative.
  b. *different* signs: subtract the smaller absolute value from the larger. The answer will have the sign of the number with the larger absolute value.

An alternate approach to obtaining the rule for addition of real numbers is to state the rule in terms of some theorems and their proofs. For instance, the fact that the sum of two negative numbers is the opposite of the sum of their absolute values can be stated as follows:

THEOREM 1.2 If a and b are any two positive numbers, then

$$(-a) + (-b) = -(a + b)$$

This may not look like what we expected. We have to remember if a and b are positive, then $-a$ and $-b$ must be negative, and the absolute values of $-a$ and $-b$ are a and b.

The justification for the truth of this theorem is simply the proof of the theorem.

Proof

$(-a) + (-b) + (a + b)$
$\quad = (-a + a) + (-b + b)$ Commutative and
                               associative properties
$\quad\quad = 0 + 0$ Opposites add to 0
$\quad\quad = 0$ 0 is the additive identity

Since $(-a) + (-b)$ and $(a + b)$ add to 0, they must be opposites. In other words, $(-a) + (-b)$ is the opposite of $(a + b)$. That is,

$$(-a) + (-b) = -(a + b)$$

Part b of our rule can also be stated as a theorem which we will not attempt to prove.

THEOREM 1.3 If a is a negative real number and b is a positive real number, then

$$\begin{aligned} a + b &= -(|a| - |b|) \quad \text{if} \quad |a| \geq |b| \\ \text{or} \quad a + b &= (|b| - |a|) \qquad \text{if} \quad |b| > |a| \end{aligned}$$

Note The point here is not to make addition of real numbers complicated. We simply want to show that there is an alternative to using the number line for adding real numbers. Remember, the goal in this section is to add positive and negative numbers as quickly and as accurately as possible. It is not really important how we arrive at our rule for addition, only that we can use it effectively.

Here are other examples of addition of real numbers:

▼ **Example 2**

a. $-2 + (-3) + (-4) = -5 + (-4)$
$$= -9$$
b. $-3 + 5 + (-7) = 2 + (-7)$
$$= -5$$
c. $-6 + (-3 + 5) + 4 = -6 + 2 + 4$
$$= -4 + 4$$
$$= 0 \qquad\qquad ▲$$

Subtraction of Real Numbers

In order to have as few rules as possible, we will not attempt to list new rules for the difference of two real numbers. We will define subtraction in terms of addition and apply the rule for addition.

DEFINITION (SUBTRACTION) If a and b are any two real numbers, then the difference of a and b is

$$\underbrace{a - b} \quad = \quad \underbrace{a + (-b)}$$

To subtract b,   add the opposite of b.

We define the process of subtracting b from a to be equivalent to adding the opposite of b to a. In short, we say, "subtraction is addition of the opposite."

Here is how it works:

▼ **Example 3**

a. $5 - 3 = 5 + (-3) = 2$ Subtracting 3 is equivalent to
adding -3.

b. $-7 - 6 = -7 + (-6) = -13$ Subtracting 6 is equivalent to
                                                          adding -6.

c. $9 - (-2) = 9 + 2 = 11$ Subtracting -2 is equivalent
                                                          to adding 2.

d. $-6 - (-5) = -6 + 5 = -1$ Subtracting -5 is equivalent
                                                          to adding 5.   ▲

The following example involves combinations of sums and differences. Generally, the differences are first changed to appropriate sums; the additions are then performed left to right.

▼   **Example 4**

a. $9 - 5 + 2 = 9 + (-5) + 2$
$$= 4 + 2$$
$$= 6$$

b. $6 - (-3) + 2 = 6 + 3 + 2$
$$= 9 + 2$$
$$= 11$$

c. $-4 - 2 - (-5) = -4 + (-2) + 5$
$$= -6 + 5$$
$$= -1$$

d. $10 - |-3| + |-2| = 10 - 3 + 2$
$$= 10 + (-3) + 2$$
$$= 9$$

e. $-8 - (-3) - |-7| = -8 - (-3) - 7$
$$= -8 + 3 + (-7)$$
$$= -12$$ ▲

Problem Set 1.5

Find each of the following sums:

1. $6 + (-2)$ 2. $11 + (-5)$
3. $-6 + 2$ 4. $-11 + 5$
5. $-6 + (-2)$ 6. $-11 + (-5)$
7. $-3 + 5 + (-7)$ 8. $-1 + 4 + (-6)$
9. $3 + (-5) - (-7)$ 10. $1 + (-4) + (-7)$
11. $-9 + (-3) + (-1)$ 12. $-10 + (-5) + (-2)$
13. $-3 + (-8 + 1) + (-4)$ 14. $-2 + (-9 + 2) + (-5)$

Find each of the following differences:

15. $7 - 3$ 16. $6 - 9$
17. $-7 - 3$ 18. $-6 - 9$

19. $-7 - (-3)$ **20.** $-6 - (-9)$
21. $7 - (-3)$ **22.** $6 - (-9)$
23. $-15 - 20$ **24.** $-11 - 15$
25. $12 - (-4)$ **26.** $5 - (-2)$
27. $-8 - (-11)$ **28.** $-4 - (-12)$

Perform the indicated operations:

29. $3 + (-2) - 6$ **30.** $8 + (-3) - 5$
31. $-4 - 3 + 8$ **32.** $-9 - 5 + 7$
33. $6 - (-2) + 11$ **34.** $8 - (-3) + 12$
35. $-8 - (-3) - 9$ **36.** $-1 - (-2) - 3$
37. $4 - (-5) - (-1)$ **38.** $7 - (-2) - (-6)$
39. $|-2| + |-3| - |-6|$ **40.** $|-5| + |-2| - |-7|$
41. $|-2| + |-3| - (-6)$ **42.** $|-5| + |-2| - (-7)$
43. $|-4| - (-2) - |-10|$ **44.** $|-2| - (-3) - |-12|$
45. $9 - (-2) - 5 - 6 + (-3)$ **46.** $11 - (-6) - 2 - 4 + (-3)$
47. $|-2| - |-3| - (-8) + |-(-1)|$
48. $|-7| - |-2| - (-4) + |-(-6)|$
49. $-14 - (-20) + |-26| - |-16|$
50. $-11 - (-14) + |-17| - |-20|$

Multiplication with whole numbers is simply a shorthand way of writing repeated addition. That is, the product $3(2)$ can be interpreted as the sum of three 2s:

$$3(2) = 2 + 2 + 2$$

Although this definition of multiplication does not hold for other kinds of numbers, such as fractions, it does, however, give us ways of interpreting products of positive and negative numbers. For example, $3(-2)$ can be evaluated as follows:

$$3(-2) = -2 + (-2) + (-2)$$
$$= -6$$

From this result it seems reasonable to say that the product of a positive and a negative number is a negative number. We can actually state this fact as a theorem and prove that it is true for all real numbers.

THEOREM 1.4 If a and b are two positive numbers, then

and   1. $a(-b) = -(ab)$
        2. $-a(b) = -(ab)$

**1.6
Multiplication,
Division, and Order
of Operation for Real
Numbers**
*Multiplication of Real
Numbers*

In words: The product of two numbers with unlike signs is negative.

Proof

In order to prove this theorem we need only recognize from our past experience that the product of two positive numbers is also a positive number. This is actually the closure property for positive numbers. Applying some of the other axioms for real numbers, we have:

$a(0) = 0$ Multiplication by 0 always produces 0.

$a[b + (-b)] = 0$ Opposites add to 0.

$ab + a(-b) = 0$ Multiplication distributes over addition.

The last line tells us ab and $a(-b)$ are opposites, since they add to zero. More specifically, $a(-b)$ is the opposite of ab or

$$a(-b) = -(ab)$$

We can prove the second part of Theorem 1.4 by a similar process. We omit the proof here. [If you would like to try it, simply start with the statement $(0)a = 0$ and follow the above procedure.]

The next theorem gives rules for finding the product of two negative numbers.

THEOREM 1.5 If a and b are two positive numbers, then

$$(-a)(-b) = ab$$

In words: The product of two negative numbers is positive.

Proof

$-a(0) = 0$ Multiplication by 0 always produces 0.

$-a[b + (-b)] = 0$ Opposites add to 0.

$-a(b) + (-a)(-b) = 0$ Multiplication distributes over addition.

We know from Theorem 1.4 that $-a(b)$ is $-(ab)$. We can therefore rewrite the last line as

$$-(ab) + (-a)(-b) = 0$$

The last line tells us that $(-a)(-b)$ is the opposite of $-(ab)$ since they add to 0, or

$$(-a)(-b) = ab$$

The results of Theorems 1.4 and 1.5 along with our previous knowledge of multiplication of positive numbers can be summarized with the following rule.

RULE   To multiply two real numbers, simply multiply their absolute values. The product is

    a. *positive* if both numbers have the same sign; that is, both are + or both are −; or

    b. *negative* if the two numbers have opposite signs; that is, one + and the other −.

The following example illustrates this rule for finding products of positive and negative numbers.

▼ **Example 1**

a. $(7)(3) = 21$
b. $(7)(-3) = -21$
c. $(-7)(3) = -21$
d. $(-7)(-3) = 21$
e. $(4)(-3)(-2) = -12(-2)$
$$= 24$$
f. $(5)(-3 \cdot 2) = 5(-6)$
$$= -30$$

▲

Division of Real Numbers

In order to have as few rules as possible in building our system of algebra, we will now define division for two real numbers in terms of multiplication.

DEFINITION   If a and b are any two real numbers, where $b \neq 0$, then

$$\frac{a}{b} = a \cdot \left(\frac{1}{b}\right)$$

Dividing a by b is equivalent to multiplying a by the reciprocal of b. In short, we say, "division is multiplication by the reciprocal."

Note   The reason for the restriction $b \neq 0$ in the above definition is that division by 0 is not defined. Dividing a number by 0 would have to be equivalent to multiplying the number by the reciprocal of 0. In Section 1.2 we found 0 to be the only real number without a reciprocal. Hence, division by 0 is not defined.

Since division is defined in terms of multiplication, the same rules hold for assigning the correct sign to a quotient as held for assigning the correct sign to a product. That is, the quotient of two numbers with like signs is positive, while the quotient of two numbers with unlike signs is negative.

▼   **Example 2**

a. $\dfrac{6}{3} = 6 \cdot \left(\dfrac{1}{3}\right) = 2$

b. $\dfrac{6}{-3} = 6 \cdot \left(-\dfrac{1}{3}\right) = -2$

c. $\dfrac{-6}{3} = -6 \cdot \left(\dfrac{1}{3}\right) = -2$

d. $\dfrac{-6}{-3} = -6 \cdot \left(\dfrac{1}{-3}\right) = 2$ ▲

The second step in the above examples is written only to show that each quotient can be written as a product. It is not actually necessary to show this step when working problems.

Order of Operation

It is important when evaluating arithmetic expressions in mathematics that each expression have only one answer in reduced form. Consider the expression

$$3 \cdot 7 + 2$$

If we find the product of 3 and 7 first, then add 2, the answer is 23. On the other hand, if we first combine the 7 and 2, then multiply by 3, we have 27. The problem seems to have two distinct answers depending on whether we multiply first or add first. To avoid this situation we will decide that multiplication in a situation like this will always be done before addition. In this case, only the first answer, 23, is correct.

Here is the complete set of rules for evaluating expressions. It is intended to avoid the type of confusion found in the illustration above:

RULE (ORDER OF OPERATION)   When evaluating a mathematical expression, we will perform the operations in the following order.

1.   Perform operations inside the innermost parentheses first if possible.

2. Then do all multiplications and divisions left to right.
3. Perform all additions and subtractions left to right.

Here are some more complicated examples using combinations of the four basic operations:

▼ **Example 3** Simplify as much as possible:

a. $\dfrac{5(-3) - 10}{-4 - 1} = \dfrac{-15 - 10}{-4 - 1}$

$= \dfrac{-25}{-5}$

$= 5$

Notice that the division rule (the line used to separate the numerator from the denominator) is treated like parentheses. It serves to group the numbers on top separately from the numbers on the bottom.

b. $3 - 5(4 - 7) - (-3) = 3 - 5(-3) + 3$
$= 3 + 15 + 3$
$= 21$

c. $\dfrac{-5(-4) + 2(-3)}{2(-1) - 5} = \dfrac{20 - 6}{-2 - 5}$

$= \dfrac{14}{-7}$

$= -2$

d. $\dfrac{3(-2 + 7) - (-5)}{4 - 2(-3)} = \dfrac{3(5) + 5}{4 + 6}$

$= \dfrac{15 + 5}{10}$

$= \dfrac{20}{10}$

$= 2$ ▲

Problem Set 1.6

Find the following products:

1. $3(-5)$ 2. $-3(5)$
3. $-3(-5)$ 4. $4(-6)$
5. $-8(3)$ 6. $-7(-6)$
7. $-5(-4)$ 8. $-4(0)$

9. $-2(-1)(-6)$
10. $-3(-2)(5)$
11. $2(-3)(4)$
12. $-2(3)(-4)$
13. $-1(-2)(-3)(4)$
14. $-3(-2)(1)(4)$
15. $-2(4)(-3)(1)$
16. $-5(6)(-3)(-2)$

Use the definition of division to write each division problem as a multiplication problem, then simplify:

17. $\dfrac{8}{-4}$

18. $\dfrac{-8}{4}$

19. $\dfrac{-8}{-4}$

20. $\dfrac{-7}{-3}$

21. $\dfrac{-7}{3}$

22. $\dfrac{7}{-3}$

23. $\dfrac{5}{20}$

24. $\dfrac{6}{30}$

25. $\dfrac{-9}{18}$

26. $\dfrac{4}{-12}$

27. $\dfrac{12}{-4}$

28. $\dfrac{-12}{-3}$

29. $\dfrac{28}{-7}$

30. $\dfrac{-28}{-7}$

Simplify each expression as much as possible:

31. $3(-4) - 2$
32. $-3(-4) - 2$
33. $5(-2) - (-3)$
34. $-8(-11) - (-1)$
35. $4(-3) - 6(-5)$
36. $-6(-3) - 5(-7)$
37. $-8(4) - (-6)(-2)$
38. $9(-1) - 4(-3)$
39. $2 - 4[3 - 5(-1)]$
40. $6 - 5[2 - 4(-8)]$
41. $(8 - 7)[4 - 7(-2)]$
42. $(6 - 9)[15 - 3(-4)]$

43. $\dfrac{6(-2) - 8}{-15 - (-10)}$

44. $\dfrac{8(-3) - 6}{-7 - (-2)}$

45. $\dfrac{3(-1) - 4(-2)}{8 - 5}$

46. $\dfrac{6(-4) - 5(-2)}{7 - 6}$

47. $8 - (-6)\left[\dfrac{2(-3) - 5(4)}{-8(6) - 4}\right]$

48. $-9 - 5\left[\dfrac{11(-1) - 9}{4(-3) - 2(5)}\right]$

49. $6 - (-3)\left[\dfrac{2 - 4(3 - 8)}{6 - 5(1 - 3)}\right]$

50. $8 - (-7)\left[\dfrac{6 - 1(6 - 10)}{4 - 3(5 - 7)}\right]$

Chapter 1
Summary and Review
Symbols

$a = b$	a is equal to b
$a \neq b$	a is not equal to b
$a < b$	a is less than b

$a \leq b$ a is less than or equal to b
$a \geq b$ a is greater than or equal to b
$a > b$ a is greater than b
$a \ngtr b$ a is not greater than b
$a \nless b$ a is not less than b
$a + b$ the sum of a and b
$a - b$ the difference of a and b
$a \cdot b$ the product of a and b
a/b the quotient of a and b

Sets

A *set* is any well-defined collection of objects or things.

The *union* of two sets A and B, written $A \cup B$, is all the elements that are in A *or* are in B, *or* are in both A and B.

The *intersection* of two sets A and B, written $A \cap B$, is the set consisting of all elements common to both A *and* B.

Set A is a *subset* of set B, written $A \subset B$, if all elements in set A are also in set B.

Special Sets

Counting numbers $= \{1, 2, 3, \ldots\}$
Whole numbers $= \{0, 1, 2, 3, \ldots\}$
Integers $= \{\ldots -3, -2, -1, 0, 1, 2, 3, \ldots\}$
Rational numbers $= \{a/b \,|\, a$ and b are integers, $b \neq 0\}$
Irrational numbers $= \{x \,|\, x$ is a nonrepeating,
    nonterminating decimal$\}$
Real numbers $= \{x \,|\, x$ is rational or x is irrational$\}$

Opposites

Any two real numbers the same distance from 0 on the number line, but in opposite directions from 0, are called *opposites* or *additive inverses*. Opposites always add to 0.

Reciprocals

Any two real numbers whose product is 1 are called *reciprocals*. Every real number has a reciprocal except 0.

Absolute Value

The *absolute value* of a real number is its distance from 0 on the number line. If $|x|$ represents the absolute value of x, then

$$|x| = \begin{cases} x & \text{if } x \geq 0 \\ -x & \text{if } x < 0 \end{cases}$$

The absolute value of a real number is never negative.

Axioms of Equality

Reflexive $a = a$
Symmetric   If $a = b$, then $b = a$.
Transitive   If $a = b$ and $b = c$, then $a = c$.
Substitution   If $a = b$, then a and b can be used in
                     place of each other whenever convenient.

Axioms of Inequality

Trichotomy   Exactly one of the following is true:
                     $a < b$, $a = b$, $a > b$
Transitive   If $a < b$ and $b < c$, then $a < c$.

Properties of Real Numbers

	For Addition	*For Multiplication*
Commutative	$a + b = b + a$	$a \cdot b = b \cdot a$
Associative	$a + (b + c) = (a + b) + c$	$a \cdot (b \cdot c) = (a \cdot b) \cdot c$
Identity	$a + 0 = a$	$a \cdot 1 = a$
Inverse	$a + (-a) = 0$	$a\left(\dfrac{1}{a}\right) = 1$

Distributive $a(b + c) = ab + ac$

Addition

To add two real numbers with

1. *the same sign:* simply add absolute values and use the common sign.
2. *different signs:* subtract the smaller absolute value from the larger absolute value. The answer has the same sign as the number with the larger absolute value.

Subtraction

If a and b are real numbers,

$$a - b = a + (-b)$$

To subtract b, add the opposite of b.

Multiplication

To multiply two real numbers simply multiply their absolute values. Like signs give a positive answer. Unlike signs give a negative answer.

Division

If a and b are real numbers and $b \neq 0$, then

$$\frac{a}{b} = a \cdot \left(\frac{1}{b}\right)$$

To divide by b, multiply by the reciprocal of b.

1. Do what is inside the parentheses first.   *Order of Operation*
2. Then perform all multiplications and divisions left to right.
3. Finally, do all additions and subtractions left to right.

---

1. Interpreting absolute value as changing the sign of the number inside the   **COMMON MISTAKES**
absolute value symbols. That is, $|-5| = +5, |+5| = -5$. To avoid this mistake,
remember, absolute value is defined as a distance and distance is always
measured in positive units.
2. Confusing $-(-5)$ with $-|-5|$. The first answer is $+5$, while the second
answer is -5.

Write each of the following in symbols: **Chapter 1**
1. Twice the sum of $3x$ and $4y$.   **Test**
2. The difference of $2a$ and $3b$ is less than their sum.

If $A = \{1, 2, 3, 4\}$, $B = \{2, 4, 6\}$, and $C = \{1, 3, 5\}$, find:
3. $(A \cup B) \cap C$ 4. $A \cap (B \cup C)$
5. $B \cap C$

Give the opposite and reciprocal of each of the following:
6. -3 7. $\frac{4}{3}$
8. $-\sqrt{5}$

Simplify each of the following:
9. $-(-3)$ 10. $|-4| + |-3|$
11. $-|-2|$

For the set $\{-5, -4.1, -3.75, -\frac{5}{6}, -\sqrt{2}, 0, \sqrt{3}, 1, 1.8, 4\}$, list all the elements
belonging to the following sets:
12. Integers   13. Rational numbers
14. Irrational numbers

Graph each of the following:
15. $\{x \mid x > 2\}$ 16. $\{x \mid x \leq -1 \text{ or } x > 5\}$
17. $\{x \mid -2 \leq x \leq 4\}$

State the property or properties that justify each of the following:
18. $4 + x = x + 4$ 19. $5(1) = 5$
20. $3(x \cdot y) = (3y) \cdot x$ 21. $(a + 1) + b = (a + b) + 1$

Simplify each of the following as much as possible:
22. $5(-4) + 1$ 23. $-4(-3) + 2$

24. $-3(5) - 4$

25. $-2(-3) - 4(-2)$

26. $5(-6) - (-4)$

27. $\dfrac{6(-3) - 2}{-6 - 2}$

28. $\dfrac{-4(-1) - (-10)}{5 - (-2)}$

29. $-4\left[\dfrac{-3 - (-6)}{2(-4) - 4}\right]$

30. $3 - 2\left[\dfrac{8(-1) - 5}{-3(2) - 4}\right]$

31. $(6 - 5)[4 - 3(2 - 1)]$

First-Degree Equations and Inequalities

To the student:

One of the best-known mathematical formulas is the formula $E = mc^2$ from Einstein's theory of relativity. Einstein viewed the universe as if everything in it were in one of two states, matter or energy. His theory states that matter and energy are constantly being transformed into one another. The amount of energy (E) that can be obtained from an object with mass m is given by the formula $E = mc^2$, where c is the speed of light. Now as far as we are concerned, the theory behind the formula is not important. What is important is that the formula $E = mc^2$ describes a certain characteristic of the universe. The universe has always had this characteristic. The formula simply gives us a way of stating it in symbols. Once this property of matter and energy has been stated in symbols we can apply any of our mathematical knowledge to it, if we need to. The idea behind all of this is that mathematics can be used to describe the world around us symbolically. Mathematics is the language of science.

In this chapter we will begin our work with equations. The information in this chapter is some of the most important information in the book. You will learn the basic steps used in solving equations. Probably the most useful tool in algebra is the ability to solve first-degree equations in one variable. The methods we develop to solve first-degree equations in one variable will be used again and again throughout the rest of the book.

In this chapter we will also consider first-degree inequalities in one variable as well as provide a section on formulas and a section on word problems.

A large part of your success in this chapter depends on how well you mastered the concepts from Chapter 1. Here is a list of the more important concepts needed to begin this chapter:

1.   You must know how to add, subtract, multiply, and divide positive and negative numbers.
2.   You should be familiar with the commutative, associative, and distributive properties.
3.   You should understand that opposites add to 0 and reciprocals multiply to 1.
4.   You must know the definition of absolute value.

**2.1
First-Degree
Equations**

In this section we will solve some first-degree equations. A first-degree equation is any equation that can be put in the form

$$ax + b = c$$

where a, b, and c are constants.

Some examples of first-degree equations are

$$5x + 3 = 2 \qquad 2x = 7 \qquad 2x + 5 = 0$$

Each is a first-degree equation because it can be put in the form $ax + b = c$. In the first equation above, $5x$, 3, and 2 are called *terms* of the equation. $5x$ is a variable term; 3 and 2 are constant terms.

DEFINITION   The *solution set* for an equation is the set of all numbers which, when used in place of the variable, make the equation a true statement.

In short, the solution set for an equation is all the numbers that make the equation true.

▼   **Example 1**   The solution set for $2x - 3 = 9$ is 6, since replacing x with 6 makes the equation a true statement.

$$
\begin{aligned}
\text{If} \qquad\qquad x &= 6 \\
\text{then} \qquad 2x - 3 &= 9 \\
\text{becomes} \qquad 2(6) - 3 &= 9 \\
12 - 3 &= 9 \\
9 &= 9 \leftarrow \text{a true statement}
\end{aligned}
$$

▲

The important thing about any equation is its solution set.

DEFINITION Two or more equations with the same solution set are called *equivalent equations*.

▼ **Example 2** The equations $2x - 5 = 9$, $x - 1 = 6$, and $x = 7$ are all equivalent equations since the solution set for each is $\{7\}$. ▲

Two equations may look completely different from one another. If they have the same solution set, they are equivalent.

The first step in solving a first-degree equation is to simplify both sides as much as possible. To do this, we use the axioms and properties developed in Chapter 1. The most useful properties here are the commutative, associative, and distributive properties.

In addition to the axioms from Chapter 1, we need two new properties—one for addition or subtraction and one for multiplication or division.

The first property states that adding the same quantity to both sides of an equation preserves equality. Or, more importantly, adding the same amount to both sides of an equation *never changes* the solution set. This property is called the *addition property of equality* and is stated in symbols as follows.

Addition Property of Equality

Let A, B, and C represent algebraic expressions.

$$\text{If} \qquad A = B$$
$$\text{then} \qquad A + C = B + C$$

The multiplication property of equality states that if we multiply both sides of an equation by the same nonzero quantity, the resulting equation will always be equivalent to the original equation. In symbols the *multiplication property of equality* is as follows.

Multiplication Property of Equality

Let A and B be any two algebraic expressions and C be any *nonzero* algebraic expression.

$$\text{If} \qquad A = B$$
$$\text{then} \qquad AC = BC$$

Since subtraction is defined in terms of addition (subtraction is addition of the opposite) and division is defined in terms of multiplication (division by a number gives the same result as multiplication by its reciprocal), we do not need to introduce separate properties for subtraction and division. The solution set for an equation will never be changed by subtracting the same amount from both sides or by dividing both sides by the same nonzero quantity.

The following examples illustrate how we use the axioms from Chapter 1 along with the addition property of equality and the multiplication property of equality to solve first-degree equations.

▼   **Example 3** Solve for x: $2x - 3 = 9$.

Solution We begin by using the addition property of equality to add $+3$, the opposite of -3, to both sides of the equation:

$$2x - 3 + 3 = 9 + 3$$
$$2x = 12$$

To get x alone on the left side, we use the multiplication property of equality and multiply both sides by $\frac{1}{2}$, the reciprocal of 2.

$$\tfrac{1}{2}(2x) = \tfrac{1}{2}(12)$$
$$x = 6$$

Since the addition and multiplication properties of equality always produce equations equivalent to the original equations, our last equation, $x = 6$, is equivalent to our first equation, $2x - 3 = 9$. The solution set is, therefore, $\{6\}$.   ▲

▼   **Example 4** Find the solution set for $3a - 5 + a - 3 = 18 - 5$.

Solution We begin by simplifying the left and right sides separately. [*Note* $3a + a = (3 + 1)a = 4a$]

$$3a - 5 + a - 3 = 18 - 5$$
$$4a - 8 = 13 \qquad \text{Simplify each side.}$$

We now proceed as in Example 1 by first adding $+8$ to both sides and then multiplying both sides by $\frac{1}{4}$:

$$4a - 8 + 8 = 13 + 8 \qquad \text{Add } +8 \text{ to both sides.}$$
$$4a = 21$$
$$\tfrac{1}{4}(4a) = \tfrac{1}{4}(21) \qquad \text{Multiply both sides by } \tfrac{1}{4}.$$
$$a = \tfrac{21}{4}$$

The solution set is $\{\tfrac{21}{4}\}$.   ▲

As the equations become more complicated, it is sometimes helpful to use the following steps as a guide in solving the equations:

Step 1.   Use the distributive property to separate terms.

Step 2.   Use the commutative and associative properties to simplify both sides as much as possible.

Step 3.   Use the addition property of equality to get all the terms containing the variable (variable terms) on one side and all other terms (constant terms) on the other side.

Step 4.   Use the multiplication property of equality to get x alone on one side of the equal sign.

Step 5.   Check your results in the original equation, if necessary.

Let's see how our steps apply by solving one last equation.

▼   **Example 5**   Solve $3(2y - 1) + y = 5y + 3$.

Solution   We begin by using the distributive property to separate terms.

$$Step\ 1 \begin{cases} 3(2y - 1) + y = 5y + 3 \\ \quad\downarrow\qquad\downarrow \\ 6y - 3\ + y = 5y + 3 \end{cases}$$ Distributive property.

Step 2 $7y - 3 = 5y + 3$ $6y + y = 7y$

$$Step\ 3 \begin{cases} 7y + (-5y) - 3 = 5y + (-5y) + 3 \\ \qquad 2y - 3 = 3 \\ \qquad 2y - 3 + 3 = 3 + 3 \\ \qquad 2y = 6 \end{cases}$$ Add $-5y$ to both sides.

Add $+3$ to both sides.

$$Step\ 4 \begin{cases} \tfrac{1}{2}(2y) = \tfrac{1}{2}(6) \\ \quad y = 3 \end{cases}$$ Multiply by $\tfrac{1}{2}$.

The solution set is $\{3\}$.

As a final note we should mention that after step 2 has been completed, there are, at most, four terms left—two variable terms and two constant terms. After step 3, there are two terms left—one variable and one constant term.   ▲

Solve each of the following equations: *Problem Set 2.1*

1. $x - 5 = 3$		**2.** $x + 2 = 7$
3. $2x - 4 = 6$		**4.** $3x - 5 = 4$
5. $4a - 1 = 7$		**6.** $3a - 5 = 10$
7. $3 - y = 10$		**8.** $5 - 2y = 11$
9. $-3 - 4x = 15$		**10.** $-8 - 5x = 23$
11. $2x + 5 = -3$		**12.** $9x + 6 = -12$
13. $-3y + 1 = 5$		**14.** $-2y + 8 = 3$

15. $-5x - 4 = 16$

16. $-6x - 5 = 11$

17. $3 - 4a = -11$

18. $8 - 2a = -13$

19. $9 + 5a = -2$

20. $3 + 7a = -7$

Simplify each side of the following equations, then find the solution set:

21. $2x - 5 = 3x + 2$

22. $5x - 1 = 4x + 3$

23. $-3a + 2 = -2a - 1$

24. $-4a - 8 = -3a + 7$

25. $2x - 3 - x = 3x + 5$

26. $3x - 5 - 2x = 2x - 3$

27. $5y - 2 + 4y = 2y + 12$

28. $7y - 3 + 2y = 7y - 9$

29. $11x - 5 + 4x - 2 = 5x$

30. $2x + 7 - 3x + 4 = -2x$

31. $3(x - 2) = 12$

32. $4(x + 1) = 12$

33. $2(3k - 5) = k$

34. $3(4k - 1) = 9k$

35. $-5(2x + 1) + 5 = 3x$

36. $-3(5x + 7) - 4 = -10x$

37. $5(y + 2) - 4(y + 1) = 3$

38. $6(y - 3) - 5(y + 2) = 8$

39. $6 - 7(m - 3) = 7$

40. $3 - 5(2m - 5) = -2$

41. $4(a - 3) + 5 = 2(3a - 1)$

42. $6(a + 4) - 7 = 2(5a + 2)$

43. $7 + 3(x + 2) = 4(x - 1)$

44. $5 + 2(3x - 4) = 7(2x - 3)$

2.2 First-Degree Inequalities

In Section 2.1 we used the addition and multiplication properties of equality as the main tools in solving first-degree equations. Solving first-degree inequalities is similar to solving equations. We need to develop only two new properties to use in solving inequalities—one property for addition and another for multiplication. As an extra step, we will also graph each of the solution sets for our inequalities.

A first-degree inequality is any inequality that can be put in the following form:

$$ax + b < c \qquad (a, b, \text{ and } c \text{ constants}, a \neq 0)$$

where the inequality symbol ($<$) can be replaced with any of the other three inequality symbols ($\leq$, $>$, or $\geq$).

Some examples of first-degree inequalities are

$$3x - 2 \geq 7 \qquad -5y < 25 \qquad 3(x - 4) > 2x$$

Each of these is a first-degree inequality because it can be put in the form $ax + b < c$. They may not be in that form to begin with, but each can be put in the correct form.

Addition Property for Inequalities

Let A, B, and C represent algebraic expressions.

$$\text{If} \qquad A < B$$
$$\text{then} \qquad A + C < B + C$$

The addition property for inequalities is very similar to the addition property for equalities. It states that adding the same amount to both sides of an inequality never changes the solution set.

Before we state the multiplication property for inequalities, we will take a look at what happens to an inequality statement when we multiply both sides by a positive number and what happens when we multiply by a negative number.

We begin by writing three true inequality statements:

$$3 < 5 \qquad -3 < 5 \qquad -5 < -3$$

We multiply both sides of each inequality by a positive number—say, 4.

$$\begin{array}{ccc} 4(3) < 4(5) & 4(-3) < 4(5) & 4(-5) < 4(-3) \\ 12 < 20 & -12 < 20 & -20 < -12 \end{array}$$

Notice in each case that the resulting inequality symbol points in the same direction as the original inequality symbol. Multiplying both sides of an inequality by a positive number preserves the *sense* of the inequality.

Let's take the same three original inequalities and multiply both sides by -4.

$$\begin{array}{ccc} 3 < 5 & -3 < 5 & -5 < -3 \\ \downarrow & \downarrow & \downarrow \\ -4(3) > -4(5) & -4(-3) > -4(5) & -4(-5) > -4(-3) \\ -12 > -20 & 12 > -20 & 20 > 12 \end{array}$$

Notice in this case that the resulting inequality symbol always points in the opposite direction from the original one. Multiplying both sides of an inequality by a negative number *reverses* the sense of the inequality. Keeping this in mind, we will now state the multiplication property for inequalities.

Multiplication Property for Inequalities

Let A, B, and C represent algebraic expressions.

$$\begin{array}{lll} \text{If} & A < B & \\ \text{then} & AC < BC & \text{if } C \text{ is positive } (C > 0) \\ \text{or} & AC > BC & \text{if } C \text{ is negative } (C < 0) \end{array}$$

The multiplication property for inequalities states that we can multiply both sides of an inequality by any nonzero number we choose. If that number happens to be *negative*, we must also *reverse* the direction of the inequality.

We have stated both properties using the "less than" ($<$) symbol. The properties also hold for the other three inequality symbols.

Since subtraction is defined as addition of the opposite and division as multiplication by the reciprocal, our two new properties hold for both subtraction and division. (Note too that multiplication or division of both sides of an inequality by a negative number *always* reverses the direction of the inequality.)

We will follow the same basic steps in solving inequalities as we did with equations. With inequalities, we will graph the solution set as an extra step.

▼   **Example 1**   Solve $3x - 5 \leq 7$.

Solution

$$3x - 5 \leq 7$$
$$3x - 5 + 5 \leq 7 + 5 \qquad \text{Add } +5 \text{ to both sides.}$$
$$3x \leq 12$$
$$\tfrac{1}{3}(3x) \leq \tfrac{1}{3}(12) \qquad \text{Multiply by } \tfrac{1}{3}.$$
$$x \leq 4$$

The solution set is $\{x \mid x \leq 4\}$, the graph of which is   ▲

▼   **Example 2**   Find the solution set for $-2y - 3 < 7$.

Solution

$$-2y - 3 < 7$$
$$-2y < 10 \qquad \text{Add } +3 \text{ to both sides.}$$
$$\downarrow$$
$$-\tfrac{1}{2}(-2y) > -\tfrac{1}{2}(10) \qquad \text{Multiply by } -\tfrac{1}{2} \text{ and reverse the}$$
$$y > -5 \qquad\qquad\quad \text{direction of the inequality symbol.}$$

The solution set is $\{y \mid y > -5\}$, the graph of which is   ▲

▼   **Example 3**   Solve $3(2x - 4) - 7x \leq -3x$.

Solution We begin by using the distributive property to separate terms. Next, simplify both sides:

$$3(2x - 4) - 7x \le -3x$$
$$6x - 12 - 7x \le -3x$$
$$-x - 12 \le -3x \qquad\qquad 6x - 7x = (6 - 7)x = -x$$
$$-12 \le -2x \qquad\qquad \text{Add } x \text{ to both sides.}$$
$$\downarrow$$
$$-\tfrac{1}{2}(-12) \ge -\tfrac{1}{2}(-2x) \qquad \text{Multiply both sides by } -\tfrac{1}{2} \text{ and}$$
$$6 \ge x \qquad\qquad\qquad \text{reverse the direction of the}$$
$$\text{inequality symbol.}$$

The solution set is $\{x \mid x \le 6\}$, and the graph is   ▲

Notice in Examples 2 and 3 that each time we multiplied both sides of the inequality by a negative number, we also reversed the direction of the inequality. If we fail to reverse the inequality in this situation, the graph of our solution set will be on the wrong side of the end point.

▼ **Example 4** Solve the continued inequality $-2 \le 5k - 7 \le 13$.

Solution We can extend our properties for addition and multiplication to cover this situation. If we add a number to the middle expression, we must add the same number to the outside expressions. If we multiply the center expression by a number, we must do the same to the outside expressions, remembering to reverse the direction of the inequality symbols if we multiply by a negative number.

$$-2 \le \quad 5k - 7 \quad \le 13$$
$$-2 + 7 \le 5k - 7 + 7 \le 13 + 7 \qquad \text{Add 7 to each expression.}$$
$$5 \le \qquad 5k \qquad \le 20$$
$$\tfrac{1}{5}(5) \le \quad \tfrac{1}{5}(5k) \quad \le \tfrac{1}{5}(20) \qquad \text{Multiply each expression by } \tfrac{1}{5}.$$
$$1 \le \qquad k \qquad \le 4$$

The graph of the solution set is   ▲

▼ **Example 5**   Solve the compound inequality $3x + 1 \leq -5$ or $2x - 3 > 7$.

Solution   We solve each half of the compound inequality separately, then graph the solution set:

$$3x + 1 \leq -5 \quad \text{or} \quad 2x - 3 > 7$$
$$3x \leq -6 \quad \text{or} \quad 2x > 10$$
$$x \leq -2 \quad \text{or} \quad x > 5$$

We graph the solution set using the methods developed in Chapter 1 for graphing compound inequalities.   ▲

Problem Set 2.2

Solve each of the following inequalities and graph each solution set:

1. $2x \leq 3$
2. $5x \geq -15$
3. $-5x \leq 25$
4. $-7x \geq 35$
5. $-6a < -12$
6. $-3a \leq -18$
7. $2x + 1 \geq 3$
8. $5x - 4 < 16$
9. $-3x + 1 > 10$
10. $-2x - 5 \leq 15$
11. $6 - m \leq 7$
12. $5 - m > -2$
13. $-3 - 4x \leq 9$
14. $-2 - 5x < 18$
15. $3 - 2y \geq -4$
16. $5 - 3y < -2$

Simplify each side first, then solve the following inequalities:

17. $2(3y + 1) \leq -10$
18. $3(2y - 4) > 0$
19. $2(x - 5) \leq -3x$
20. $5(2x - 1) < 9x$
21. $-(a + 1) - 4a \leq 2a - 8$
22. $-(a - 2) - 5a \leq 3a + 7$
23. $4x - 5 + 3x - 2 \leq 2x + 3$
24. $3x + 7 - 2x + 1 \leq -2x - 4$
25. $2t - 3(5 - t) < 0$
26. $3t - 4(2t - 5) < 0$
27. $5 - 7(2a + 3) \geq -2$
28. $3 - 4(3a - 1) > 1$
29. $-3(x + 5) \leq -2(x - 1)$
30. $-4(2x + 1) \leq -3(x + 2)$
31. $5(y + 3) + 4 < 6y - 1 - 5y$
32. $4(y - 1) + 2 \geq 3y + 8 - 2y$
33. $3(x + 1) - 2(x - 2) \leq 0$
34. $5(x + 4) - 3(x + 7) \leq 5$

Solve the following continued inequalities. Be sure to graph the solution sets.

35. $2 \leq m - 5 \leq 7$
36. $-3 < m + 1 \leq 5$
37. $-8 < -5x + 2 < 3$
38. $-4 \leq -3x - 1 \leq 5$
39. $-6 < 2a + 2 < 6$
40. $-6 < 5a - 4 < 6$
41. $5 \leq -3a - 7 \leq 11$
42. $1 \leq -4a + 1 \leq 3$

Graph the solution sets for the following compound inequalities:

43. $x + 5 \leq -2$ or $2x + 1 > 7$
44. $3x + 2 < -3$ or $x - 5 \geq 3$
45. $5y + 1 \leq -4$ or $2y - 4 \geq 6$
46. $7y - 5 \leq 2$ or $3y - 6 \geq 3$
47. $2x + 5 > 3x - 1$ or $x - 4 < 2x + 6$
48. $3x - 1 < 2x + 4$ or $5x - 2 > 3x + 4$
49. $3(a + 1) < 2(a - 5)$ or $2(a - 1) \geq a + 2$
50. $5(a + 3) < 2(a - 4)$ or $4(a - 6) \geq a + 3$

In Chapter 1 we defined the absolute value of x, $|x|$, to be the distance between x and 0 on the number line. The absolute value of a number measures its distance from 0.

In this section we will solve some first-degree equations involving absolute value. In order to solve these equations, we must rely heavily on the definition of absolute value.

**2.3
Equations with
Absolute Value**

▼ **Example 1** Solve for x: $|x| = 5$.

Solution Using the definition of absolute value, we can read the equation as, "The distance between x and 0 on the number line is 5." If x is 5 units from 0, then x can be 5 or -5.

$$\text{If } |x| = 5, \quad \text{then } x = 5 \quad \text{or} \quad x = -5$$

The solution set is $\{5, -5\}$. ▲

In general, then, we can see that any equation of the form $|a| = b$ is equivalent to the equations $a = b$ or $a = -b$.

▼ **Example 2** Solve $|2a - 1| = 7$.

Solution We can read this equation as "$2a - 1$ is 7 units from 0 on the number line." The quantity $2a - 1$ must be equal to 7 or -7.

$$|2a - 1| = 7$$
$$2a - 1 = 7 \quad \text{or} \quad 2a - 1 = -7$$

We have transformed our absolute value equation into two first-degree equations that do not involve absolute value. We can solve each equation using the method in Section 2.1:

$$2a - 1 = 7 \quad \text{or} \quad 2a - 1 = -7$$
$$2a = 8 \quad \text{or} \quad 2a = -6 \qquad \text{Add } +1 \text{ to both sides.}$$
$$a = 4 \quad \text{or} \quad a = -3 \qquad \text{Multiply by } \tfrac{1}{2}.$$

Our solution set is $\{4, -3\}$.   ▲

To check our solutions, we put them into the original absolute value equation:

When	$a = 4$	When	$a = -3$
the equation	$\lvert 2a - 1 \rvert = 7$	the equation	$\lvert 2a - 1 \rvert = 7$
becomes	$\lvert 2(4) - 1 \rvert = 7$	becomes	$\lvert 2(-3) - 1 \rvert = 7$
	$\lvert 7 \rvert = 7$		$\lvert -7 \rvert = 7$
	$7 = 7$		$7 = 7$

▼   **Example 3**   Solve $\lvert 3y - 5 \rvert = 2$.

Solution   Proceeding as in Example 1, we have

$$\lvert 3y - 5 \rvert = 2$$
$$3y - 5 = 2 \quad \text{or} \quad 3y - 5 = -2$$
$$3y = 7 \quad \text{or} \quad 3y = 3 \qquad \text{Add } +5 \text{ to both sides.}$$
$$y = \tfrac{7}{3} \quad \text{or} \quad y = 1 \qquad \text{Multiply by } \tfrac{1}{3}.$$

The solution set is $\{\tfrac{7}{3}, 1\}$.   ▲

▼   **Example 4**   Solve $\lvert 5x - 3 \rvert + 5 = 12$.

Solution   In order to use the definition of absolute value to solve this equation, we must isolate the absolute value on the left side of the equal sign. We can delete the $+5$ from the left side by adding -5 to both sides of the equation.

$$\lvert 5x - 3 \rvert + 5 + (-5) = 12 + (-5)$$
$$\lvert 5x - 3 \rvert = 7$$

Now that the equation is in the correct form, we can see that $5x - 3$ is 7 or -7. We continue as in Examples 2 and 3:

$$5x - 3 = 7 \quad \text{or} \quad 5x - 3 = -7$$
$$5x = 10 \quad \text{or} \quad 5x = -4 \qquad \text{Add } +3 \text{ to both sides.}$$
$$x = 2 \quad \text{or} \quad x = -\tfrac{4}{5} \qquad \text{Multiply by } \tfrac{1}{5}.$$

The solution set is $\{2, -\tfrac{4}{5}\}$.   ▲

Notice in each of the last three examples that one of the solutions comes out negative. This does not contradict our statement in Chapter 1 that absolute value is always a positive quantity, or zero. The quantity inside the absolute value symbols can be negative. When we take its absolute value, the *result* is always positive.

▼   **Example 5**   Solve $|3a - 6| = -4$.

Solution   The solution set is ∅ because the left side cannot be negative and the right side is negative. No matter what we try to substitute for the variable a, the quantity $|3a - 6|$ will always be positive, or zero. It can never be -4.   ▲

As a last case in absolute value equations, let's consider the situation where we have two quantities whose absolute values are equal.

Consider the statement $|x| = |y|$. What can we say about x and y? We know they are equal in absolute value. By the definition of absolute value, they are the same distance from 0 on the number line. They must be equal to each other or opposites of each other. In symbols we write

$$|x| = |y| \quad \Leftrightarrow \quad x = y \quad \text{or} \quad x = -y$$

$$\uparrow \qquad\qquad\qquad \uparrow \qquad\qquad \uparrow$$

Equal in   Equals   or   Opposites
absolute value

▼   **Example 6**   Solve $|3a + 2| = |2a + 3|$.

Solution   The quantities $(3a + 2)$ and $(2a + 3)$ have equal absolute values. They are, therefore, the same distance from 0 on the number line. They must be equals or opposites.

$$|3a + 2| = |2a + 3|$$

Equals	*Opposites*
$3a + 2 = 2a + 3$ or	$3a + 2 = -(2a + 3)$
$a + 2 = 3$	$3a + 2 = -2a - 3$
$a = 1$	$5a + 2 = -3$
	$5a = -5$
	$a = -1$

The solution set is $\{1, -1\}$.

It makes no difference in the outcome of the problem if we take the opposite of the first or second expression. It is very important, once we have decided which one to take the opposite of, that we take the opposite of both its terms and not just the first term. That is, the opposite of $2a + 3$ is $-(2a + 3)$, which we can think of as $-1(2a + 3)$. Distributing the -1 across *both* terms, we have

$$-1(2a + 3) = -1(2a) + (-1)(3)$$
$$= -2a - 3 \qquad \blacktriangle$$

Problem Set 2.3

Use the definition of absolute value to solve each of the following problems:

1. $|x| = 4$ 2. $|x| = 7$
3. $|a| = 2$ 4. $|a| = 5$
5. $|x| = -3$ 6. $|x| = -4$
7. $|a| + 2 = 3$ 8. $|a| - 5 = 2$
9. $|y| + 4 = 3$ 10. $|y| + 3 = 1$
11. $|x| - 2 = 4$ 12. $|x| - 5 = 3$
13. $|x - 2| = 5$ 14. $|x + 1| = 2$
15. $|a - 4| = 1$ 16. $|a + 2| = 7$
17. $|3 - x| = 1$ 18. $|4 - x| = 2$
19. $|2x + 1| = -3$ 20. $|2x - 5| = -5$
21. $|3a + 1| = 5$ 22. $|2x - 3| - 4 = 3$
23. $|3x + 4| + 1 = 7$ 24. $|5x - 3| - 4 = 3$
25. $|2y - 3| + 4 = 3$ 26. $|7y - 8| + 9 = 1$
27. $|1 + 2(x - 1)| = 7$ 28. $|3 + 4(x - 2)| = 5$
29. $|2(k + 4) - 3| = 1$ 30. $|3(k - 2) + 1| = 4$

Solve the following equations by the method used in Example 6:

31. $|3a + 1| = |2a - 4|$ 32. $|5a + 2| = |4a + 7|$
33. $|6x - 2| = |3x + 1|$ 34. $|x - 5| = |2x + 1|$
35. $|y - 2| = |y + 3|$ 36. $|y - 5| = |y - 4|$
37. $|3x - 1| = |3x + 1|$ 38. $|5x - 8| = |5x + 8|$
39. $|3 - m| = |m + 4|$ 40. $|5 - m| = |m + 8|$
41. $|3 - x| = |4 + 5x|$ 42. $|7 - x| = |8 - 2x|$

**2.4
Inequalities Involving
Absolute Value**

In this section we will again apply the definition of absolute value to solve inequalities involving absolute value. Again, the absolute value of x, which is $|x|$, represents the distance that x is from 0 on the number line. We will begin by considering three absolute value expressions and their English translations:

Expression	In Words
$|x| = 7$	x is exactly 7 units from 0 on the number line.
$|a| < 5$	a is less than 5 units from 0 on the number line.
$|y| \geq 4$	y is greater than or equal to 4 units from 0 on the number line.

Once we have translated the expression into words, we can use the translation to graph the original equation or inequality. The graph is then used to write a final equation or inequality that does not involve absolute value.

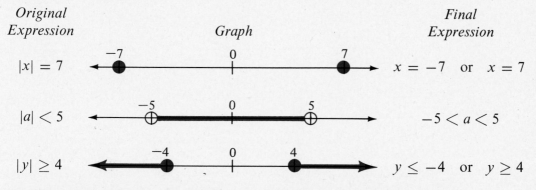

Original Expression	Graph	Final Expression
$|x| = 7$		$x = -7$ or $x = 7$
$|a| < 5$		$-5 < a < 5$
$|y| \geq 4$		$y \leq -4$ or $y \geq 4$

Although we will not always write out the English translation of an absolute value inequality, it is important that we understand the translation. Our second expression, $|a| < 5$, means a is within 5 units of 0 on the number line. The graph of this relationship is

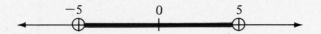

which can be written with the following continued inequality:

$$-5 < a < 5$$

We can follow this same kind of reasoning to solve more complicated absolute value inequalities.

▼ **Example 1** Graph the solution set $|2x - 5| < 3$.

Solution The absolute value of $2x - 5$ is the distance that $2x - 5$ is from 0 on the number line. We can translate the inequality as, "$2x - 5$ is less than 3 units from 0 on the number line." That is, $2x - 5$ must appear between -3 and 3 on the number line.

A picture of this relationship is

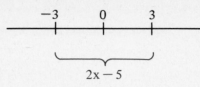

Using the picture, we can write an inequality without absolute value that describes the situation:

$$-3 < 2x - 5 < 3$$

Next, we solve the continued inequality by first adding $+5$ to all three members and then multiplying all three by $\frac{1}{2}$:

$$-3 < 2x - 5 < 3$$
$$2 < 2x < 8 \qquad \text{Add } +5 \text{ to all three members.}$$
$$1 < x < 4 \qquad \text{Multiply each member by } \tfrac{1}{2}.$$

The graph of the solution set is

We can see from the solution that in order for the absolute value of $2x - 5$ to be within 3 units of 0 on the number line, x must be between 1 and 4.   ▲

▼ **Example 2** Solve and graph $|3a + 7| \leq 4$.

Solution We can read the inequality as, "The distance between $3a + 7$ and 0 is less than or equal to 4." Or, "$3a + 7$ is within 4 units of 0 on the number line." This relationship can be written without absolute value as

$$-4 \leq 3a + 7 \leq 4$$

Solving as usual, we have

$$-4 \leq 3a + 7 \leq 4$$
$$-11 \leq 3a \leq -3 \qquad \text{Add } -7 \text{ to all three members.}$$
$$\frac{-11}{3} \leq a \leq -1 \qquad \text{Multiply each by } \tfrac{1}{3}.$$

▲

We can see from Examples 1 and 2 that in order to solve an inequality involving absolute value, we must be able to write an equivalent expression that does not involve absolute value.

▼   **Example 3**   Solve $|x - 3| > 5$.

Solution   We interpret the absolute value inequality to mean that $x - 3$ is more than 5 units from 0 on the number line. The quantity $x - 3$ must be either above $+5$ or below -5. Here is a picture of the relationship:

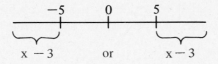

An inequality without absolute value that also describes this situation is

$$x - 3 < -5 \quad \text{or} \quad x - 3 > 5$$

Adding $+3$ to both sides of each inequality we have

$$x < -2 \quad \text{or} \quad x > 8$$

the graph of which is

From our results we see that in order for the absolute value of $x - 3$ to be more than 5 units from 0, x must be below -2 or above $+8$.   ▲

▼   **Example 4**   Graph the solution set $|4t - 3| \geq 9$.

Solution   The quantity $4t - 3$ is greater than or equal to 9 units from 0. It must be either above $+9$ or below -9.

$$
\begin{aligned}
4t - 3 &\leq -9 \quad \text{or} \quad 4t - 3 \geq 9 \\
4t &\leq -6 \quad \text{or} \qquad\quad 4t \geq 12 \qquad \text{Add } +3. \\
t &\leq -\tfrac{6}{4} \quad \text{or} \qquad\quad t \geq \tfrac{12}{4} \qquad \text{Multiply by } \tfrac{1}{4}. \\
t &\leq -\tfrac{3}{2} \quad \text{or} \qquad\quad t \geq 3
\end{aligned}
$$

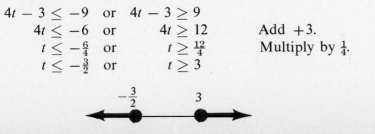

Since absolute value always results in a nonnegative quantity, we sometimes come across special solution sets when a negative number appears on the right side of an absolute value inequality.

▼ **Example 5** Solve $|7y - 1| < -2$.

Solution The *left* side is never negative because it is an absolute value. The *right* side is negative. We have a positive quantity less than a negative quantity, which is impossible. The solution set is the empty set, $\emptyset$. There is no real number to substitute for y to make the above inequality a true statement.   ▲

▼ **Example 6** Solve $|6x + 2| > -5$.

Solution This is the opposite case from that in Example 5. No matter what real number we use for x on the *left* side, the result will always be positive, or zero. The *right* side is negative. We have a positive quantity greater than a negative quantity. Every real number we choose for x gives us a true statement. The solution set is the set of all real numbers.   ▲

Problem Set 2.4

Solve each of the following inequalities using the definition of absolute value. Graph the solution set in each case.

1. $|x| < 3$ 2. $|x| \leq 7$
3. $|x| \geq 2$ 4. $|x| > 4$
5. $|a| < 1$ 6. $|a| \leq 4$
7. $|x| + 2 < 5$ 8. $|x| - 3 < -1$
9. $|t| - 3 > 4$ 10. $|t| + 5 > 8$
11. $|y| < -5$ 12. $|y| > -3$
13. $|x| \geq -2$ 14. $|x| \leq -4$
15. $|x - 3| < 7$ 16. $|x + 4| < 2$
17. $|a + 5| \geq 4$ 18. $|a - 6| \geq 3$
19. $|a - 1| < -3$ 20. $|a + 2| \geq -5$
21. $|2x - 3| \leq 5$ 22. $|2x + 7| \leq 3$
23. $|2x - 4| < 6$ 24. $|2x + 6| < 2$
25. $|3y + 9| \geq 6$ 26. $|5y - 1| \geq 4$
27. $|6x - 7| > 3$ 28. $|4x + 2| > 5$
29. $|2k + 3| \geq 7$ 30. $|2k - 5| \geq 3$
31. $|x - 3| + 2 < 6$ (*Hint:* Isolate $x - 3$ on the left side)
32. $|x + 4| - 3 < -1$ 33. $|2a + 1| + 4 \geq 7$

wa

34. $|2a - 6| - 1 \geq 2$ **35.** $|3x + 5| - 8 < 5$
36. $|6x - 1| - 4 \leq 2$ **37.** $|3(x - 5)| \leq -8$
38. $|4(2x - 1)| \geq -1$ **39.** $|3(2a + 1)| \leq 6$
40. $|2(4a - 5)| \leq 8$

A formula in mathematics is an equation which contains more than one
variable. There are probably some formulas which are already familiar
to you—for example, the formula for the area (A) of a rectangle with
length l and width w:

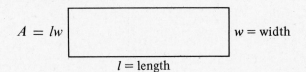

$A = lw$ $w = \text{width}$

$l = \text{length}$

There are some others with which you are probably not as familiar—for
example, the formula for the surface area (s) of a closed cylinder with a
given height h and radius r:

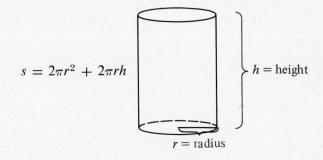

$s = 2\pi r^2 + 2\pi rh$ $h = \text{height}$

$r = \text{radius}$

Formulas are very common in the application of algebra to other
disciplines. They are found in chemistry, physics, biology, and business,
among others.

There are generally two main types of problems associated with
formulas. We can solve for one of the variables in a formula if we are
given numerical replacements for the other variables. Or we can solve
for one of the variables in a formula without being given replacements
for the other variables.

To illustrate the first case, let's consider a fairly simple example
involving the formula for the area of a triangle.

▼ **Example 1** Using the formula $A = \frac{1}{2}bh$, find h if $b = 6$ in. (inches)
and $A = 15$ sq in. (square inches).

Solution We have been given the formula for the area (A) of a triangle with base b and height h:

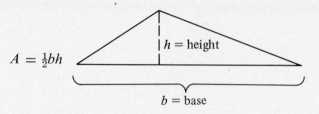

$$A = \tfrac{1}{2}bh$$

When $A = 15$ and $b = 6$, we have

$$15 = \tfrac{1}{2}(6)h$$
$$15 = 3h$$
$$5 = h$$

The height is $h = 5$ in.   ▲

Let's look at a slightly more difficult example.

▼   **Example 2**   Given the formula $A = \tfrac{1}{2}(b + B)h$, find B when $A = 20$ sq ft (square feet), $b = 3$ ft (feet), and $h = 4$ ft.

Solution The formula is for the area of a trapezoid with bases b and B and height h:

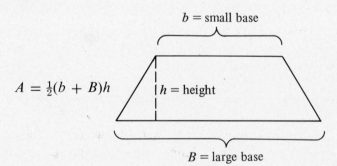

$$A = \tfrac{1}{2}(b + B)h$$

When $A = 20$, $b = 3$, and $h = 4$, we have

$$20 = \tfrac{1}{2}(3 + B)4$$

Multiplication is commutative, so we can multiply the $\tfrac{1}{2}$ and the 4:

$$20 = 2(3 + B)$$
$$10 = 3 + B \qquad \text{Multiply both sides by } \tfrac{1}{2}.$$
$$7 = B \qquad\quad \text{Add } -3 \text{ to both sides.}$$

The larger base is $B = 7$ ft.   ▲

We will now consider examples where we are given a formula and are asked to solve for one of the variables without having replacements for the others.

▼ **Example 3** Given the formula $P = 2w + 2l$, solve for w.

Solution The formula represents the relationship between the perimeter P (the distance around the outside), the length l, and the width w of a rectangle.

To solve for w we must isolate it on one side of the equation. We can accomplish this if we delete the $2l$ term and the coefficient 2 from the right side of the equation.

To begin, we add $-2l$ to both sides:

$$P + (-2l) = 2w + 2l + (-2l)$$
$$P - 2l = 2w$$

To delete the 2 from the right side, we can multiply both sides by $\frac{1}{2}$:

$$\tfrac{1}{2}(P - 2l) = \tfrac{1}{2}(2w)$$
$$\frac{P - 2l}{2} = w$$

We know we are finished because w appears alone on the right side of the equal sign and does not appear on the left side.

The two formulas

$$P = 2l + 2w \quad \text{and} \quad w = \frac{P - 2l}{2}$$

give the relationship between P, l, and w. They look different, but they both say the same thing about P, l, and w. The first formula gives P in terms of l and w, and the second formula gives w in terms of P and l.   ▲

▼ **Example 4** Solve the formula $s = 2\pi rh + \pi r^2$ for h.

Solution To isolate h, we first add $-\pi r^2$ to both sides and then multiply both sides by $1/2\pi r$:

$$s = 2\pi rh + \pi r^2$$

$$s + (-\pi r^2) = 2\pi rh + \pi r^2 + (-\pi r^2) \qquad \text{Add } -\pi r^2$$
$$\text{to both sides.}$$

$$s - \pi r^2 = 2\pi rh$$

$$\frac{1}{2\pi r}(s - \pi r^2) = \frac{1}{2\pi r}(2\pi rh) \qquad \text{Multiply by } \frac{1}{2\pi r}.$$

$$\frac{s - \pi r^2}{2\pi r} = h$$

$$\text{or} \qquad h = \frac{s - \pi r^2}{2\pi r}$$

▲

Problem Set 2.5

Solve each of the following formulas for the variable that does not have a numerical replacement:

1. $A = lw$; $A = 30$, $l = 5$
2. $A = lw$; $A = 16$, $w = 2$
3. $A = \frac{1}{2}bh$; $A = 2$, $b = 2$
4. $A = \frac{1}{2}bh$; $A = 10$, $b = 10$
5. $I = prt$; $I = 200$, $p = 1,000$, $t = .05$
6. $I = prt$; $I = 100$, $p = 2,000$, $t = 2$
7. $P = 2l + 2w$; $P = 16$, $l = 3$
8. $P = 2l + 2w$; $P = 40$, $w = 5$
9. $A = 2\pi r^2 + 2\pi rh$; $r = 1$, $\pi = 3.14$, $h = 2$
10. $A = \frac{1}{2}(b + B)h$; $A = 7$, $b = 2$, $B = 3$
11. $A = \frac{1}{2}(b + B)h$; $A = 9$, $b = 3$, $h = 6$
12. $A = \frac{1}{2}(b + B)h$; $A = 15$, $B = 2$, $h = 8$
13. $y = mx + b$; $y = 18$, $b = 2$, $m = 2$
14. $y = mx + b$; $y = 30$, $m = 2$, $x = 5$
15. $s = \frac{1}{2}(a + b + c)$; $s = 12$, $a = 1$, $b = 2$
16. $s = \frac{1}{2}(a + b + c)$; $s = 15$, $a = 8$, $c = 4$

Solve each of the following formulas for the indicated variables:

17. $A = lw$ for l
18. $A = \frac{1}{2}bh$ for b
19. $I = prt$ for t
20. $I = prt$ for r
21. $E = mc^2$ for m
22. $C = 2\pi r$ for r
23. $PV = nRT$ for T
24. $PV = nRT$ for R
25. $c^2 = a^2 + b^2$ for b^2
26. $c^2 = a^2 + b^2$ for a^2
27. $y = mx + b$ for b
28. $y = mx + b$ for x
29. $A = \frac{1}{2}(b + B)h$ for h
30. $A = \frac{1}{2}(b + B)h$ for b
31. $s = \frac{1}{2}(a + b + c)$ for c
32. $s = \frac{1}{2}(a + b + c)$ for b

There are a number of word problems whose solutions depend on solving first-degree equations in one variable. We will begin our study of word problems by considering some simple problems stated in words. Admittedly, the problems are a bit contrived. That is, the problems themselves are not that useful. They are not the kind of problems we would find in physics or chemistry; however, they are good practice. Ultimately, all the problems solved with algebra are stated in words. We are on our way to solving some of these problems. We need to practice on some simpler ones first.

Here are some general steps we will follow in solving word problems:

Step 1. Let x represent the quantity asked for in the problem.
Step 2. Write expressions, using the variable x, that represent any other unknown quantities in the problem.
Step 3. Write an equation, in x, that describes the situation.
Step 4. Solve the equation found in step 3.
Step 5. Check the solution in the original words of the problem.

Step 3 is usually the most difficult step. Step 3 is really what word problems are all about—translating a problem stated in words into an algebraic equation.

As an aid in simplifying step 3, we will look at a number of phrases written in English and the equivalent mathematical expression.

English Phrase	*Algebraic Expression*
The sum of a and b	$a + b$
The difference of a and b	$a - b$
The product of a and b	$a \cdot b$
The quotient of a and b	a/b

The word *sum* always indicates addition. The word *difference* always implies subtraction. *Product* indicates multiplication, and *quotient* means division.

Let's continue our list:

English Phrase	*Algebraic Expression*
4 more than x	$4 + x$
Twice the sum of a and 5	$2(a + 5)$
The sum of twice a and 5	$2a + 5$
8 decreased by y	$8 - y$
3 less than m	$m - 3$
7 times the difference of x and 2	$7(x - 2)$

We could extend the list even more. For every English sentence involving a relationship with numbers, there is an associated mathematical expression.

▼ **Example 1**   Twice the sum of a number and 3 is 16. Find the number.

Solution

Step 1.   Let x = the number asked for.
Step 2.   Twice the sum of x and 3 = $2(x + 3)$.
Step 3.   An equation that describes the situation is

$$2(x + 3) = 16$$

(The word *is* always translates to =.)
Step 4.   Solving the equation, we have

$$2(x + 3) = 16$$
$$2x + 6 = 16$$
$$2x = 10$$
$$x = 5$$

Step 5.   Checking $x = 5$ in the original problem, we see that twice the sum of 5 and 3 is twice 8 or 16.   ▲

▼ **Example 2**   The sum of two consecutive even integers is 12 more than their difference. Find the two integers.

Solution

Step 1.   Let x = the smaller integer.
Step 2.   The next consecutive even integer after x is $x + 2$. The sum of the two integers is $x + (x + 2)$, or $2x + 2$. Their difference is $(x + 2) - x$, or simply 2.
Step 3.   An equation that describes the situation is

$$2x + 2 = 2 + 12$$
$$2x = 12$$
$$x = 6$$

The two integers are $x = 6$ and $x + 2 = 8$.
Step 4.   Checking the original problem, we see that the sum of 6 and 8 is 14. Their difference is 2, and 14 is 12 more than 2. The answers check.   ▲

▼ **Example 3** A rectangle is 3 times as long as it is wide. The perimeter is 40 feet. Find the dimensions.

Solution

Step 1. Let x = the width of the rectangle.
Step 2. The length is 3 times the width, or $3x$.

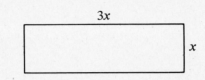

Step 3. The perimeter of a rectangle is twice the width plus twice the length:
$$2x + 2(3x) = 40$$

Step 4. Solve:
$$8x = 40$$
$$x = 5$$

The width is 5 feet. The length is 15 feet.
Step 5. Since the sum of twice 5 plus twice 15 is 40, the solutions check in the original problem. ▲

In all problems that involve geometric figures (rectangles, triangles, squares, etc.), it is helpful to draw the figure and label the dimensions.

▼ **Example 4** Diane is 4 years older than JoAnn. In 6 years the sum of their ages will be 68. What are their ages now?

Solution

Step 1. Let x = JoAnn's age now.
Step 2. With age problems like this, it is usually helpful to use a table like the following:

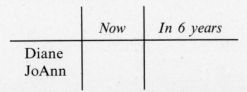

	Now	*In 6 years*
Diane		
JoAnn		

There is one column for each of the different times

given ("Now" and "In 6 years") and a row for each of the people mentioned (Diane and JoAnn).

To fill in the table, we use JoAnn's age as x. In 6 years she will be $x + 6$ years old. Diane is 4 years older than JoAnn or $x + 4$ years old. In 6 years she will be $x + 10$ years old.

	Now	In 6 years
Diane	$x + 4$	$x + 10$
JoAnn	x	$x + 6$

Step 3.   In 6 years the sum of their ages will be 68.

$$(x + 6) + (x + 10) = 68$$

Step 4.   Solve:

$$2x + 16 = 68$$
$$2x = 52$$
$$x = 26$$

JoAnn is 26 and Diane is 30.

Step 5.   In 6 years JoAnn will be 32 and Diane will be 36. The sum of their ages will then be $32 + 36 = 68$. The solutions check in the original problem.   ▲

Problem Set 2.6

Solve each of the following word problems. Be sure to show the equation used in each case.

1. A number increased by 2 is 5 less than twice the number. Find the number.
2. Twice a number decreased by 3 is equal to the number. Find the number.
3. Three times the sum of a number and 4 is 3. Find the number.
4. Five times the difference of a number and 3 is 10. Find the number.
5. Twice the sum of 2 times a number and 1 is the same as 3 times the difference of the number and 5. Find the number.
6. The sum of 3 times a number and 2 is the same as the difference of the number and 4. Find the number.
7. If 5 times a number is increased by 2, the result is 8 more than 3 times the number. Find the number.
8. If 6 times a number is decreased by 5, the result is 7 more than 4 times the number decreased by 5. Find the number.

9. The sum of two consecutive even integers is 16 more than their difference. Find the two integers.

10. The sum of two consecutive odd integers is 14 more than their difference. Find the two integers.

11. The sum of two consecutive integers is 1 less than 3 times the smaller. Find the two integers.

12. The sum of two consecutive integers is 5 less than 3 times the larger. Find the integers.

13. If twice the smaller of two consecutive integers is added to the larger, the result is 7. Find the smaller one.

14. If twice the larger of two consecutive integers is added to the smaller, the result is 23. Find the smaller one.

15. If the larger of two consecutive odd integers is subtracted from twice the smaller, the result is 5. Find the two integers.

16. If the smaller of two consecutive even integers is subtracted from twice the larger, the result is 12. Find the two integers.

17. A rectangle is twice as long as it is wide. The perimeter is 60 feet. Find the dimensions.

18. The length of a rectangle is 5 times the width. The perimeter is 48 inches. Find the dimensions.

19. A square has a perimeter of 28 feet. Find the length of the side.

20. A square has a perimeter of 36 centimeters. Find the length of the side.

21. A triangle has a perimeter of 23 inches. The medium side is 3 more than the smallest side, and the longest side is twice the shortest side. Find the shortest side.

22. The longest side of a triangle is 3 times the shortest side. While the medium side is twice the shortest side, the perimeter is 18 meters. Find the dimensions.

23. The length of a rectangle is 3 less than twice the width. The perimeter is 18 meters. Find the width.

24. The length of a rectangle is one more than twice the width. The perimeter is 20 feet. Find the dimensions.

25. Patrick is 4 years older than Amy. In 10 years the sum of their ages will be 36. How old are they now?

26. Mr. Lloyd is 2 years older than Mrs. Lloyd. Five years ago the sum of their ages was 44. How old are they now?

27. John is twice as old as Bill. In 5 years John will be $1\frac{1}{2}$ times as old as Bill. How old is each boy now?

28. Bob is 6 years older than his little brother Ron. In 4 years he will be twice as old as his little brother. How old are they now?

29. Jane is 3 times as old as Kate. In 5 years Jane's age will be 2 less than twice Kate's. How old are the girls now?

30. Carol is 2 years older than Mike. In 4 years her age will be 8 less than twice his. How old are they now?

**Chapter 2
Summary and Review**

*Addition Property of
Equality*

For algebraic expressions A, B, and C,

$$\text{if} \qquad A = B$$
$$\text{then} \qquad A + C = B + C$$

This property states that we can add the same quantity to both sides of an equation without changing the solution set.

*Multiplication
Property of Equality*

For algebraic expressions A, B, and C,

$$\text{if} \qquad A = B$$
$$\text{then} \qquad AC = BC, \quad C \neq 0$$

Multiplying both sides of an equation by the same nonzero quantity never changes the solution set.

*Solving a First-Degree
Equation in One
Variable*

Step 1. Use the distributive property to separate terms.

Step 2. Simplify the left and right sides of the equation separately whenever possible.

Step 3. Use the addition property of equality to write all terms containing the variable on one side of the equal sign and all remaining terms on the other.

Step 4. Use the multiplication property of equality in order to get x alone on one side of the equal sign.

Step 5. Check your solution in the original equation, if necessary.

*Addition Property for
Inequalities*

For expressions A, B, and C,

$$\text{if} \qquad A < B$$
$$\text{then} \qquad A + C < B + C$$

Adding the same quantity to both sides of an inequality never changes the solution set.

*Multiplication
Property for
Inequalities*

For expressions A, B, and C,

$$\text{if} \qquad A < B$$
$$\text{then} \qquad AC < BC \quad \text{if} \quad C > 0, C \text{ is positive}$$
$$\text{or} \qquad AC > BC \quad \text{if} \quad C < 0, C \text{ is negative}$$

We can multiply both sides of an inequality by the same nonzero number without changing the solution set as long as each time we multiply by a negative number we also reverse the direction of the inequality symbol.

Follow the same first four steps used to solve a first-degree equation— using, of course, the addition and multiplication properties for inequalities. As a fifth step, graph the solution set.

Solving a First-Degree Inequality in One Variable

Use the definition of absolute value to write an equivalent equation or inequality that does not involve absolute value.

Absolute Value Equations and Inequalities

$$|x| = 5 \quad \text{is equivalent to} \quad x = 5 \quad \text{or} \quad x = -5$$
$$|x| < 5 \quad \text{is equivalent to} \quad -5 < x < 5$$
$$|x| > 5 \quad \text{is equivalent to} \quad x < -5 \quad \text{or} \quad x > 5$$

A formula in algebra is an equation involving more than one variable. To solve a formula for one of its variables, simply isolate that variable on one side of the equation.

Formulas

Step 1. Let x represent the quantity asked for.

Step 2. If possible, write all other unknown quantities in terms of x.

Step 3. Write an equation, using x, that describes the situation.

Step 4. Solve the equation in step 3.

Step 5. Check the solution from step 4 with the original words of the problem.

Solving a Problem Stated in Words

A very common mistake in solving inequalities is to forget to reverse the direction of the inequality symbol when multiplying both sides by a negative number. When this mistake occurs, the graph of the solution set is always to the wrong side of the end point.

COMMON MISTAKE

Solve the following equations using the steps outlined in Section 2.1:

Chapter 2 Test

1. $x - 5 = 7$ **2.** $3y = -4$

3. $3a - 6 = 11$ **4.** $5(x - 1) - 2(2x + 3) = 5x - 4$

Solve the following inequalities by the method developed in Section 2.2:

5. $-5t \leq 30$ **6.** $8 - 2x \geq 6$

7. $4x - 5 < 2x + 7$ **8.** $3(2y + 4) \geq 5(y - 8)$

Solve the following equations and inequalities using the definition of absolute value:

9. $|x - 4| = 2$ **10.** $|2a + 7| = 5$

11. $|6x - 1| > 7$ **12.** $|3x - 5| - 4 \leq 3$

13. Solve the formula $A = 2l + 2w$ for w.

14. Solve the formula $A = \frac{1}{2}h(b + B)$ for B.

15. Find two consecutive even integers whose sum is 18. (Be sure to write an equation that describes the situation.)

16. Patrick is 4 years older than Amy. In 5 years he will be twice as old as she is now. Find their ages now.

Exponents and Polynomials 3

To the student:

There are many expressions and equations in mathematics that involve exponents. For example, in genetics, the phenotypic variance P^2 of a given trait is the sum of the genetic variance (G^2), the environmental variance (E^2), and the variance due to the interaction between them (I^2), or $P^2 = G^2 + E^2 + I^2$. The 2's in the equation are exponents. Before we get to the point where we can work with equations like this, we must review the properties of exponents and practice working with expressions that involve exponents.

We will begin the chapter by reviewing the properties of exponents. The exponents encountered will be integer exponents, both positive and negative. We will then turn our attention to a special kind of expression that involves exponents. These expressions are known as polynomials. We will give some definitions and notation on polynomials, and then apply the four basic operations—addition, subtraction, multiplication, and division—to polynomials. The chapter will end with two sections on factoring polynomials, with the emphasis on factoring trinomials.

In addition to being able to add, subtract, multiply, and divide real numbers, your success in this chapter depends on your ability to understand and apply the distributive property. The fact that multiplication distributes over addition allows us to add and multiply polynomials. Understanding addition and multiplication of polynomials allows us to understand subtraction and factoring of polynomials. Actually, we could probably rename this chapter "Exponents and Applications of the Distributive Property." Understanding the distributive property is the key to understanding this chapter.

3.1
Properties of
Exponents

In Chapter 1 we showed multiplication by positive integers as repeated addition. That is, $3 \cdot 4 = 4 + 4 + 4$. In this section we will define positive integer exponents to be shorthand notation for repeated multiplication.

Consider the expression 3^4, which reads "3 to the fourth power" or just "3 to the fourth." The 3 is called the *base* and the 4 is called the *exponent*. The exponent gives us the number of times the base is used as a factor in the expansion of the expression. For example,

$$3^4 = \underbrace{3 \cdot 3 \cdot 3 \cdot 3}_{4 \text{ factors}} = 81$$

The expression 3^4 is said to be in *exponential form*, while $3 \cdot 3 \cdot 3 \cdot 3$ is said to be in *expanded form*.

Here are some more examples of expressions with integer exponents and their expansions.

▼   **Example 1**

a. $(-4)^3 = (-4)(-4)(-4) = -64$ Base -4, exponent 3
b. $(-5)^2 = (-5)(-5) = 25$ Base -5, exponent 2
c. $-5^2 = -5 \cdot 5 = -25$ Base 5, exponent 2
d. $\left(\frac{2}{3}\right)^4 = \frac{2}{3} \cdot \frac{2}{3} \cdot \frac{2}{3} \cdot \frac{2}{3} = \frac{16}{81}$ Base $\frac{2}{3}$, exponent 4
e. $(x + y)^2 = (x + y)(x + y)$ Base $(x + y)$, exponent 2
f. $x^6 = x \cdot x \cdot x \cdot x \cdot x \cdot x$ Base x, exponent 6   ▲

In this section we are concerned with the simplification of products and quotients involving more than one base and more than one exponent. Actually, almost all of the expressions can be simplified by applying the definition of positive integer exponents, that is, by writing all expressions in expanded form and then simplifying by using ordinary arithmetic. This process, however, is usually very time-consuming. We can shorten the process considerably by making some generalizations about exponents and applying the generalizations whenever it is convenient.

Let's expand the product $x^3 \cdot x^4$.

$$x^3 \cdot x^4 = (x \cdot x \cdot x)(x \cdot x \cdot x \cdot x)$$
$$= (x \cdot x \cdot x \cdot x \cdot x \cdot x \cdot x)$$
$$= x^7 \qquad \textit{Notice: } 3 + 4 = 7$$

We can generalize this result into the first property of exponents.

PROPERTY 1   If a is a real number, and r and s are positive integers, then

$$a^r \cdot a^s = a^{r+s}$$

The product of two expressions with the same base is equivalent to the base raised to the sum of the exponents from the original two expressions. Here is the justification for Property 1:

$$a^r \cdot a^s = \underbrace{(a \cdot a \cdot a \cdots a)}_{r \text{ factors}}\underbrace{(a \cdot a \cdot a \cdots a)}_{s \text{ factors}}$$

$$= \underbrace{a \cdot a \cdot a \cdots a}_{(r+s) \text{ factors}}$$

$$= a^{r+s}$$

Expressions of the form $(a^r)^s$ occur frequently with exponents. Here is an example:

$$(5^3)^2 = 5^3 \cdot 5^3$$
$$= 5^6 \qquad \textit{Notice: } 3 \cdot 2 = 6$$

Generalizing this result we have a second property of exponents:

PROPERTY 2   If a is a real number and r and s are positive integers, then

$$(a^r)^s = a^{r \cdot s}$$

An expression with an exponent, raised to another power, is the same as the base from the original expression raised to the product of the powers. Here is the proof of Property 2 for exponents:

$$(a^r)^s = \underbrace{a^r \cdot a^r \cdot a^r \cdots a^r}_{s \text{ factors}}$$

$$= \underbrace{\underbrace{(a \cdot a \cdot a \cdots a)}_{r \text{ factors}}\underbrace{(a \cdot a \cdot a \cdots a)}_{r \text{ factors}} \cdots \underbrace{(a \cdot a \cdot a \cdots a)}_{r \text{ factors}}}_{s \text{ factors}} \quad \text{(each of which has } r \text{ factors)}$$

$$= \underbrace{a \cdot a \cdot a \cdots a}_{r \cdot s \text{ factors}}$$

$$= a^{r \cdot s}$$

A third property of exponents arises when we have the product of two or more numbers raised to an integer power. For example:

$$(3x)^4 = (3x)(3x)(3x)(3x)$$
$$= (3 \cdot 3 \cdot 3 \cdot 3)(x \cdot x \cdot x \cdot x)$$
$$= 3^4 \cdot x^4 \qquad \textit{Notice: } \text{The exponent 4 dis-}$$
$$\qquad\qquad\qquad \text{tributes over the product } 3x.$$
$$= 81x^4$$

PROPERTY 3 If a and b are any two real numbers, and r is a positive integer, then

$$(ab)^r = a^r \cdot b^r$$

Here's why:

$$(ab)^r = \underbrace{(ab)(ab)(ab) \cdots (ab)}_{r \text{ factors}}$$
$$= \underbrace{(a \cdot a \cdot a \cdots a)}_{r \text{ factors}}\underbrace{(b \cdot b \cdot b \cdots b)}_{r \text{ factors}}$$
$$= a^r \cdot b^r$$

Here are some examples that use combinations of the first three properties of exponents to simplify expressions involving exponents:

▼ **Example 2**

a. $(-3x^2)(5x^4) = -3(5)(x^2 \cdot x^4)$ Commutative, associative
                                        properties
             $= -15x^6$ Property 1 for exponents

b. $(-2x^2)^3(4x^5) = (-2)^3(x^2)^3(4x^5)$ Property 3
              $= -8x^6 \cdot (4x^5)$ Property 2
              $= (-8 \cdot 4)(x^6 \cdot x^5)$ Commutative and associative
              $= -32x^{11}$ Property 1

c. $(x^2)^4(x^2y^3)^2(y^4)^3 = x^8 \cdot x^4 \cdot y^6 \cdot y^{12}$ Properties 2 and 3
              $= x^{12}y^{18}$ Property 1   ▲

Property 3 was stated for products. Since division is defined in terms of multiplication, we can expect a similar property involving quotients.

PROPERTY 4 If a and b are any two real numbers with $b \neq 0$, and r is a positive integer, then

$$\left(\frac{a}{b}\right)^r = \frac{a^r}{b^r}$$

Proof

$$\left(\frac{a}{b}\right)^r = \underbrace{\left(\frac{a}{b}\right)\left(\frac{a}{b}\right)\left(\frac{a}{b}\right)\cdots\left(\frac{a}{b}\right)}_{r \text{ factors}}$$

$$= \frac{a \cdot a \cdot a \cdots a}{b \cdot b \cdot b \cdots b} \begin{array}{l}\leftarrow r \text{ factors} \\ \leftarrow r \text{ factors}\end{array}$$

$$= \frac{a^r}{b^r}$$

We will now consider some examples of the quotient of two expressions involving exponents in which the bases are the same. Since multiplication with the same base resulted in addition of exponents, and since division is to multiplication as subtraction is to addition, it seems reasonable to expect division with the same base to result in subtraction of exponents.

Let's begin with an example in which the exponent in the numerator is larger than the exponent in the denominator:

$$\frac{6^5}{6^3} = \frac{6 \cdot 6 \cdot 6 \cdot 6 \cdot 6}{6 \cdot 6 \cdot 6}$$

Dividing out the 6's common to the numerator and denominator, we have

$$\frac{6^5}{6^3} = \frac{\cancel{6} \cdot \cancel{6} \cdot \cancel{6} \cdot 6 \cdot 6}{\cancel{6} \cdot \cancel{6} \cdot \cancel{6}}$$

$$= 6 \cdot 6$$

$$= 6^2 \qquad \textit{Notice: } 5 - 3 = 2$$

If we were simply to subtract the exponent in the denominator from the exponent in the numerator, we would obtain the correct result, which is 6^2.

Let's consider an example where the exponent in the denominator is larger than the exponent in the numerator:

$$\frac{5^3}{5^7} = \frac{\cancel{5} \cdot \cancel{5} \cdot \cancel{5}}{\cancel{5} \cdot \cancel{5} \cdot \cancel{5} \cdot 5 \cdot 5 \cdot 5 \cdot 5}$$

$$= \frac{1}{5 \cdot 5 \cdot 5 \cdot 5}$$

$$= \frac{1}{5^4} \qquad \textit{Notice: } 7 - 3 = 4$$

Here again, subtraction of exponents gives the correct result. In this case, however, we must subtract the exponent in the numerator from the exponent in the denominator.

We can simplify our generalization of the above results with the following definition:

DEFINITION (PROPERTY 5) If a is any nonzero real number and r is a positive integer, then

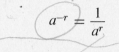

$$a^{-r} = \frac{1}{a^r}$$

This is the definition for negative exponents. It says negative exponents indicate reciprocals.

The following examples illustrate the definition for negative exponents:

▼   **Example 3**

a. $5^{-2} = \dfrac{1}{5^2} = \dfrac{1}{25}$

b. $(-2)^{-3} = \dfrac{1}{(-2)^3} = \dfrac{1}{-8} = -\dfrac{1}{8}$

c. $\left(\dfrac{3}{4}\right)^{-2} = \left(\dfrac{1}{\frac{3}{4}}\right)^2 = \dfrac{1}{\frac{9}{16}} = \dfrac{16}{9}$ ▲

Consider again the expression $5^3/5^7$. We will simplify by subtracting the exponent in the denominator from the exponent in the numerator and applying the definition for negative exponents to the result:

$$\frac{5^3}{5^7} = 5^{3-7}$$

$$= 5^{-4}$$

$$= \frac{1}{5^4}$$

The result is the same as before. Our definition for negative exponents does not contradict what we have done previously.

We can use our definition of negative exponents as Property 5 for exponents and summarize the above results into Property 6:

PROPERTY 6   If a is any real number, and r and s are any two integers, then

$$\frac{a^r}{a^s} = a^{r-s}$$

Notice we have specified r and s to be any integers instead of only positive integers as was the case in our previous properties. Our definition of negative exponents is such that all the above properties hold for all integer exponents, whether positive integers or negative.

▼   **Example 4**

a. $\dfrac{7^8}{7^3} = 7^{8-3} = 7^5$

b. $\dfrac{9^4}{9^{10}} = 9^{4-10} = 9^{-6} = \dfrac{1}{9^6}$

c. $\dfrac{x^2}{x^{18}} = x^{2-18} = x^{-16} = \dfrac{1}{x^{16}}$

d. $\dfrac{3^6}{3^{-8}} = 3^{6-(-8)} = 3^{14}$

e. $\dfrac{x^{-5}}{x^{-7}} = x^{-5-(-7)} = x^2$

f. $\dfrac{(2x^2)^4}{(2^2x^3)^5} = \dfrac{2^4 \cdot x^8}{2^{10}x^{15}}$ Properties 2 and 3

$\qquad\qquad = 2^{-6}x^{-7}$ Property 6

$\qquad\qquad = \dfrac{1}{2^6} \cdot \dfrac{1}{x^7}$ Property 5

$\qquad\qquad = \dfrac{1}{2^6x^7}$

g. $3^{-2} + 2^{-3} = \dfrac{1}{3^2} + \dfrac{1}{2^3}$ Property 5

$\qquad\qquad = \dfrac{1}{9} + \dfrac{1}{8}$

$\qquad\qquad = \dfrac{8}{72} + \dfrac{9}{72}$

$\qquad\qquad = \dfrac{17}{72}$

▲

Finally, we will consider 0 and 1 as exponents. The number 1 is easy to deal with since we can use the original definition for exponents:

$$a^1 = \underbrace{a}_{\text{1 factor}}$$

For 0 as an exponent, consider the expression $3^4/3^4$. Since $3^4 = 81$, we have

$$\frac{3^4}{3^4} = \frac{81}{81} = 1$$

On the other hand, since we have the quotient of two expressions with the same base, we can subtract exponents.

$$\frac{3^4}{3^4} = 3^{4-4} = 3^0$$

Hence, 3^0 must be the same as 1.

Summarizing these results, we have our last property for exponents:

PROPERTY 7 If a is any real number, then

$$\text{and} \quad \begin{aligned} a^1 &= a \\ a^0 &= 1 \quad \text{(as long as } a \neq 0\text{)} \end{aligned}$$

▼   **Example 5**

a. $(2x^2y^4)^0 = 1$
b. $(2x^2y^4)^1 = 2x^2y^4$ ▲

We will end this section with two examples that use many of the properties of exponents. There are a number of different ways to proceed on problems like these. You should use the method that works best for you.

▼   **Example 6**

a. $\dfrac{(2x^2)^4(-3x^{-2})^2}{4x^{-6}(9x^{-3})} = \dfrac{16x^8 \cdot 9x^{-4}}{4x^{-6} \cdot 9x^{-3}}$

$$= \frac{4x^4}{x^{-9}} \qquad \text{Property 1}$$

$$= 4x^{13} \qquad \text{Property 6}$$

b. $\left(\dfrac{2x^0y^4z^{-3}}{8x^5y^{-3}z^{-2}}\right)^{-2} = \left(\dfrac{x^{-5}\cdot y^7\cdot z^{-1}}{4}\right)^{-2}$ Property 6

$\quad\quad = \dfrac{x^{10}y^{-14}z^2}{4^{-2}}$ Properties 2 and 3

$\quad\quad = \dfrac{x^{10}\left(\dfrac{1}{y^{14}}\right)z^2}{\frac{1}{16}}$ Property 5

$\quad\quad = \dfrac{16x^{10}z^2}{y^{14}}$ ▲

Evaluate each of the following: ***Problem Set 3.1***

1. 4^2 2. $(-4)^2$
3. -4^2 4. $-(-4)^2$
5. -3^3 6. $(-3)^3$
7. 2^5 8. 2^4
9. $(\frac{1}{2})^3$ 10. $(\frac{3}{4})^2$
11. $(-\frac{5}{6})^2$ 12. $(-\frac{7}{8})^2$
13. $3^0 - 2^0 - 1^0$ 14. $(4 - 2 - 1)^0$

Use the properties of exponents to simplify each of the following as much as possible:

15. $x^5\cdot x^4$ 16. $x^6\cdot x^3$
17. $a^4\cdot a^6$ 18. $a^3\cdot a^2$
19. $(x^3)^2$ 20. $(x^4)^3$
21. $(-2x^2)^3$ 22. $(-3x^4)^3$
23. $-3x^2(2x^4)$ 24. $5x^7(-4x^6)$
25. $6x^2(-3x^4)(2x^5)$ 26. $(5x^3)(-7x^4)(-2x^6)$
27. $(2x^2)^2(3x^3)^4$ 28. $(3x^5)^4(-2x^6)^2$
29. $(-3x)^4(2x^3)^2(-x^6)^4$ 30. $(5x^6)^2(-2x^3)^2(-3x^7)^2$

Write each of the following with positive exponents. Then simplify as much as possible.

31. 3^{-2} 32. $(-5)^{-2}$
33. 2^{-5} 34. $(-2)^{-5}$
35. $(\frac{3}{4})^{-2}$ 36. $(-\frac{5}{6})^{-1}$
37. $3^{-2} + 2^{-1}$ 38. $4^{-2} + 2^{-3}$
39. $4^0 + 3^{-1} + 2^{-3}$ 40. $5^0 - 4^{-1} + 2^{-4}$

Use the properties of exponents to simplify each expression as much as possible. All answers should have positive exponents. Assume all variables are nonnegative.

41. $\dfrac{3^4}{3^6}$

42. $\dfrac{-2^7}{-2^3}$

43. $\dfrac{2x^{-5}}{x^{-6}}$

44. $\dfrac{x^{-3}}{3x^{-4}}$

✳45. $\dfrac{3^5 \cdot 3^2}{3^6 \cdot 3^5}$

46. $\dfrac{2^7 \cdot 2^3}{2^6 \cdot 2^8}$

47. $2^5 \cdot 2^{-4}$

48. $5^{-3} \cdot 5^2 \cdot 5^{-4}$

✳49. $6^0 \cdot 6^{-3} \cdot 6^7$

50. $10^4 \cdot 10^{-7} \cdot 10^{-3}$

51. $\dfrac{(2x^2)^3(-3x^4)^{-1}}{8x^{-3}}$

52. $\dfrac{(3x)^4(-2x^2)^{-3}}{9x^{-2}}$

53. $\dfrac{(5x^2)(-3x^4)^2}{18x^{-3} \cdot 10x^0}$

54. $\dfrac{(4x^{-3})(6x^2)^2}{(3x^5)^2(16x^{-2})}$

55. $\left(\dfrac{2x^{-3}y^0}{4x^6y^{-5}}\right)^{-2}$

56. $\left(\dfrac{2x^6y^4z^0}{8x^{-3}y^0z^{-5}}\right)^{-1}$

57. $\left(\dfrac{ab^{-3}c^{-2}}{a^{-3}b^0c^{-5}}\right)^{-1}$

58. $\left(\dfrac{a^3b^2c^1}{a^{-1}b^{-2}c^{-3}}\right)^{-2}$

59. $x^m \cdot x^{1-m}$

60. $x^{m-5} \cdot x^{5-m}$

61. $x^0 \cdot x^1 \cdot x^{-1}$

62. $\dfrac{x^{3m} \cdot x^{m+1}}{x^{2m}}$

63. $\dfrac{x^{2r} \cdot x^{r+3}}{x^{3r}}$

64. $\dfrac{x^{6r+1} \cdot x^{1-r}}{x^{r+2}}$

**3.2
Polynomials, Sums,
and Differences**

We begin this section with the definition of the basic unit, a term, around which polynomials are defined. Once we have listed all the terminology associated with polynomials, we will show how the distributive property is used to find sums and differences of polynomials.

*Polynomials in
General*

DEFINITION   A *term* or *monomial* is a constant or the product of a constant and one or more variables raised to whole-number exponents.

The following are monomials or terms:

$$-16, \qquad 3x^2y, \qquad -\tfrac{2}{5}a^3b^2c, \qquad xy^2z$$

The numerical part of each monomial is called the *numerical coefficient*, or just *coefficient* for short. For the above terms the coefficients are -16, 3, $-\tfrac{2}{5}$, and 1. Notice that the coefficient for xy^2z is understood to be 1.

DEFINITION A *polynomial* is any finite sum of terms. Since subtraction can be written in terms of addition, finite differences are also included in this definition.

The following are polynomials:

$$2x^2 - 6x + 3, \qquad -5x^2y + 2xy^2, \qquad 4a - 5b + 6c + 7d$$

Polynomials can be classified further according to the number of terms present. If a polynomial consists of two terms, it is said to be a *binomial*. If it has three terms, it is called a *trinomial*. And as stated above, a polynomial with only one term is said to be a *monomial*.

DEFINITION The *degree* of a polynomial with one variable is the highest power to which the variable is raised in any one term.

▼ **Example 1**

a. $6x^2 + 2x - 1$ A trinomial of degree 2
b. $5x - 3$ A binomial of degree 1
c. $7x^6 - 5x^3 + 2x - 4$ A polynomial of degree 6
d. $-7x^4$ A monomial of degree 4
e. 15 A monomial of degree 0   ▲

Polynomials in one variable are usually written in decreasing powers of the variable. When this is the case, the coefficient of the first term is called the *leading coefficient*. In part a of Example 1 above, the leading coefficient is 6. In part b it is 5. The leading coefficient in part c is 7.

Addition and Subtraction of Polynomials

Combining polynomials to find sums and differences is a very simple process. For the most part it involves applying the distributive property to combine similar terms.

DEFINITION Two or more terms which differ only in their numerical coefficients are called *similar* or *like* terms. Since similar terms differ only in their coefficients, they have identical variable parts—that is, the same variables raised to the same power. For example, $3x^2$ and $-5x^2$ are similar terms. So are $15x^2y^3z$, $-27x^2y^3z$ and $\frac{3}{4}x^2y^3z$.

We can use the distributive property to combine the similar terms $6x^2$ and $9x^2$ as follows:

$$6x^2 + 9x^2 = (6 + 9)x^2 \qquad \text{Distributive property.}$$
$$= 15x^2 \qquad\qquad \text{The sum of 6 and 9 is 15.}$$

The distributive property can also be used to combine more than two like terms.

▼ **Example 2** Combine $7x^2y + 4x^2y - 10x^2y + 2x^2y$.

Solution

$$7x^2y + 4x^2y - 10x^2y + 2x^2y = (7 + 4 - 10 + 2)x^2y \qquad \text{Distributive property}$$
$$= 3x^2y \qquad\qquad\qquad \text{Addition} \quad \blacktriangle$$

To add two polynomials, we simply apply the commutative and associative properties to group similar terms and then use the distributive property as we have in the example above.

▼ **Example 3** Add $5x^2 - 4x + 2$ and $3x^2 + 9x - 6$.

Solution

$$(5x^2 - 4x + 2) + (3x^2 + 9x - 6)$$
$$= (5x^2 + 3x^2) + (-4x + 9x) + (2 - 6) \qquad \text{Commutative and associative properties}$$
$$= (5 + 3)x^2 + (-4 + 9)x + (2 - 6) \qquad \text{Distributive property}$$
$$= 8x^2 + 5x + (-4)$$
$$= 8x^2 + 5x - 4 \qquad\qquad\qquad\qquad\qquad\qquad \blacktriangle$$

In actual practice it is not necessary to show all the steps shown in Example 3. It is important to understand that addition of polynomials is equivalent to combining similar terms. We add similar terms by combining coefficients.

Sometimes it is convenient to add polynomials vertically in columns.

▼ **Example 4** Find the sum of $-8x^3 + 7x^2 - 6x + 5$ and $10x^3 + 3x^2 - 2x - 6$.

Solution We can add the two polynomials using the method of Example 3, or we can arrange similar terms in columns and add vertically. Using the column method, we have:

$$
\begin{array}{r}
-8x^3 + 7x^2 - 6x + 5 \\
\underline{10x^3 + 3x^2 - 2x - 6} \\
2x^3 + 10x^2 - 8x - 1
\end{array}
$$
▲

It is important to notice that no matter which method is used to combine two polynomials, the variable part of the combined terms never changes. It is sometimes tempting to say $7x^2 + 3x^2 = 10x^4$, which is incorrect. Remember, we combine similar terms by using the distributive property: $7x^2 + 3x^2 = (7 + 3)x^2 = 10x^2$. The variable part common to each term is unchanged.

To find the difference of two polynomials, we need to use the fact that the opposite of a sum is the sum of the opposites. That is,

$$-(a + b) = -a + (-b)$$

One way to remember this is to observe that $-(a + b)$ is equivalent to $-1(a + b) = (-1)a + (-1)b = -a + (-b)$.

If there is a negative sign directly preceding the parentheses surrounding a polynomial, we may remove the parentheses and preceding negative sign by changing the sign of each term within the parentheses.

For example:

$$
\begin{aligned}
-(3x + 4) &= -3x + (-4) = -3x - 4 \\
-(5x^2 - 6x + 9) &= -5x^2 + 6x - 9 \\
-(-x^2 + 7x - 3) &= x^2 - 7x + 3
\end{aligned}
$$

To find the difference of two or more polynomials, we simply apply this principle and proceed as we did when finding sums.

▼ **Example 5**

$$
\begin{aligned}
&(9x^2 - 3x + 5) - (4x^2 + 2x - 3) \\
&= 9x^2 - 3x + 5 + (-4x^2) + (-2x) + 3 \qquad \text{The opposite of} \\
&\qquad\qquad\qquad\qquad\qquad\qquad\qquad\quad \text{a sum is the sum} \\
&\qquad\qquad\qquad\qquad\qquad\qquad\qquad\quad \text{of the opposites.} \\
&= (9x^2 - 4x^2) + (-3x - 2x) + (5 + 3) \qquad \text{Commutative and} \\
&\qquad\qquad\qquad\qquad\qquad\qquad\qquad\qquad\quad \text{associative} \\
&\qquad\qquad\qquad\qquad\qquad\qquad\qquad\qquad\quad \text{properties} \\
&= 5x^2 - 5x + 8 \qquad\qquad\qquad\qquad\qquad\quad \text{Combine similar} \\
&\qquad\qquad\qquad\qquad\qquad\qquad\qquad\qquad\quad \text{terms.} \qquad ▲
\end{aligned}
$$

▼   **Example 6**

$(2x^3 + 5x^2 + 3) - (4x^2 - 2x - 7) - (6x^3 - 3x + 1)$
$$= (2x^3 + 5x^2 + 3) + (-4x^2 + 2x + 7) + (-6x^3 + 3x - 1)$$
$$= 2x^3 + 5x^2 + 3 - 4x^2 + 2x + 7 - 6x^3 + 3x - 1$$
$$= (2x^3 - 6x^3) + (5x^2 - 4x^2) + (2x + 3x) + (3 + 7 - 1)$$
$$= -4x^3 + x^2 + 5x + 9 \qquad ▲$$

▼   **Example 7**   Subtract $4x^2 - 9x + 1$ from $-3x^2 + 5x - 2$.

   Solution

$$(-3x^2 + 5x - 2) - (4x^2 - 9x + 1)$$
$$= (-3x^2 + 5x - 2) + (-4x^2 + 9x - 1)$$
$$= -3x^2 + 5x - 2 - 4x^2 + 9x - 1$$
$$= (-3x^2 - 4x^2) + (5x + 9x) + (-2 - 1)$$
$$= -7x^2 + 14x - 3 \qquad ▲$$

When one set of grouping symbols is contained within another, it is best to begin the process of simplification with the innermost grouping symbol and work out from there.

▼   **Example 8**   Simplify $4x - 3[2 - (3x + 4)]$

   Solution   Removing the innermost parentheses first, we have

$$4x - 3[2 - (3x + 4)] = 4x - 3(2 - 3x - 4)$$
$$= 4x - 3(-3x - 2)$$
$$= 4x + 9x + 6$$
$$= 13x + 6 \qquad ▲$$

▼   **Example 9**

$$(2x + 3) - [(3x + 1) - (x - 7)] = (2x + 3) - (3x + 1 - x + 7)$$
$$= (2x + 3) - (2x + 8)$$
$$= -5 \qquad ▲$$

There is a special type of substitution problem that arises when working with polynomials. Many times polynomials are designated with the notation $P(x)$. Also quite common are $Q(x)$ and $R(x)$.

▼   **Example 10** If $P(x) = 5x^2 - 3x + 7$, find $P(2)$.

Solution   This is just a substitution problem. $P(x)$ is $5x^2 - 3x + 7$ and we are asked for $P(2)$. We simply substitute 2 for x in the expression $5x^2 - 3x + 7$:

$$\text{If} \quad P(x) = 5x^2 - 3x + 7$$
$$\text{then} \quad P(2) = 5(2)^2 - 3(2) + 7$$
$$= 5 \cdot 4 - 3(2) + 7$$
$$= 20 - 6 + 7$$
$$= 21 \qquad \blacktriangle$$

▼ **Example 11** If $Q(x) = 2x^3 - 3x^2 + 4x - 5$, find $Q(-1)$ and $Q(0)$.

Solution

$$\text{If} \quad Q(x) = 2x^3 - 3x^2 + 4x - 5$$
$$\text{then} \quad Q(-1) = 2(-1)^3 - 3(-1)^2 + 4(-1) - 5$$
$$= 2(-1) - 3(1) + 4(-1) - 5$$
$$= -2 - 3 - 4 - 5$$
$$= -14$$
$$\text{and} \quad Q(0) = 2(0)^3 - 3(0)^2 + 4(0) - 5$$
$$= 0 - 0 + 0 - 5$$
$$= -5 \qquad \blacktriangle$$

We will do more work with notation like $P(x)$ and $Q(x)$ in Chapter 10 when we work with relations and functions.

Identify those of the following that are monomials, binomials, or trinomials. Give the degree of each and name the leading coefficient.

Problem Set 3.2

1. $5x^2 - 3x + 2$
2. $2x^2 + 4x - 1$
3. $3x - 5$
4. $5y + 3$
5. $8a^2 + 3a - 5$
6. $9a^2 - 8a - 4$
7. $4x^3 - 6x^2 + 5x - 3$
8. $9x^4 + 4x^3 - 2x^2 + x$
9. $-\frac{3}{4}$
10. -16
11. $4x - 5 + 6x^3$
12. $9x + 2 + 3x^3$

Simplify each of the following by combining similar terms:

13. $(4x + 2) + (3x - 1)$
14. $(8x - 5) + (-5x + 4)$
15. $(3 - 2x) + (5x - 1) + (4 - 3x)$
16. $(11 - x) + (3 - 5x) + (4x - 8)$
17. $(9a + 1) - (4a - 6)$

18. $(8a - 4) - (-2a + 8)$
19. $(5x^2 - 6x + 1) - (4x^2 + 7x - 2)$
20. $(11x^2 - 8x) - (4x^2 - 2x - 7)$
21. $(6x^2 - 4x - 2) - (3x^2 + 7x) + (4x - 1)$
22. $(8x^2 - 6x) - (3x^2 + 2x + 1) - (6x^2 + 3)$
23. $(y^3 - 2y^2 - 3y + 4) - (2y^3 - y^2 + y - 3)$
24. $(8y^3 - 3y^2 + 7y + 2) - (-4y^3 + 6y^2 - 5y - 8)$
25. $(5x^3 - 4x^2) - (3x + 4) + (5x^2 - 7) - (3x^3 + 6)$
26. $(x^3 - x) - (x^2 + x) + (x^3 - 1) - (-3x + 2)$
27. $(8x^2 - 2xy + y^2) - (7x^2 - 4xy - 9y^2)$
28. $(2x^2 - 5xy + y^2) + (-3x^2 + 4xy - 5)$
29. $(3a^3 + 2a^2b + ab^2 - b^3) - (6a^3 - 4a^2b + 6ab^2 - b^3)$
30. $(a^3 - 3a^2b + 3ab^2 - b^3) - (a^3 + 3a^2b + 3ab^2 + b^3)$
31. Subtract $9x^2 - 6x$ from $-11x^2 + 4x$.
32. Subtract $6x + 2$ from $3 - 5x$.
33. Find the sum of $x^2 - 6xy + y^2$ and $2x^2 - 6xy - y^2$.
34. Find the sum of $9x^3 - 6x^2 + 2$ and $3x^2 - 5x + 4$.
35. Subtract $-8x^5 - 4x^3 + 6$ from $9x^5 - 4x^3 - 6$.
36. Subtract $4x^4 - 3x^3 - 2x^2$ from $2x^4 + 3x^3 + 4x^2$.
37. Find   the   sum   of $11a^2 + 3ab + 2b^2$, $9a^2 - 2ab + b^2$,   and $-6a^2 - 3ab + 5b^2$.
38. Find the sum of $a^2 - ab - b^2$, $a^2 + ab - b^2$, and $a^2 + 2ab + b^2$.

Simplify each of the following. Begin by working on the innermost parentheses first.

39. $-[2 - (4 - x)]$
40. $-[-3 - (x - 6)]$
41. $-5[-(x - 3) - (x + 2)]$
42. $-6[(2x - 5) - 3(8x - 2)]$
43. $4x - 5[3 - (x - 4)]$
44. $x - 7[3x - (2 - x)]$
45. $-(3x - 4y) - [(4x + 2y) - (3x + 7y)]$
46. $(8x - y) - [-(2x + y) - (-3x - 6y)]$
47. $4a - \{3a + 2[a - 5(a + 1) + 4]\}$
48. $6a - \{-2a - 6[2a + 3(a - 1) - 6]\}$

Use the process given in Examples 9 and 10 to work the following problems:

49. If $P(x) = 2x^2 - 3x - 4$, find $P(2)$.
50. If $P(x) = 4x^2 + 3x - 2$, find $P(-1)$.
51. If $Q(x) = x^2 - 6x + 5$, find $Q(0)$.
52. If $Q(x) = 4x^2 - 4x + 4$, find $Q(0)$.
53. If $P(x) = 5x^2 - 2x + 3$, find $P(-1)$, $P(0)$, and $P(2)$.
54. If $P(x) = 3x^2 - 6x - 6$, find $P(-1)$, $P(0)$, and $P(2)$.

55. If $R(x) = x^3 - x^2 + x - 1$, find $R(-2)$, $R(0)$, and $R(3)$.

56. If $R(x) = x^3 + x^2 + x + 1$, find $R(-3)$, $R(0)$, and $R(1)$.

57. If $P(x) = 2x^3 - 6x^2 + 3$, find $P(-2)$, $P(-1)$, and $P(0)$.

58. If $P(x) = -3x^3 - 2x + 1$, find $P(-5)$, $P(0)$, and $P(4)$.

59. If an object is thrown straight up into the air with a velocity of 128 feet/second, then its height H above the ground t seconds later is given by the equation

$$H(t) = -16t^2 + 128t$$

We can let $H(1)$ represent its height after 1 second, $H(2)$ its height after 2 seconds, and so on. Find the height of the object after 1 second, after 2 seconds, and after 3 seconds.

The distributive property is the key to multiplying polynomials. The simplest type of multiplication occurs when we multiply a polynomial by a monomial.

3.3 Multiplication of Polynomials

▼ **Example 1** Find the product of $4x^3$ and $5x^2 - 3x + 1$.

Solution

$4x^3(5x^2 - 3x + 1)$
$= 4x^3(5x^2) + 4x^3(-3x) + 4x^3(1)$ Distributive property
$= 20x^5 - 12x^4 + 4x^3$

Notice we multiply coefficients and add exponents. ▲

▼ **Example 2** Multiply $-2a^2b(4a^3 - 6a^2b + ab^2 - 3b^3)$.

Solution

$-2a^2b(4a^3 - 6a^2b + ab^2 - 3b^3)$
$= (-2a^2b)(4a^3) + (-2a^2b)(-6a^2b) + (-2a^2b)(ab^2)$
$\quad + (-2a^2b)(-3b^3)$
$= -8a^5b + 12a^4b^2 - 2a^3b^3 + 6a^2b^4$ ▲

The distributive property can also be applied to multiply a polynomial by a polynomial. Let's consider the case where both polynomials have two terms:

▼ **Example 3** Multiply $2x - 3$ and $x + 5$.

Solution

$(2x - 3)(x + 5)$
$= (2x - 3)x + (2x - 3)5$ Distributive property
$= 2x(x) + (-3)x + 2x(5) + (-3)5$ Distributive property
$= 2x^2 - 3x + 10x - 15$
$= 2x^2 + 7x - 15$ Combine like terms.

Notice the third line in this example. It consists of all possible products of terms in the first binomial and those of the second binomial. We can generalize this into a rule for multiplying two polynomials.   ▲

RULE To multiply two polynomials, multiply each term in the first polynomial by each term in the second polynomial.

Multiplying polynomials can be accomplished by a method that looks very similar to long multiplication with whole numbers. We line up the polynomials vertically and then apply our rule for multiplication of polynomials. Here's how it looks using the same two binomials used in the last example:

$$
\begin{array}{r}
2x - 3 \\
x + 5 \\
\hline
10x - 15 \\
2x^2 - 3x \quad\ \\
\hline
2x^2 + 7x - 15
\end{array}
$$

Multiply $+5$ times $2x - 3$.
Multiply x times $2x - 3$.
Add in columns.

The vertical method of multiplying polynomials does not directly show the use of the distributive property. It is, however, very useful since it always gives the correct result and is easy to remember.

▼ **Example 4** Multiply $(2x - 3y)$ and $(3x^2 - xy + 4y^2)$ vertically.

Solution

$$
\begin{array}{r}
3x^2 - xy + 4y^2 \\
2x - 3y \\
\hline
-9x^2y + 3xy^2 - 12y^3 \\
6x^3 - 2x^2y + 8xy^2 \quad\ \\
\hline
6x^3 - 11x^2y + 11xy^2 - 12y^3
\end{array}
$$

Multiply $(3x^2 - xy + 4y^2)$ by $-3y$.
Multiply $(3x^2 - xy + 4y^2)$ by $2x$.
Add similar terms.   ▲

▼   **Example 5**   Multiply $(x - 3)(2x + 1)(3x - 4)$.

Solution   By the associative property we can first find the product of any two of the binomials and then multiply the result by the other binomial.

We begin by multiplying the first two binomials:

$$
\begin{array}{r}
2x + 1 \\
x - 3 \\
\hline
-6x - 3 \\
2x^2 + x \\
\hline
2x^2 - 5x - 3
\end{array}
$$

We now multiply this product by the last binomial, $3x - 4$:

$$
\begin{array}{r}
2x^2 - 5x - 3 \\
3x - 4 \\
\hline
-8x^2 + 20x + 12 \\
6x^3 - 15x^2 - 9x \\
\hline
6x^3 - 23x^2 + 11x + 12
\end{array}
$$

The product is

$$(x - 3)(2x + 1)(3x - 4) = 6x^3 - 23x^2 + 11x + 12 \qquad ▲$$

The product of two binomials occurs very frequently in algebra. Since this type of product is so common, we have a special method of multiplication that applies only to products of binomials.

Consider the product of $(2x - 5)$ and $(3x - 2)$. Distributing $(3x - 2)$ over $2x$ and -5 we have

$$
\begin{aligned}
(2x - 5)(3x - 2) &= (2x)(3x - 2) + (-5)(3x - 2) \\
&= (2x)(3x) + (2x)(-2) + (-5)(3x) + (-5)(-2) \\
&= 6x^2 - 4x - 15x + 10 \\
&= 6x^2 - 19x + 10
\end{aligned}
$$

Looking closely at the second and third lines we notice the following relationships:

1. $6x^2$ comes from multiplying the *first* terms in each binomial:

$$(2x - 5)(3x - 2) \qquad 2x(3x) = 6x^2 \qquad First \text{ terms}$$

2. $-4x$ comes from multiplying the *outside* terms in the product:

$$(2x - 5)(3x - 2) \qquad 2x(-2) = -4x \qquad Outside \text{ terms}$$

3. $-15x$ comes from multiplying the *inside* terms in the product:

$$(2x - 5)(3x - 2) \qquad -5(3x) = -15x \qquad Inside \text{ terms}$$

4. 10 comes from multiplying the *last* two terms in the product:

$$(2x - 5)(3x - 2) \qquad -5(-2) = 10 \qquad Last \text{ terms}$$

Once we know where the terms in the answer come from, we can reduce the number of steps used in finding the product:

$$(2x - 5)(3x - 2) = 6x^2 - 4x - 15x + 10$$
$$ \text{First} \quad \text{Outside} \quad \text{Inside} \quad \text{Last}$$
$$= 6x^2 - 19x + 10$$

This method is called the FOIL Method—*F*irst-*O*utside-*I*nside-*L*ast. The FOIL method does not show the properties used in multiplying two binomials. It is simply a way of finding products of binomials quickly. Remember, the FOIL method only applies to products of two binomials. The vertical method applies to all products of polynomials with two or more terms.

▼ **Example 6** Multiply $(4a - 5b)(3a + 2b)$.

Solution

$$(4a - 5b)(3a + 2b) = 12a^2 + 8ab - 15ab - 10b^2$$
$$ \updownarrow \qquad \updownarrow \qquad \updownarrow \qquad \updownarrow$$
$$ \text{F} \qquad \text{O} \qquad \text{I} \qquad \text{L}$$
$$= 12a^2 - 7ab - 10b^2 \qquad\qquad ▲$$

▼ **Example 7** Find $(4x - 6)^2$.

Solution

$$(4x - 6)^2 = (4x - 6)(4x - 6)$$
$$= 16x^2 - 24x - 24x + 36$$
$$ \text{F} \qquad \text{O} \qquad \text{I} \qquad \text{L}$$
$$= 16x^2 - 48x + 36 \qquad\qquad ▲$$

The last example is the square of a binomial. This type of product occurs frequently enough in algebra that we have a special formula for it.

Here are the formulas for binomial squares:

$$(a + b)^2 = (a + b)(a + b) = a^2 + ab + ab + b^2 = a^2 + 2ab + b^2$$

$$(a - b)^2 = (a - b)(a - b) = a^2 - ab - ab + b^2 = a^2 - 2ab + b^2$$

Observing the results in both cases we have the following rule:

RULE   The square of a binomial is the sum of the square of the first term, twice the product of the two terms, and the square of the last term. Or:

$$(a + b)^2 = \quad a^2 \quad + \quad 2ab \quad + \quad b^2$$

	Square of first term	Twice the product of the two terms	Square of last term

$$(a - b)^2 = \quad a^2 \quad - \quad 2ab \quad + \quad b^2$$

▼   **Example 8**

a. $(x + y)^2 = \quad x^2 \quad + \quad 2xy \quad + \quad y^2 \quad = x^2 + 2xy + y^2$
b. $(x + 7)^2 = \quad x^2 \quad + \quad 2(x)(7) \quad + \quad 7^2 \quad = x^2 + 14x + 49$
c. $(3x - 5)^2 = \quad (3x)^2 \quad + 2(3x)(-5) + \quad (-5)^2 \quad = 9x^2 - 30x + 25$
d. $(4x - 2y)^2 = \quad (4x)^2 \quad + 2(4x)(-2y) + \quad (-2y)^2 \quad = 16x^2 - 16xy + 4y^2$

	First term squared	Twice their product	Last term squared	Answer
				▲

Note   From the above rule and examples it should be obvious that $(a + b)^2 \neq a^2 + b^2$. That is, the square of a sum is not the same as the sum of the squares.

Another frequently occurring kind of product is found when multiplying two binomials which differ only in the sign between their terms.

▼   **Example 9**   Multiply $(3x - 5)$ and $(3x + 5)$.

Solution

$$(3x - 5)(3x + 5) = 9x^2 + 15x - 15x - 25 \qquad \text{Two middle terms}$$
$$\text{add to } 0.$$

$$= 9x^2 - 25 \qquad \blacktriangle$$

The outside and inside products in Example 9 are opposites and therefore add to 0.
Here it is in general:

$$(a - b)(a + b) = a^2 + ab - ab + b^2 \qquad \text{Two middle terms}$$
$$\text{add to } 0.$$

$$= a^2 - b^2$$

RULE To multiply two binomials which differ only in the sign between their two terms, simply subtract the square of the second term from the square of the first term:

$$(a - b)(a + b) = a^2 - b^2$$

The expression $a^2 - b^2$ is called the *difference of two squares*.

Once we memorize and understand this rule, we can multiply binomials of this form with a minimum of work.

▼ **Example 10**

a. $(x - 5)(x + 5) = x^2 - 25$
b. $(2a - 3)(2a + 3) = 4a^2 - 9$
c. $(x^2 + 4)(x^2 - 4) = x^4 - 16$
d. $(3x + 7y)(3x - 7y) = 9x^2 - 49y^2$
e. $(x^3 - 2a)(x^3 + 2a) = x^6 - 4a^2$ 　　　　　　　　　▲

When working through the problems in the problem set, pay particular attention to products of binomials. We want to be aware of the relationship between the terms in the answers and the terms in the original binomials. In the next section we will do factoring. Since factoring is actually the reverse of multiplication, the more familiar we are with products, the easier factoring will be.

Problem Set 3.3

Multiply the following by applying the distributive property:

1. $2x(6x^2 - 5x + 4)$ 　　　　　　　2. $-3x(5x^2 - 6x - 4)$
3. $-3a^2(a^3 - 6a^2 + 7)$ 　　　　　4. $4a^3(3a^2 - a + 1)$

5. $2a^2b(a^3 - ab + b^3)$
6. $-5a^2b^2(8a^2 - 2ab + b^2)$
7. $-4x^2y^3(7x^2 - 3xy + 6y^2)$
8. $-3x^3y^2(6x^2 - 3xy + 4y^2)$
9. $3r^3s^2(r^3 - 2r^2s + 3rs^2 + s^3)$
10. $-5r^2s^3(2r^3 + 3r^2s - 4rs^2 + 5s^3)$

Multiply the following vertically:

11. $(x - 5)(x + 3)$
12. $(x + 4)(x + 6)$
13. $(2x - 3)(3x - 5)$
14. $(3x + 4)(2x - 5)$
15. $(x + 3)(x^2 + 6x + 5)$
16. $(x - 2)(x^2 - 5x + 7)$
17. $(3a + 5)(2a^3 - 3a^2 + a)$
18. $(2a - 3)(3a^2 - 5a + 1)$
19. $(a - b)(a^2 + ab + b^2)$
20. $(a + b)(a^2 - ab + b^2)$
21. $(2x + y)(4x^2 - 2xy + y^2)$
22. $(x - 3y)(x^2 + 3xy + 9y^2)$
23. $(2a - 3b)(a^2 + ab + b^2)$
24. $(5a - 2b)(a^2 - ab - b^2)$
25. $(3x - 4y)(6x^2 + 3xy + 4y^2)$
26. $(2x - 6y)(7x^2 - 6xy + 3y^2)$
27. $2x^2(x - 5)(3x - 7)$
28. $-5x^3(3x - 2)(x + 4)$
29. $(x - 2)(2x + 3)(3x - 4)$
30. $(x + 5)(2x - 6)(x - 3)$

Multiply the following using the FOIL method:

31. $(x - 2)(x + 3)$
32. $(x + 2)(x - 3)$
33. $(x - 2)(x - 3)$
34. $(x + 2)(x + 3)$
35. $(2a + 3)(3a + 2)$
36. $(5a - 4)(2a + 1)$
37. $(3x - 5)(2x + 4)$
38. $(x - 7)(3x + 6)$
39. $(5x - 4)(x - 5)$
40. $(7x - 5)(3x + 1)$
41. $(4a + 1)(5a + 1)$
42. $(3a - 1)(2a - 1)$
43. $(5x - 6y)(4x + 3y)$
44. $(6x - 5y)(2x - 3y)$
45. $(2x - 3y)(4x - 5y)$
46. $(6x - 2y)(3x + y)$
47. $(4a + b)(7a - 2b)$
48. $(3a - 4b)(6a + b)$

Find the following special products:

49. $(x + 2)^2$
50. $(x - 5)^2$
51. $(2a - 3)^2$
52. $(3a + 2)^2$
53. $(5x + 2y)^2$
54. $(3x - 4y)^2$
55. $(x - 4)(x + 4)$
56. $(x + 5)(x - 5)$
57. $(2a + 3b)(2a - 3b)$
58. $(6a - 1)(6a + 1)$
59. $(3r + 7s)(3r - 7s)$
60. $(5r - 2s)(5r + 2s)$
61. $(5x - 4y)(5x + 4y)$
62. $(4x - 5y)(4x + 5y)$

Find the following products:

63. $(x - 2)^3$
64. $(x + 4)^3$
65. $3(x - 1)(x - 2)(x - 3)$
66. $2(x + 1)(x + 2)(x + 3)$
67. $(x^N + 3)(x^N - 2)$
68. $(x^N + 4)(x^N - 1)$
69. $(x^{2N} - 3)(x^{2N} + 3)$
70. $(x^{3N} + 4)(x^{3N} - 4)$
71. The flower color (either red, pink, or white) of a certain species of sweet pea plant is due to a single gene. If p is the proportion of the dominant form of the gene in a population, and q is the proportion of the recessive

form of this gene in the same population, then the proportion of plants in the next generation that have pink flowers is given by the middle term in the expansion of $(p + q)^2$. If $p = \frac{1}{4}$ and $q = \frac{3}{4}$, find the proportion of the next generation that will have pink flowers.

3.4
Division of
Polynomials

We begin this section by considering division of a polynomial by a monomial. This is the simplest kind of polynomial division. The rest of the section is devoted to division of a polynomial by a polynomial. This kind of division is very similar to long division with whole numbers.

Dividing a Polynomial
by a Monomial

To divide a polynomial by a monomial we use the definition of division and apply the distributive property. The following example illustrates the procedure.

▼   **Example 1**

$$\frac{10x^5 - 15x^4 + 20x^3}{5x^2}$$

$$= (10x^5 - 15x^4 + 20x^3) \cdot \frac{1}{5x^2}$$ Dividing by $5x^2$ is the same as multiplying by $1/5x^2$.

$$= 10x^5 \cdot \frac{1}{5x^2} - 15x^4 \cdot \frac{1}{5x^2} + 20x^3 \cdot \frac{1}{5x^2}$$ Distributive property.

$$= \frac{10x^5}{5x^2} - \frac{15x^4}{5x^2} + \frac{20x^3}{5x^2}$$ Multiplying by $1/5x^2$ is the same as dividing by $5x^2$.

$$= 2x^3 - 3x^2 + 4x$$ Divide coefficients, subtract exponents.

Notice that division of a polynomial by a monomial is accomplished by dividing each term of the polynomial by the monomial. The first two steps are usually not shown in a problem like this. They are part of Example 1 to justify distributing $5x^2$ under all three terms of the polynomial $10x^5 - 15x^4 + 20x^3$.   ▲

Here are some more examples of this kind of division:

▼ **Example 2**

a. $\dfrac{8x^3y^5 - 16x^2y^2 + 4x^4y^3}{-2x^2y} = \dfrac{8x^3y^5}{-2x^2y} + \dfrac{-16x^2y^2}{-2x^2y} + \dfrac{4x^4y^3}{-2x^2y}$

$$= -4xy^4 + 8y - 2x^2y^2$$

b. $\dfrac{10a^4b^2 + 8ab^3 - 12a^3b + 6ab}{4a^2b^2} = \dfrac{10a^4b^2}{4a^2b^2} + \dfrac{8ab^3}{4a^2b^2} - \dfrac{12a^3b}{4a^2b^2} + \dfrac{6ab}{4a^2b^2}$

$$= \dfrac{5a^2}{2} + \dfrac{2b}{a} - \dfrac{3a}{b} + \dfrac{3}{2ab} \qquad ▲$$

Notice in part b of Example 2 that the result is not a polynomial because of the last three terms. If we were to write each as a product, some of the variables would have negative exponents. For example, the second term would be

$$\frac{2b}{a} = 2a^{-1}b$$

Since division of a polynomial by a polynomial is very similar to long division with whole numbers, we begin with a detailed example of whole-number division:

Dividing a Polynomial by a Polynomial

▼ **Example 3**   Divide: $25\overline{)4628}$.

Solution

$$
\begin{array}{r}
1 \\
25\overline{)4628} \\
\underline{25} \\
21
\end{array}
$$
  ⟵ Estimate: 25 into 46
  ⟵ Multiply: $1 \times 25 = 25$
  ⟵ Subtract: $46 - 25 = 21$

$$
\begin{array}{r}
1 \\
25\overline{)4628} \\
\underline{25\downarrow} \\
212
\end{array}
$$
  ⟵ Bring down the 2.

These are the four basic steps in long division: estimate, multiply, subtract, and bring down the next term. To complete the problem we simply perform the same four steps again.

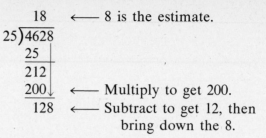

$$
\begin{array}{r}
18 \\
25\overline{)4628} \\
25 \\
\overline{212} \\
200 \\
\overline{128}
\end{array}
$$

18 ⟵ 8 is the estimate.

200 ⟵ Multiply to get 200.

128 ⟵ Subtract to get 12, then bring down the 8.

One more time:

$$
\begin{array}{r}
185 \\
25\overline{)4628} \\
25 \\
\overline{212} \\
200 \\
\overline{128} \\
125 \\
\overline{3}
\end{array}
$$

185 ⟵ 5 is the estimate.

125 ⟵ Multiply to get 125.

3 ⟵ Subtract to get 3.

Since 3 is less than 25, we have our answer:

$$\frac{4628}{25} = 185 + \frac{3}{25}$$

To check our answer, we multiply 185 by 25, then add 3 to the result:

$$25(185) + 3 = 4625 + 3 = 4628 \qquad \blacktriangle$$

Note You may realize when looking over this last example that you don't have a very good idea why you proceed as you do with the steps in long division. What you do know is the process always works. We are going to approach the explanation for division of two polynomials with this in mind. That is, we won't always be sure why the steps we use are important, only that they always produce the correct result.

The method used to divide a polynomial by a polynomial is very similar to the method used to divide the numbers in Example 3. Both use the same four basic steps: estimate, multiply, subtract, and bring down the next term.

▼ **Example 4** Divide: $\dfrac{2x^2 - 7x + 9}{x - 2}$.

Solution

$$\begin{array}{r} 2x \\ x-2\overline{)\ 2x^2-7x+9} \end{array}$$ ⟵ Estimate: $2x^2 \div x = 2x$.

$$\begin{array}{r} \cancel{-}\ \cancel{+} \\ \cancel{}2x^2\ \cancel{-}\ 4x \\ \hline -3x \end{array}$$ ⟵ Multiply: $2x(x-2) = 2x^2 - 4x$
               ⟵ Subtract: $(2x^2 - 7x) - (2x^2 - 4x) = -3x$

$$\begin{array}{r} 2x \\ x-2\overline{)\ 2x^2-7x+9} \\ \cancel{-}\ \cancel{+} \downarrow \\ \cancel{}2x^2\ \cancel{-}\ 4x \\ \hline -3x+9 \end{array}$$ ⟵ Bring down the 9.

      Notice we change the signs on $2x^2 - 4x$ and add in the subtraction step. Subtracting a polynomial is equivalent to adding its opposite.

      We repeat the four steps again:

$$\begin{array}{r} 2x\ -3 \\ x-2\overline{)\ 2x^2-7x+9} \end{array}$$ ⟵ -3 is the estimate: $-3x \div x = -3$

$$\begin{array}{r} \cancel{-}\ \cancel{+} \\ \cancel{}2x^2\ \cancel{-}\ 4x \\ \hline -3x+9 \\ +\ \ - \\ \cancel{-}\ 3x\ \cancel{+}\ 6 \\ \hline 3 \end{array}$$ ⟵ Multiply: $-3(x-2) = -3x + 6$
               ⟵ Subtract: $(-3x + 9) - (-3x + 6) = 3$

      Since we have no other term to bring down, we have our answer:

$$\frac{2x^2 - 7x + 9}{x - 2} = 2x - 3 + \frac{3}{x - 2}$$

      To check we multiply $(2x - 3)(x - 2)$ to get $2x^2 - 7x + 6$; then, adding the remainder 3 to this result, we have $2x^2 - 7x + 9$. ▲

      In setting up a division problem involving two polynomials there are two things to remember: (1) both polynomials should be in decreasing powers of the variable, and (2) neither should skip any powers from the highest power down to the constant term. If there are any missing terms, they can be filled in using a coefficient of 0.

▼   **Example 5**   Divide: $2x - 4 \overline{)4x^3 - 6x - 11}$.

Solution   Since the first polynomial is missing a term in x^2, we can fill it in with $0x^2$:

$$4x^3 - 6x - 11 = 4x^3 + 0x^2 - 6x - 11$$

Adding $0x^2$ does not change our original problem.

$$
\begin{array}{r}
2x^2 + 4x\ + 5 \\
2x - 4 \overline{)\ 4x^3 + 0x^2 -\ \ \ 6x - 11} \\
\end{array}
$$

Notice: adding the $0x^2$ term gives us a column in which to write $+8x^2$.

$$
\begin{array}{r}
 \\
\underset{\,}{\;\;\;\;\;\; - + } \\
\cancel{+}4x^3 \cancel{+}\ 8x^2 \\
\hline
+ 8x^2 - \ \ \ 6x \\
- + \\
\cancel{+}\ 8x^2 \cancel{+}\ 16x \\
\hline
+ 10x - 11 \\
- + \\
\cancel{+}\ 10x \cancel{+}\ 20 \\
\hline
+9
\end{array}
$$

$$\frac{4x^3 - 6x - 11}{2x - 4} = 2x^2 + 4x + 5 + \frac{9}{2x - 4}$$

▲

▼   **Example 6**   Divide: $x + 3 \overline{)5 + 8x^2 + 2x^3}$.

Solution   To begin we must write the first polynomial in descending powers of the variable and add a $0x$ term, to keep a column for terms in x.

Here is the complete problem:

$$
\begin{array}{r}
2x^2 + 2x - 6 \\
x + 3 \overline{)\ 2x^3 + 8x^2 + 0x + \ \ 5} \\
- - \\
\cancel{+}2x^3 \cancel{+}\ 6x^2 \\
\hline
2x^2 + 0x \\
- - \\
\cancel{+}\ 2x^2 \cancel{+}\ 6x \\
\hline
- 6x + \ \ 5 \\
+ + \\
\cancel{+}\ 6x \cancel{+}\ 18 \\
\hline
23
\end{array}
$$

$$\frac{2x^3 + 8x^2 + 5}{x + 3} = 2x^2 + 2x - 6 + \frac{23}{x + 3}$$ ▲

▼ **Example 7**   Divide $\dfrac{x^2 - 6xy - 7y^2}{x + y}$.

Solution

$$
\begin{array}{r}
x \ \ - 7y \\
x + y\overline{) \ x^2 - 6xy - 7y^2} \\
\underline{-\quad\ \ -\quad\quad} \\
\not{x^2} \not{+}\ xy \\
\underline{- 7xy - 7y^2} \\
+\qquad + \\
\underline{\not{+} 7xy \not{+} 7y^2} \\
0
\end{array}
$$

In this case the remainder is 0 and we have

$$\frac{x^2 - 6xy - 7y^2}{x + y} = x - 7y$$

which is easy to check since

$$(x + y)(x - 7y) = x^2 - 6xy - 7y^2$$ ▲

Find the following quotients: *Problem Set 3.4*

1. $\dfrac{4x^3 - 8x^2 + 6x}{2x}$

2. $\dfrac{6x^3 + 12x^2 - 9x}{3x}$

3. $\dfrac{10x^4 + 15x^3 - 20x^2}{-5x^2}$

4. $\dfrac{12x^5 - 18x^4 - 6x^3}{6x^3}$

5. $\dfrac{8y^5 + 10y^3 - 6y}{4y^3}$

6. $\dfrac{6y^4 - 3y^3 + 18y^2}{9y^2}$

7. $\dfrac{5x^3 - 8x^2 - 6x}{-2x^2}$

8. $\dfrac{-9x^5 + 10x^3 - 12x}{-6x^4}$

9. $\dfrac{28a^3b^5 + 42a^4b^3}{7a^2b^2}$

10. $\dfrac{a^2b + ab^2}{ab}$

11. $\dfrac{10x^3y^2 - 20x^2y^3 - 30x^3y^3}{-10x^2y}$

12. $\dfrac{9x^4y^4 + 18x^3y^4 - 27x^2y^4}{-9xy^3}$

13. $\dfrac{8x^3 - 6x^2y + 10xy^2 + 4y^3}{2xy}$

14. $\dfrac{6x^3 + 12x^2y - 9xy^2 - 15y^3}{3xy}$

15. $\dfrac{a^{9N} - a^{6N}}{a^{3N}}$ **16.** $\dfrac{a^{5r} + a^{4r} + a^{3r}}{a^{2r}}$

17. $\dfrac{24x^{6M} - 18x^{3M} + 12x^{M}}{6x^{M}}$ **18.** $\dfrac{10x^{2M} - 25x^{3M} - 30x^{5M}}{5x^{2M}}$

Do each division problem using the long division method:

19. $\dfrac{x^2 - 5x - 7}{x + 2}$ **20.** $\dfrac{x^2 + 4x - 8}{x - 3}$

21. $\dfrac{6x^2 + 7x - 18}{3x - 4}$ **22.** $\dfrac{8x^2 - 26x - 9}{2x - 7}$

23. $\dfrac{2x^3 - 3x^2 - 4x + 5}{x + 1}$ **24.** $\dfrac{3x^3 - 5x^2 + 2x - 1}{x - 2}$

25. $\dfrac{2y^3 - 9y^2 - 17y + 12}{2y - 3}$ **26.** $\dfrac{3y^3 - 19y^2 + 17y + 4}{3y - 4}$

27. $\dfrac{8x^2 - 6xy + 9y^2}{2x - 3y}$ **28.** $\dfrac{6x^2 - 13xy - 8y^2}{3x + y}$

29. $\dfrac{2x^3 - 11x - 9x^2 - 5}{2x^2 - 3x + 2}$ **30.** $\dfrac{6x^3 + 5 + 5x - 11x^2}{3x^2 - x + 1}$

31. $\dfrac{6y^3 - 8y + 5}{2y - 4}$ **32.** $\dfrac{9y^3 - 6y^2 + 8}{3y - 3}$

33. $\dfrac{x^4 - 2x + 5}{x - 7}$ **34.** $\dfrac{x^4 + x^3 - 1}{x + 2}$

35. $\dfrac{y^4 - 16}{y - 2}$ **36.** $\dfrac{y^4 - 81}{y - 3}$

37. $\dfrac{x^3 - y^3}{x - y}$ **38.** $\dfrac{x^3 + y^3}{x + y}$

39. $\dfrac{x^4 + x^3 - 3x^2 - x + 2}{x^2 + 3x + 2}$ **40.** $\dfrac{2x^4 + x^3 + 3x - 3}{2x^2 - x + 3}$

**3.5
The Greatest
Common Factor and
Factoring by
Grouping**

In this section we will consider two basic types of factoring: factoring out the greatest common factor, and factoring by grouping. Both types of factoring rely heavily on the distributive property.

In general, factoring is the reverse of multiplication. The following diagram illustrates the relationship between factoring and multiplication:

Multiplication

Factors $3 \cdot 7 = 21$ Product

Factoring

Reading from left to right we say the product of 3 and 7 is 21. Reading in the other direction, from right to left, we say 21 factors into 3 times 7. Or, 3 and 7 are factors of 21.

DEFINITION   The *greatest common factor* for a polynomial is the largest monomial that divides (is a factor of) each term of the polynomial.

The greatest common factor for the polynomial $25x^5 + 20x^4 - 30x^3$ is $5x^3$ since it is the largest monomial that is a factor of each term. We can apply the distributive property and write

$$25x^5 + 20x^4 - 30x^3 = 5x^3(5x^2) + 5x^3(4x) + 5x^3(-6)$$
$$= 5x^3(5x^2 + 4x - 6)$$

The last line is written in factored form.

Once we recognize the greatest common factor for a polynomial, we apply the distributive property and factor it from each term. We rewrite the original polynomial as the product of its greatest common factor and the polynomial that remains after the greatest common factor has been factored from each term.

▼   **Example 1**   Factor the greatest common factor from

$$16a^5b^4 - 24a^2b^5 - 8a^3b^3$$

Solution   The largest monomial that divides each term is $8a^2b^3$. We write each term of the original polynomial in terms of $8a^2b^3$ and apply the distributive property to write the polynomial in factored form:

$16a^5b^4 - 24a^2b^5 - 8a^3b^3$
$$= 8a^2b^3(2a^3b) + 8a^2b^3(-3b^2) + 8a^2b^3(-a)$$
$$= 8a^2b^3(2a^3b - 3b^2 - a)   \blacktriangle$$

Note   The term *largest monomial* as used here refers to the monomial with the largest integer exponents whose coefficient has the greatest absolute value. We could have factored the polynomial in Example 1 correctly by taking out $-8a^2b^3$. We usually keep the coefficient of the greatest common factor positive. However, it is not incorrect (and is sometimes useful) to have the coefficient negative.

Here are some further examples of factoring out the greatest common factor:

▼ **Example 2**

a. $16x^3y^4z^5 + 10x^4y^3z^2 - 12x^2y^3z^4$
$$= 2x^2y^3z^2(8xyz^3) + 2x^2y^3z^2(5x^2) - 2x^2y^3z^2(6z^2)$$
$$= 2x^2y^3z^2(8xyz^3 + 5x^2 - 6z^2)$$

b. $5(a + b)^3 - 10(a + b)^2 + 5(a + b)$
$$= 5(a + b)(a + b)^2 - 5(a + b)2(a + b) + 5(a + b)(1)$$
$$= 5(a + b)[(a + b)^2 - 2(a + b) + 1] \qquad \blacktriangle$$

The second step in the two parts of the above example is not necessary. It is shown here simply to emphasize use of the distributive property.

Factoring by Grouping

Many polynomials have no greatest common factor other than the number 1. Some of these can be factored using the distributive property if those terms with a common factor are grouped together.

For example, the polynomial $5x + 5y + x^2 + xy$ can be factored by noticing that the first two terms have a 5 in common, whereas the last two have an x in common.

Applying the distributive property we have

$$5x + 5y + x^2 + xy = 5(x + y) + x(x + y)$$

This last expression can be thought of as having two terms, $5(x + y)$ and $x(x + y)$, each of which has a common factor $(x + y)$. We apply the distributive property again to factor $(x + y)$ from each term:

$$\underline{5(x + y) + x(x + y)}$$
$$= (5 + x)(x + y)$$

▼ **Example 3**   Factor $a^2b^2 + b^2 + 8a^2 + 8$.

Solution The first two terms have b^2 in common; the last two have 8 in common.

$$a^2b^2 + b^2 + 8a^2 + 8 = b^2(a^2 + 1) + 8(a^2 + 1)$$
$$= (b^2 + 8)(a^2 + 1) \qquad \blacktriangle$$

▼ **Example 4**   Factor $15 - 5y^4 - 3x^3 + x^3y^4$.

Solution Let's try factoring a 5 from the first two terms and an x^3 from the last two terms:

$$15 - 5y^4 - 3x^3 + x^3y^4 = 5(3 - y^4) + x^3(-3 + y^4)$$

Now, $3 - y^4$ and $-3 + y^4$ are not equal and we cannot factor further. Notice, however, that we can factor $-x^3$ instead of x^3 from the last two terms and obtain the desired result:

$$15 - 5y^4 - 3x^3 + x^3y^4 = 5(3 - y^4) - x^3(3 - y^4)$$
$$= (5 - x^3)(3 - y^4) \quad \blacktriangle$$

The type of factoring shown in these last two examples is called factoring by grouping. With some practice it becomes fairly mechanical.

Factor the greatest common factor from each of the following. (The answers in the back of the book all show greatest common factors whose coefficients are positive.)

Problem Set 3.5

1. $10x^3 - 15x^2$
2. $12x^5 + 18x^7$
3. $9y^6 + 18y^3$
4. $24y^4 - 8y^2$
5. $9a^2b - 6ab^2$
6. $30a^3b^4 + 20a^4b^3$
7. $21xy^4 + 7x^2y^2$
8. $14x^6y^3 - 6x^2y^4$
9. $8a^3 - 6a^2 + 10a$
10. $27a^5 - 9a^2 + 18a^4$
11. $22x^4 - 33x^2 + 11x^6$
12. $-35x^2 + 21x^3 - 14x^2$
13. $10x^3y^2 + 20x^3y^3 - 30x^2y^3$
14. $6x^4y - 9xy^4 + 18x^3y^3$
15. $-x^2y + xy^2 - x^2y^2$
16. $-x^3y^2 - x^2y^3 - x^2y^2$
17. $4x^3y^2z - 8x^2y^2z^2 + 6xy^2z^3$
18. $7x^4y^3z^2 - 21x^2y^2z^2 - 14x^2y^3z^4$
19. $20a^2b^2c^2 - 30ab^2c + 25a^2bc^2$
20. $8a^3bc^5 - 48a^2b^4c + 16ab^3c^5$
21. $5x(a - 2b) - 3y(a - 2b)$
22. $3a(x - y) - 7b(x - y)$
23. $3x^2(x + y)^2 - 6y^2(x + y)^2$
24. $10x^3(2x - 3y) - 15x^2(2x - 3y)$
25. $11(a - b)^4 + 22(a - b)^3 + 33(a - b)^2$
26. $25(a + 2b)^5 - 30(a + 2b)^4 + 20(a + 2b)^3$

Factor each of the following by grouping:

27. $3xy + 3y + 2ax + 2a$
28. $5xy^2 + 5y^2 + 3ax + 3a$
29. $x^2y + x + 3xy + 3$
30. $x^3y^3 + 2x^3 + 5x^2y^3 + 10x^2$
31. $a + b + 5ax + 5bx$
32. $a + b + 7ax + 7bx$
33. $3x - 2y + 6x - 4y$
34. $4x - 5y + 12x - 15y$
35. $3xy^2 - 6y^2 + 4x - 8$
36. $8x^2y - 4x^2 + 6y - 3$
37. $2xy^3 - 8y^3 + x - 4$
38. $8x^2y^4 - 4y^4 + 14x^2 - 7$
39. $x^2 - ax - bx + ab$
40. $ax - x^2 - bx + ab$
41. $ab + 5a - b - 5$
42. $x^2 - xy - ax + ay$
43. $a^4b^2 + a^4 - 5b^2 - 5$
44. $2a^2 - bc^2 - a^2b + 2c^2$
45. $4x^2y + 5y - 8x^2 - 10$
46. $x^3y^2 + 8y^2b^2 - x^3 - 8b^2$

3.6
Factoring Trinomials

Factoring trinomials is probably the most common type of factoring found in algebra. We begin this section by considering trinomials which have a leading coefficient of 1. The remainder of the section is concerned with trinomials with leading coefficients other than 1. The more familiar we are with multiplication of binomials, the easier factoring trinomials will be.

In Section 3.3 we multiplied binomials:

$$(x - 2)(x + 3) = x^2 + x - 6$$
$$(x + 5)(x + 2) = x^2 + 7x + 10$$

In each case the product of two binomials is a trinomial. The first term in the resulting trinomial is obtained by multiplying the first term in each binomial. The middle term comes from adding the product of the two inside terms and the two outside terms. The last term is the product of the last term in each binomial.

In general,

$$(x + a)(x + b) = x^2 + ax + bx + ab$$
$$= x^2 + (a + b)x + ab$$

Writing this as a factoring problem we have

$$x^2 + (a + b)x + ab = (x + a)(x + b)$$

To factor a trinomial with a leading coefficient of 1, we simply find the two numbers a and b whose sum is the coefficient of the middle term and whose product is the constant term.

▼   **Example 1**   Factor $x^2 + 5x + 6$.

Solution   The leading coefficient is 1. We need two numbers whose sum is 5 and whose product is 6. The numbers are 2 and 3. (6 and 1 do not work because their sum is not 5.)

$$x^2 + 5x + 6 = (x + 2)(x + 3)$$

To check our work we simply multiply:

$$(x + 2)(x + 3) = x^2 + 3x + 2x + 6$$
$$= x^2 + 5x + 6$$

▲

▼   **Example 2**   Factor $x^2 + 2x - 15$.

Solution Again the leading coefficient is 1. We need two integers whose product is -15 and whose sum is $+2$. The integers are $+5$ and -3.

$$x^2 + 2x - 15 = (x + 5)(x - 3) \qquad \blacktriangle$$

If a trinomial is factorable, then its factors are unique. For instance, in the example above we found factors of $x + 5$ and $x - 3$. These are the only two factors for $x^2 + 2x - 15$. There is no other pair of binomials whose product is $x^2 + 2x - 15$.

▼ **Example 3** Factor $x^2 - xy - 12y^2$.

Solution We need two numbers whose product is $-12y^2$ and whose sum is $-y$. The numbers are $-4y$ and $3y$:

$$x^2 - xy - 12y^2 = (x - 4y)(x + 3y)$$

Checking this result gives

$$(x - 4y)(x + 3y) = x^2 + 3xy - 4xy - 12y^2$$
$$= x^2 - xy - 12y^2 \qquad \blacktriangle$$

▼ **Example 4** Factor $x^2 - 8x + 6$.

Solution Since there is no pair of integers whose product is 6 and whose sum is 8, the trinomial $x^2 - 8x + 6$ is not factorable. We say it is a *prime polynomial*. $\qquad \blacktriangle$

▼ **Example 5** Factor $3x^4 - 15x^3y - 18x^2y^2$.

Solution The leading coefficient is not 1. However, each term is divisible by $3x^2$. Factoring this out to begin with we have

$$3x^4 - 15x^3y - 18x^2y^2 = 3x^2(x^2 - 5xy - 6y^2)$$

Factoring the resulting trinomial as in the examples above gives

$$3x^2(x^2 - 5xy - 6y^2) = 3x^2(x - 6y)(x + y) \qquad \blacktriangle$$

Note As a general rule, it is best to factor out the greatest common factor first.

We want to turn our attention now to trinomials with leading coefficients other than 1 and with no greatest common factor other than 1.

Suppose we want to factor $3x^2 - x - 2$. The factors will be a pair of binomials. The product of the first terms will be $3x^2$ and the product of the last terms will be -2. We can list all the possible factors along with their products as follows.

Possible factors	First term	Middle term	Last term
$(x + 2)(3x - 1)$	$3x^2$	$+5x$	-2
$(x - 2)(3x + 1)$	$3x^2$	$-5x$	-2
$(x + 1)(3x - 2)$	$3x^2$	$+x$	-2
$(x - 1)(3x + 2)$	$3x^2$	$-x$	-2

From the last line we see that the factors of $3x^2 - x - 2$ are $(x - 1)(3x + 2)$. That is,

$$3x^2 - x - 2 = (x - 1)(3x + 2)$$

Unlike the first examples in this section, there is no straightforward way of factoring trinomials with leading coefficients other than 1, when the greatest common factor is 1. We must use trial and error or list all the possible factors. In either case the idea is this: look only at pairs of binomials whose products give the correct first and last terms, then look for combinations that will give the correct middle term.

▼ **Example 6** Factor $2x^2 + 13xy + 15y^2$.

Solution Listing all possible factors the product of whose first terms is $2x^2$ and the product of whose last terms is $+15y^2$ yields

Possible factors	Middle term of product
$(2x - 5y)(x - 3y)$	$-11xy$
$(2x - 3y)(x - 5y)$	$-13xy$
$(2x + 5y)(x + 3y)$	$+11xy$
$(2x + 3y)(x + 5y)$	$+13xy$

The last line has the correct middle term:

$$2x^2 + 13xy + 15y^2 = (2x + 3y)(x + 5y)$$

Actually, we did not need to check the first two pairs of possible factors in the above list. All the signs in the trinomial

$2x^2 + 13xy + 15y^2$ are positive. The binomial factors must then be of the form $(ax + b)(cx + d)$, where a, b, c, and d are all positive.

▲

There are other ways to reduce the number of possible factors to consider. For example, if we were to factor the trinomial $2x^2 - 11x + 12$, we would not have to consider the pair of possible factors $(2x - 4)(x - 3)$. If the original trinomial has no greatest common factor other than 1, then neither of its binomial factors will either. The trinomial $2x^2 - 11x + 12$ has a greatest common factor of 1, but the possible factor $2x - 4$ has a greatest common factor of 2: $2x - 4 = 2(x - 2)$. Therefore, we do not need to consider $2x - 4$ as a possible factor.

▼ **Example 7** Factor $18x^3y + 3x^2y^2 - 36xy^3$.

Solution First factor out the greatest common factor $3xy$. Then factor the remaining trinomial:

$$18x^3y + 3x^2y^2 - 36xy^3 = 3xy(6x^2 + xy - 12y^2)$$
$$= 3xy(3x - 4y)(2x + 3y) \qquad ▲$$

▼ **Example 8** Factor $12x^4 + 17x^2 + 6$.

Solution This is a trinomial in x^2:

$$12x^4 + 17x^2 + 6 = (4x^2 + 3)(3x^2 + 2)$$

We could have made the substitution $y = x^2$ to begin with, in order to simplify the trinomial, and then factor as

$$12y^2 + 17y + 6 = (4y + 3)(3y + 2)$$

Again using $y = x^2$, we have:

$$(4y + 3)(3y + 2) = (4x^2 + 3)(3x^2 + 2) \qquad ▲$$

Factor each of the following trinomials:

Problem Set 3.6

1. $x^2 + 7x + 12$ 2. $x^2 - 7x + 12$
3. $x^2 - x - 12$ 4. $x^2 + x - 12$
5. $y^2 + y - 6$ 6. $y^2 - y - 6$
7. $x^2 - 6x - 16$ 8. $x^2 + 2x - 3$
9. $x^2 + 8x + 12$ 10. $x^2 - 2x - 15$

11. $3a^2 - 21a + 30$ 12. $3a^2 - 3a - 6$
13. $4x^3 - 16x^2 - 20x$ 14. $2x^3 - 14x^2 + 20x$
15. $x^2 + 3xy + 2y^2$ 16. $x^2 - 5xy - 24y^2$
17. $a^2 + 3ab - 18b^2$ 18. $a^2 - 8ab - 9b^2$
19. $3x^2 - 6xy - 9y^2$ 20. $5x^2 + 45xy + 100y^2$
21. $2x^5 + 4x^4y + 4x^3y^2$ 22. $3x^4 - 18x^3y + 27x^2y^2$

Factor each of the following:

23. $2x^2 + 7x - 15$ 24. $2x^2 - 7x - 15$
25. $2x^2 + x - 15$ 26. $2x^2 - x - 15$
27. $2x^2 + 13x + 15$ 28. $2x^2 - 13x + 15$
29. $6a^2 + 7a + 2$ 30. $6a^2 - 7a + 2$
31. $4y^2 - y - 3$ 32. $6y^2 + 5y - 6$
33. $6x^2 - x - 2$ 34. $3x^2 + 2x - 5$
35. $4a^2 - 12a + 9$ 36. $12a^2 + 7a - 12$
37. $4x^2 - 11xy - 3y^2$ 38. $3x^2 + 19xy - 14y^2$
39. $10x^2 - 3xy - 18y^2$ 40. $9x^2 + 9xy - 10y^2$
41. $18a^2 + 3ab - 28b^2$ 42. $6a^2 - 7ab - 5b^2$
43. $4x^2 + 2x - 6$ 44. $35x^2 - 60x - 20$
45. $9x^4 + 9x^3 - 10x^2$ 46. $4x^5 + 7x^4 - 2x^3$
47. $12a^4 - 14a^3 - 6a^2$ 48. $13a^5 + 39a^4 + 26a^3$
49. $3x^4 + 10x^2 + 3$ 50. $6x^4 - x^2 - 7$
51. $20a^4 + 37a^2 + 15$ 52. $20a^4 + 13a^2 - 15$
53. $12x^4 + 3x^2 - 9$ 54. $30a^4 - 4a^2 - 2$

**3.7
Special Factoring**

In this section we will end our study of factoring by considering some special formulas. Some of the formulas are familiar, others are not. In any case, the formulas will work best if they are memorized.

**Perfect Square
Trinomials**

We previously listed some special products found in multiplying polynomials. Two of the formulas looked like this:

$$(a + b)^2 = a^2 + 2ab + b^2$$
$$(a - b)^2 = a^2 - 2ab + b^2$$

If we exchange the left and right sides of each formula we have two special formulas for factoring:

$$a^2 + 2ab + b^2 = (a + b)^2$$
$$a^2 - 2ab + b^2 = (a - b)^2$$

The left side of each formula is called a *perfect square trinomial*. The right sides are binomial squares. Perfect square trinomials can always be factored using the usual methods for factoring trinomials. However, if we notice that the first and last terms of a trinomial are perfect squares, it is wise to see if the trinomial factors as a binomial square before attempting to factor by the usual method.

▼   **Example 1**   Factor $x^2 - 6x + 9$.

Solution   Since the first and last terms are perfect squares, we attempt to factor according to the formulas above:

$$x^2 - 6x + 9 = (x - 3)^2$$

If we expand $(x - 3)^2$, we have $x^2 - 6x + 9$, indicating we have factored correctly.   ▲

▼   **Example 2**   Factor $8x^2 - 24xy + 18y^2$.

Solution   We begin by factoring the greatest common factor 2 from each term. We then proceed as in Example 1:

$$8x^2 - 24xy + 18y^2 = 2(4x^2 - 12xy + 9y^2)$$
$$= 2(2x - 3y)^2 \qquad ▲$$

Recall the formula that results in the difference of two squares: $(a - b)(a + b) = a^2 - b^2$. Writing this as a factoring formula we have

The Difference of Two Squares

$$a^2 - b^2 = (a - b)(a + b)$$

▼   **Example 3**   Factor $16x^4 - 81y^4$.

Solution   The first and last terms are perfect squares. We factor according to the formula above:

$$16x^4 - 81y^4 = (4x^2)^2 - (9y^2)^2$$
$$= (4x^2 - 9y^2)(4x^2 + 9y^2)$$

Notice that the first factor is also the difference of two squares. Factoring completely we have

$$16x^4 - 81y^4 = (2x - 3y)(2x + 3y)(4x^2 + 9y^2) \qquad ▲$$

Note   The sum of two squares never factors into the product of two binomials. That is, if we were to attempt to factor $(4x^2 + 9y^2)$ in the last example, we would be unable to find two binomials (or any other polynomials) whose product was $4x^2 + 9y^2$. The factors do not exist as polynomials.

Here is another example of the difference of two squares:

▼ **Example 4** Factor $(x - y)^2 - (x + y)^2$.

Solution If we want to simplify matters somewhat, we can substitute a for $x - y$ and b for $x + y$, and then factor using the formula for the difference of two squares. Or we can factor directly:

$$
\begin{aligned}
(x - y)^2 - (x + y)^2 &= [(x - y) + (x + y)][(x - y) - (x + y)] \\
&= (x - y + x + y)(x - y - x - y) \\
&= (2x)(-2y) \\
&= -4xy
\end{aligned}
$$
▲

The Sum and Difference of Two Cubes

Here are the formulas for factoring the sum and difference of two cubes:

$$
\begin{aligned}
a^3 + b^3 &= (a + b)(a^2 - ab + b^2) \\
a^3 - b^3 &= (a - b)(a^2 + ab + b^2)
\end{aligned}
$$

Since these formulas are unfamiliar, it is important that we verify them.

▼ **Example 5** Verify the two formulas given above.

Solution We verify the formulas by multiplying the right sides and comparing the results with the left sides:

$$
\begin{array}{r}
a^2 - ab + b^2 \\
a + b \\
\hline
a^2b - ab^2 + b^3 \\
a^3 - a^2b + ab^2 \qquad\quad \\
\hline
a^3 \qquad\qquad\qquad + b^3
\end{array}
$$

The first formula is correct.

$$
\begin{array}{r}
a^2 + ab + b^2 \\
a - b \\
\hline
- a^2b - ab^2 - b^3 \\
a^3 + a^2b + ab^2 \qquad\quad \\
\hline
a^3 \qquad\qquad\qquad - b^3
\end{array}
$$

The second formula is correct. ▲

Here are some examples using the formulas for factoring the sum and difference of two cubes:

▼ **Example 6** Factor $x^3 - 8$.

1. When subtracting one polynomial from another it is common to forget to   COMMON MISTAKES
add the opposite of each term in the second polynomial. For example:

$$(6x - 5) - (3x + 4) = 6x - 5 - 3x + 4 \qquad \text{Mistake}$$
$$= 3x - 1$$

This mistake occurs if the negative sign outside the second set of parentheses is
not distributed over all terms inside the parentheses. To avoid this mistake,
remember: The opposite of a sum is the sum of the opposite, or,

$$-(3x + 4) = -3x + (-4)$$

2. Interpreting the square of a sum to be the sum of the squares. That is,

$$(x + y)^2 = x^2 + y^2$$

This can easily be shown as false by trying a couple of numbers for x and y. If
$x = 4$ and $y = 3$, we have

$$(4 + 3)^2 = 4^2 + 3^2$$
$$7^2 = 16 + 9$$
$$49 = 25$$

There has obviously been a mistake. The correct formula for $(a + b)^2$ is

$$(a + b)^2 = a^2 + 2ab + b^2$$

Simplify: **Chapter 3**
1. $3^0 + 2^0 - 5x^0$ **2.** 2^{-5} **Test**

3. $(3x^2 y)^3 (2x^3 y^4)^2$ **4.** $\dfrac{(5ab^3)^{-2}(10a^4 b^{-3})}{(a^{-4}b^3)^4(2a^{-2}b^2)^{-3}}$

5. $(3x^3 - 4x^2 - 6) - (x^2 + 8x - 2)$
6. $3 - 4[2x - 3(x + 6)]$

7. If $P(x) = 4x^2 - 3x + 2$, find $P(-3)$, $P(0)$, and $P(2)$.

Multiply the following:
8. $(2x - 5)(x^2 + 4x - 3)$ **10.** $(4a - 3b)^2$
9. $(3y - 7)(2y + 5)$ **11.** $(6y - 1)(6y + 1)$

Perform the following divisions:
12. $(30x^5 y^3 - 15x^2 y^2 + 20x^3 y^5) \div 10xy^3$
13. $(2x^3 - 6x^2 + 10) \div (2x - 1)$

Factor the following completely:
14. $12x^2 + 26x - 10$ **15.** $16a^4 - 81y^4$
16. $7ax^2 - 14ay - b^2 x^2 + 2b^2 y$ **17.** $x^3 + 27$
18. $4a^5 b - 24a^4 b^2 - 64a^3 b^3$

4 Rational Expressions

To the student:

This chapter is mostly concerned with simplifying a certain kind of algebraic expression. The expressions are called rational expressions because they are to algebra what rational numbers are to arithmetic. Most of the work we will do with rational expressions parallels the work you have done in previous math classes with fractions.

Once we have learned to add, subtract, multiply, and divide rational expressions, we will turn our attention to equations involving rational expressions. Equations of this type are used to describe a number of concepts in science, medicine, and other fields. For example, in electronics, it is a well-known fact that if R is the equivalent resistance of two resistors, R_1 and R_2, connected in parallel, then

$$\frac{1}{R} = \frac{1}{R_1} + \frac{1}{R_2}$$

This formula involves three fractions or rational expressions. If two of the resistances are known, we can solve for the third using the methods we will develop in this chapter.

The single most important tool needed for success in this chapter is factoring. Almost every problem encountered in this chapter involves factoring at one point or another. You may be able to understand all the theory and steps involved in solving the problems, but unless you can factor the polynomials in the problems, you will be unable to work any of them. Essentially, this chapter is a review of the properties of fractions and an exercise in factoring.

We will begin this section with the definition of a rational expression. We will then develop the two basic properties associated with rational expressions, and end the section by applying one of the properties to reduce rational expressions to lowest terms.

Recall from Chapter 1 that a *rational number* is any number that can be expressed as the ratio of two integers:

$$\text{Rational numbers} = \left\{ \frac{a}{b} \,\middle|\, a \text{ and } b \text{ are integers, } b \neq 0 \right\}$$

A rational expression is defined similarly as any expression that can be written as the ratio of two polynomials:

$$\text{Rational expressions} = \left\{ \frac{P}{Q} \,\middle|\, P \text{ and } Q \text{ are polynomials, } Q \neq 0 \right\}$$

Some examples of rational expressions are

$$\frac{2x - 3}{x + 5} \qquad \frac{x^2 - 5x - 6}{x^2 - 1} \qquad \frac{a - b}{b - a}$$

Note A polynomial can be considered as a rational expression since it can be thought of as the ratio of itself to 1. (You see, any real number is a polynomial of degree 0. That is, $5 = 5x^0$, $1 = 1x^0$, $2 = 2x^0$, etc.)

The basic properties associated with rational expressions are equivalent to the properties of fractions. It is important when working with fractions that we are able to change the form of a fraction (as when reducing to lowest terms or writing with a common denominator) without changing the value of the fraction. That is, if a problem involves the fraction $\frac{1}{2}$, we may change the form to $\frac{3}{6}$ or $\frac{9}{18}$ without changing the value. The same principle holds for rational expressions. We may change the form of a rational expression, but the resulting expression must be equivalent to the original expression.

The two procedures that may change the form of a fraction but will never change its value are multiplying the numerator and denominator by the same nonzero number and dividing the numerator and denominator by the same nonzero number. That is, if a, b, and c are real numbers with $b \neq 0$ and $c \neq 0$, then

$$\frac{a}{b} = \frac{ac}{bc} \quad \text{and} \quad \frac{a}{b} = \frac{a/c}{b/c}$$

For rational expressions, multiplying the numerator and denominator by the same nonzero expression may change the form of the rational expression, but it will always produce an expression equivalent to the original one. The same is true when dividing the numerator and denominator by the same nonzero quantity.

If P, Q, and K are polynomials with $Q \neq 0$ and $K \neq 0$, then

$$\frac{P}{Q} = \frac{PK}{QK} \quad \text{and} \quad \frac{P}{Q} = \frac{P/K}{Q/K}$$

The two statements above are equivalent since division is defined as multiplication by the reciprocal. We choose to state them separately for clarity.

Reducing to Lowest Terms

The fraction $\frac{6}{8}$ can be written in lowest terms as $\frac{3}{4}$. The process is shown below:

$$\frac{6}{8} = \frac{3 \cdot 2}{4 \cdot 2} = \frac{\dfrac{3 \cdot 2}{2}}{\dfrac{4 \cdot 2}{2}} = \frac{3}{4}$$

Reducing $\frac{6}{8}$ to $\frac{3}{4}$ involves dividing the numerator and denominator by 2, the factor they have in common. Before dividing out the common factor 2, we must notice that the common factor *is* 2! (This may not be obvious since we are very familiar with the numbers 6 and 8 and therefore do not have to put much thought into finding what number divides both of them.)

We reduce rational expressions to lowest terms by first factoring the numerator and denominator and then dividing both numerator and denominator by any factors they have in common.

▼   **Example 1**   Reduce $\dfrac{x^2 - 9}{x - 3}$ to lowest terms.

Solution   Factoring, we have

$$\frac{x^2 - 9}{x - 3} = \frac{(x + 3)(x - 3)}{x - 3}$$

The numerator and denominator have the factor $x - 3$ in common.

Dividing the numerator and denominator by $x - 3$, we have

$$\frac{(x + 3)(x - 3)}{x - 3} = \frac{x + 3}{1} = x + 3 \qquad \blacktriangle$$

Note The lines drawn through the $(x - 3)$ in the numerator and denominator indicate that we have divided through by $(x - 3)$. As the problems become more involved these lines will help keep track of which factors have been divided out and which have not. The lines do not indicate some new property of rational expressions. The only way to reduce a rational expression to lowest terms is to divide the numerator and denominator by the factors they have in common.

Here are some other examples of reducing rational expressions to lowest terms:

▼ **Example 2** Reduce to lowest terms.

a.
$$\frac{y^2 - 5y - 6}{y^2 - 1} = \frac{(y - 6)(y + 1)}{(y - 1)(y + 1)} \qquad \text{Factor numerator and denominator.}$$

$$= \frac{y - 6}{y - 1} \qquad \text{Divide out common factor } (y + 1).$$

b.
$$\frac{2a^3 - 16}{4a^2 - 12a + 8} = \frac{2(a^3 - 8)}{4(a^2 - 3a + 2)}$$

$$= \frac{2(a - 2)(a^2 + 2a + 4)}{4(a - 2)(a - 1)} \qquad \left. \begin{array}{l} \text{Factor numerator} \\ \text{and denominator.} \end{array} \right.$$

$$= \frac{a^2 + 2a + 4}{2(a - 1)} \qquad \text{Divide out common factor } 2(a - 2).$$

c.
$$\frac{x^2 - 3x + ax - 3a}{x^2 - ax - 3x + 3a} = \frac{x(x - 3) + a(x - 3)}{x(x - a) - 3(x - a)} \qquad \left. \begin{array}{l} \text{Factor numerator} \\ \text{and denominator.} \end{array} \right.$$

$$= \frac{(x - 3)(x + a)}{(x - 3)(x - a)}$$

$$= \frac{x + a}{x - a} \qquad \begin{array}{l} \text{Divide out common} \\ \text{factor } (x - 3). \end{array} \blacktriangle$$

The answer to part c in Example 2 is $(x + a)/(x - a)$. The problem cannot be reduced further. It is a fairly common mistake to attempt to divide out an x or an a in this last expression. Remember, we can divide out only the factors common to the numerator and denominator of a rational expression. For the last expression in Example 2, part c, neither the numerator nor the denominator can be factored further; x is not a factor of the numerator or the denominator and neither is a. The expression is in lowest terms.

The next example involves what we may call a trick. It is a way of changing one of the factors that may not be obvious the first time we see it. This is very useful in some situations.

▼   **Example 3**   Reduce to lowest terms: $\dfrac{x^2 - 25}{5 - x}$.

Solution   We begin by factoring the numerator:
$$\frac{x^2 - 25}{5 - x} = \frac{(x - 5)(x + 5)}{5 - x}$$

The factors $(x - 5)$ and $(5 - x)$ are similar but are not exactly the same. We can reverse the order of either by factoring -1 from them. That is:

$$5 - x = -1(-5 + x) = -1(x - 5).$$

$$\frac{(x - 5)(x + 5)}{5 - x} = \frac{(x - 5)(x + 5)}{-1(x - 5)}$$
$$= \frac{x + 5}{-1}$$
$$= -(x + 5) \qquad\qquad ▲$$

Sometimes we can apply the trick before we actually factor the polynomials in our rational expression.

▼   **Example 4**   Reduce to lowest terms: $\dfrac{x^2 - 6xy + 9y^2}{9y^2 - x^2}$.

Solution   We begin by factoring -1 from the denominator to reverse the order of the terms in $9y^2 - x^2$:

$$\frac{x^2 - 6xy + 9y^2}{9y^2 - x^2} = \frac{x^2 - 6xy + 9y^2}{-1(x^2 - 9y^2)}$$
$$= \frac{(x - 3y)(x - 3y)}{-1(x - 3y)(x + 3y)}$$
$$= -\frac{x - 3y}{x + 3y} \qquad\qquad ▲$$

Reduce each fraction to lowest terms:

1. $\dfrac{8}{24}$

2. $\dfrac{13}{39}$

3. $-\dfrac{12}{36}$

4. $-\dfrac{45}{60}$

5. $\dfrac{9x^3}{3x}$

6. $\dfrac{14x^5}{7x^2}$

7. $\dfrac{2a^2b^3}{4a^2}$

8. $\dfrac{3a^3b^2}{6b^2}$

9. $-\dfrac{24x^3y^5}{16x^4y^2}$

10. $-\dfrac{36x^6y^8}{24x^3y^9}$

11. $\dfrac{144a^2b^3c^4}{56a^4b^3c^2}$

12. $\dfrac{108a^5b^2c^5}{27a^2b^5c^2}$

Reduce each rational expression to lowest terms:

13. $\dfrac{x^2 - 16}{6x + 24}$

14. $\dfrac{5x + 25}{x^2 - 25}$

15. $\dfrac{12x - 9y}{3x^2 + 3xy}$

16. $\dfrac{x^3 - xy^2}{4x + 4y}$

17. $\dfrac{a^4 - 81}{a - 3}$

18. $\dfrac{a + 4}{a^2 - 16}$

19. $\dfrac{y^2 - y - 12}{y - 4}$

20. $\dfrac{y^2 + 7y + 10}{y + 5}$

21. $\dfrac{a^2 - 4a - 12}{a^2 + 8a + 12}$

22. $\dfrac{a^2 - 7a + 12}{a^2 - 9a + 20}$

23. $\dfrac{4y^2 - 9}{2y^2 - y - 3}$

24. $\dfrac{9y^2 - 1}{3y^2 - 10y + 3}$

25. $\dfrac{x^2 + x - 6}{x^2 + 2x - 3}$

26. $\dfrac{x^2 + 10x + 25}{x^2 - 25}$

27. $\dfrac{(x - 3)^2(x + 2)}{(x + 2)^2(x - 3)}$

28. $\dfrac{(x - 4)^3(x + 3)}{(x + 3)^2(x - 4)}$

29. $\dfrac{a^3 + b^3}{a^2 - b^2}$

30. $\dfrac{a^2 - b^2}{a^3 - b^3}$

31. $\dfrac{6x^2 + 7xy - 3y^2}{6x^2 + xy - y^2}$

32. $\dfrac{4x^2 - y^2}{4x^2 - 8xy - 5y^2}$

33. $\dfrac{ax + 2x + 3a + 6}{ay + 2y - 4a - 8}$

34. $\dfrac{ax - x - 5a + 5}{ax + x - 5a - 5}$

35. $\dfrac{x^2 + bx - 3x - 3b}{x^2 - 2bx - 3x + 6b}$

36. $\dfrac{x^2 - 3ax - 2x + 6a}{x^2 - 3ax + 2x - 6a}$

Refer to Examples 3 and 4 in this section and reduce the following to lowest terms:

37. $\dfrac{x-4}{4-x}$

38. $\dfrac{6-x}{x-6}$

39. $\dfrac{y^2-36}{6-y}$

40. $\dfrac{1-y}{y^2-1}$

41. $\dfrac{1-9a^2}{9a^2-6a+1}$

42. $\dfrac{1-a^2}{a^2-2a+1}$

43. $\dfrac{6-5x-x^2}{x^2+5x-6}$

44. $\dfrac{x^2-5x+6}{6-x-x^2}$

Explain the mistake made in each of the following problems:

45. $\dfrac{\cancel{x^2}-\overset{3}{\cancel{9}}}{\cancel{x}-\cancel{3}}=x-3$

46. $\dfrac{\cancel{x^2}-6\cancel{x}+\overset{3}{\cancel{9}}}{\cancel{x^2}+\cancel{x}-\underset{2}{\cancel{6}}}=\dfrac{3}{2}$

47. $\dfrac{\cancel{x}+y}{\cancel{x}}=y$

**4.2
Multiplication and
Division of Rational
Expressions**

In Section 4.1 we found the process of reducing rational expressions to lowest terms to be the same process used in reducing fractions to lowest terms. The similarity also holds for the process of multiplication or division of rational expressions.

Multiplication with fractions is the simplest of the four basic operations. To multiply two fractions we simply multiply numerators and multiply denominators. That is, if a, b, c, and d are real numbers, with $b \neq 0$ and $d \neq 0$, then

$$\frac{a}{b} \times \frac{c}{d} = \frac{ac}{bd}$$

▼ **Example 1**   Multiply $\frac{6}{7} \times \frac{14}{18}$.

Solution

$$\frac{6}{7} \times \frac{14}{18} = \frac{6(14)}{7(18)} \qquad \text{Multiply numerators and denominators.}$$

$$= \frac{\cancel{2}\cdot\cancel{3}(2\cdot\cancel{7})}{\cancel{7}(\cancel{2}\cdot\cancel{3}\cdot 3)} \qquad \text{Factor.}$$

$$= \frac{2}{3} \qquad \text{Divide out common factors.}$$ ▲

Division, as always, is defined in terms of multiplication. To divide by a fraction we simply multiply by its reciprocal. Or, if a, b, c, and d are real numbers, with $b \neq 0$, $c \neq 0$, and $d \neq 0$, then

$$\frac{a}{b} \div \frac{c}{d} = \frac{a}{b} \times \frac{d}{c} = \frac{ad}{bc}$$

▼ **Example 2**   Divide $\frac{6}{8} \div \frac{3}{5}$.

Solution

$$\frac{6}{8} \div \frac{3}{5} = \frac{6}{8} \times \frac{5}{3}$$ Write division in terms of multiplication.

$$= \frac{6(5)}{8(3)}$$ Multiply numerators and denominators.

$$= \frac{2 \cdot 3(5)}{2 \cdot 2 \cdot 2(3)}$$ Factor.

$$= \frac{5}{4}$$ Divide out common factors.   ▲

To multiply two rational expressions, we apply the same process used in Example 1. If P, Q, R, and S are polynomials, with $Q \neq 0$ and $S \neq 0$, then

$$\frac{P}{Q} \times \frac{R}{S} = \frac{P \cdot R}{Q \cdot S}$$

The product of two rational expressions is the product of their numerators over the product of their denominators.

▼ **Example 3**   Multiply $\dfrac{x - 3}{x^2 - 4} \cdot \dfrac{x + 2}{x^2 - 6x + 9}$.

Solution   We begin by multiplying numerators and denominators. We then factor all polynomials and divide out factors common to the numerator and denominator:

$$\frac{x - 3}{x^2 - 4} \cdot \frac{x + 2}{x^2 - 6x + 9}$$

$$= \frac{(x - 3)(x + 2)}{(x^2 - 4)(x^2 - 6x + 9)}$$ Multiply.

$$= \frac{(x-3)(x+2)}{(x+2)(x-2)(x-3)(x-3)} \qquad \text{Factor.}$$

$$= \frac{1}{(x-2)(x-3)} \qquad \begin{array}{l} \text{Divide out} \\ \text{common factors.} \end{array} \quad \blacktriangle$$

The first two steps can be combined to save time. We can perform the multiplication and factoring steps together.

▼ **Example 4**   Multiply $\dfrac{2y^2 - 4y}{2y^2 - 2} \cdot \dfrac{y^2 - 2y - 3}{y^2 - 5y + 6}$.

Solution

$$\frac{2y^2 - 4y}{2y^2 - 2} \cdot \frac{y^2 - 2y - 3}{y^2 - 5y + 6} = \frac{2y(y-2)(y-3)(y+1)}{2(y+1)(y-1)(y-3)(y-2)}$$

$$= \frac{y}{y-1} \qquad \blacktriangle$$

Notice in both of the above examples that we did not actually multiply the polynomials as we did in Chapter 3. It would be senseless to do that since we would then have to factor each of the resulting products to reduce them to lowest terms.

The quotient of two rational expressions is the product of the first and the reciprocal of the second. That is, we find the quotient of two rational expressions the same way we find the quotient of two fractions.

▼ **Example 5**   Divide $\dfrac{x^2 - y^2}{x^2 - 2xy + y^2} \div \dfrac{x^3 + y^3}{x^3 - x^2y}$.

Solution   We begin by writing the problem as the product of the first and the reciprocal of the second and then proceed as in the previous two examples:

$$\frac{x^2 - y^2}{x^2 - 2xy + y^2} \div \frac{x^3 + y^3}{x^3 - x^2y}$$

$$= \frac{x^2 - y^2}{x^2 - 2xy + y^2} \cdot \frac{x^3 - x^2y}{x^3 + y^3} \qquad \begin{array}{l} \text{Multiply by the} \\ \text{reciprocal of the} \\ \text{divisor.} \end{array}$$

$$= \frac{(x-y)(x+y)(x^2)(x-y)}{(x-y)(x-y)(x+y)(x^2-xy+y^2)}$$ Factor and multiply.

$$= \frac{x^2}{x^2-xy+y^2}$$ Divide out common factors. ▲

Here are some more examples of multiplication and division with rational expressions:

▼ **Example 6** Perform the indicated operations.

a. $\dfrac{a^2-8a+15}{a+4} \cdot \dfrac{a+2}{a^2-5a+6} \div \dfrac{a^2-3a-10}{a^2+2a-8}$

$$= \frac{(a^2-8a+15)(a+2)(a^2+2a-8)}{(a+4)(a^2-5a+6)(a^2-3a-10)}$$ Change division to multiplication by using the reciprocal.

$$= \frac{(a-5)(a-3)(a+2)(a+4)(a-2)}{(a+4)(a-3)(a-2)(a-5)(a+2)}$$ Factor.

$$= 1$$ Divide out common factors.

b. $\dfrac{xa+xb+ya+yb}{xa-xb-ya+yb} \cdot \dfrac{xa+xb-ya-yb}{xa-xb+ya-yb}$

$$= \frac{x(a+b)+y(a+b)}{x(a-b)-y(a-b)} \cdot \frac{x(a+b)-y(a+b)}{x(a-b)+y(a-b)}$$

Factor by grouping.

$$= \frac{(x+y)(a+b)(x-y)(a+b)}{(x-y)(a-b)(x+y)(a-b)}$$

$$= \frac{(a+b)^2}{(a-b)^2}$$ ▲

Perform the indicated operations involving fractions: *Problem Set 4.2*

1. $\frac{2}{9} \cdot \frac{3}{4}$ 2. $\frac{5}{6} \cdot \frac{7}{8}$

3. $\frac{3}{4} \div \frac{1}{3}$ 4. $\frac{3}{8} \div \frac{5}{4}$

5. $\frac{3}{7} \cdot \frac{14}{24} \div \frac{1}{2}$ 6. $\frac{6}{5} \cdot \frac{10}{36} \div \frac{3}{4}$

7. $\dfrac{10x^2}{5y^2} \cdot \dfrac{15y^3}{2x^4}$ 8. $\dfrac{8x^3}{7y^4} \cdot \dfrac{14y^6}{16x^2}$

9. $\dfrac{11a^2b}{5ab^2} \div \dfrac{22a^3b^2}{10ab^4}$ 10. $\dfrac{8ab^3}{9a^2b} \div \dfrac{16a^2b^2}{18ab^3}$

11. $\dfrac{6x^2}{5y^3} \cdot \dfrac{11z^2}{2x^2} \div \dfrac{33z^5}{10y^8}$ **12.** $\dfrac{4x^3}{7y^2} \cdot \dfrac{6z^5}{5x^6} \div \dfrac{24z^2}{35x^6}$

Perform the indicated operations. Be sure to write all answers in lowest terms.

13. $\dfrac{x^2 - 9}{x^2 - 4} \cdot \dfrac{x - 2}{x - 3}$ **14.** $\dfrac{x^2 - 16}{x^2 - 25} \cdot \dfrac{x - 5}{x - 4}$

15. $\dfrac{y^2 - 1}{y + 2} \cdot \dfrac{y^2 + 5y + 6}{y^2 + 2y - 3}$ **16.** $\dfrac{y - 1}{y^2 - y - 6} \cdot \dfrac{y^2 + 5y + 6}{y^2 - 1}$

17. $\dfrac{3x - 12}{x^2 - 4} \cdot \dfrac{x^2 + 6x + 8}{x - 4}$ **18.** $\dfrac{x^2 + 5x + 1}{4x - 4} \cdot \dfrac{x - 1}{x^2 + 5x + 1}$

19. $\dfrac{a^2 - 5a + 6}{a^2 - 2a - 3} \div \dfrac{a - 5}{a^2 + 3a + 2}$

20. $\dfrac{a^2 + 7a + 12}{a - 5} \div \dfrac{a^2 + 9a + 18}{a^2 - 7a + 10}$

21. $\dfrac{2x^2 - 5x - 12}{4x^2 + 8x + 3} \div \dfrac{x^2 - 16}{2x^2 + 7x + 3}$

22. $\dfrac{x^2 - 2x + 1}{3x^2 + 7x - 20} \div \dfrac{x^2 + 3x - 4}{3x^2 - 2x - 5}$

23. $\dfrac{x^2 + 5x + 6}{x + 1} \cdot \dfrac{x^2 - 1}{x^2 + 7x + 10} \div \dfrac{x^2 + 2x - 3}{x + 5}$

24. $\dfrac{2x^2 - x - 1}{x^2 - 2x - 15} \cdot \dfrac{x - 5}{4x^2 - 1} \div \dfrac{x - 1}{x^2 - 9}$

25. $\dfrac{xy - 2x + 3y - 6}{xy + 2x - 4y - 8} \cdot \dfrac{xy + x - 4y - 4}{xy - x + 3y - 3}$

26. $\dfrac{x^3 - 8}{x^4 - 16} \cdot \dfrac{x^2 + 4}{x^2 + 2x + 4}$

27. $\dfrac{a^2 - 16}{a^2 - 8a + 16} \cdot \dfrac{a^2 - 9a + 20}{a^2 - 7a + 12} \div \dfrac{a^2 - 25}{a^2 - 6a + 9}$

28. $\dfrac{a^2 - 6a + 9}{a^2 - 4} \cdot \dfrac{a^2 - 5a + 6}{(a - 3)^2} \div \dfrac{a^2 - 9}{a^2 - a - 6}$

29. $\dfrac{xy - 2x + 3y - 6}{xy + 2x - 4y - 8} \cdot \dfrac{xy + x - 4y - 4}{xy - x + 3y - 3}$

30. $\dfrac{ax + bx + 2a + 2b}{ax - 3a + bx - 3b} \cdot \dfrac{ax - bx - 3a + 3b}{ax - bx - 2a + 2b}$

31. $\dfrac{xy - y + 4x - 4}{xy - 3y + 4x - 12} \div \dfrac{xy + 2x + y + 2}{xy - 3y + 2x - 6}$

32. $\dfrac{xb - 2b + 3x - 6}{xb + 3b + 3x + 9} \cdot \dfrac{xb - 2b - 2x + 4}{xb + 3b - 2x - 6}$

We begin this section by considering two examples of addition with fractions. Since subtraction is defined as addition of the opposite, we will not need to show a separate example for subtraction.

To add two fractions with the same denominator, we simply add numerators and use the common denominator.

▼   **Example 1**   Add $\frac{4}{9} + \frac{2}{9}$.

Solution   Since both fractions have the same denominator, we simply combine numerators and reduce to lowest terms:

$$\frac{4}{9} + \frac{2}{9} = \frac{4+2}{9}$$

$$= \frac{6}{9}$$

$$= \frac{2}{3} \qquad\qquad ▲$$

The main step in adding fractions is actually another application of the distributive property, even if we don't show it. To show the use of the distributive property we must first observe that $\frac{4}{9} = 4(\frac{1}{9})$ and $\frac{2}{9} = 2(\frac{1}{9})$, since division by 9 is equivalent to multiplication by $\frac{1}{9}$. Here is the solution to Example 1 again showing the use of the distributive property:

$$\frac{4}{9} + \frac{2}{9} = 4\left(\frac{1}{9}\right) + 2\left(\frac{1}{9}\right)$$

$$= (4 + 2)\left(\frac{1}{9}\right) \qquad \text{Distributive property}$$

$$= 6\left(\frac{1}{9}\right)$$

$$= \frac{6}{9}$$

$$= \frac{2}{3}$$

Adding fractions is always done by using the distributive property, even if it is not shown. The distributive property is the reason we begin all addition (or subtraction) problems involving fractions by making sure all fractions have the same denominator.

DEFINITION   The *least common denominator* (LCD) for a set of denominators is the smallest quantity divisible by each of the denominators.

The first step in adding two fractions is to find a common denominator. Once we have a common denominator we write each fraction again (without changing the value of either one) as a fraction with the common denominator. The procedure then follows that of Example 1 from this section.

Example 2 is a step-by-step solution to an addition problem involving fractions with unlike denominators. The purpose behind spending so much time here working with fractions is that addition of rational expressions follows the same procedure. If we understand addition of fractions, addition of rational expressions will follow naturally.

▼   **Example 2**   Add $\frac{3}{14} + \frac{7}{30}$.

Solution

Step 1.   Find the least common denominator.
To do this we first factor both denominators into prime factors.

$$\text{Factor 14:} \quad 14 = 2 \cdot 7$$
$$\text{Factor 30:} \quad 30 = 2 \cdot 3 \cdot 5$$

Since the LCD must be divisible by 14, it must have factors of $2 \cdot 7$. It must also be divisible by 30 and therefore have factors of $2 \cdot 3 \cdot 5$. We do not need to repeat the 2 that appears in both the factors of 14 and those of 30. Therefore,

$$\text{LCD} = 2 \cdot 3 \cdot 5 \cdot 7 = 210$$

Step 2.   Change to equivalent fractions.
Since we want each fraction to have a denominator of 210 and at the same time keep its original value, we multiply each by 1 in the appropriate form.
Change $\frac{3}{14}$ to a fraction with denominator 210:

$$\frac{3}{14} \cdot \frac{15}{15} = \frac{45}{210}$$

Change $\frac{7}{30}$ to a fraction with denominator 210:

$$\frac{7}{30} \cdot \frac{7}{7} = \frac{49}{210}$$

Step 3. Add numerators of equivalent fractions found in step 2:

$$\frac{45}{210} + \frac{49}{210} = \frac{94}{210}$$

Step 4. Reduce to lowest terms if necessary:

$$\frac{94}{210} = \frac{47}{105} \qquad \blacktriangle$$

The main idea in adding fractions is to write each fraction again with the LCD for a denominator. In doing so, we must be sure not to change the value of either of the original fractions.

When adding rational expressions we follow the same steps used in Example 2 to add fractions.

▼ **Example 3** Add $\dfrac{x}{x^2 - 1} + \dfrac{1}{x^2 - 1}$.

Solution Since the denominators are the same, we simply add numerators:

$$\frac{x}{x^2 - 1} + \frac{1}{x^2 - 1} = \frac{x + 1}{x^2 - 1} \qquad \text{Add numerators.}$$

$$= \frac{\cancel{x + 1}}{(x - 1)(\cancel{x + 1})} \qquad \text{Factor denominator.}$$

$$= \frac{1}{x - 1} \qquad \begin{array}{l}\text{Divide out common} \\ \text{factor } x + 1. \end{array} \qquad \blacktriangle$$

▼ **Example 4** Add $\dfrac{5}{x^2 - 2x - 3} + \dfrac{4}{x^2 - 9}$.

Solution In this case the denominators are not the same, so we follow the steps from Example 2 of this section.

Step 1. Factor each denominator and build the LCD from the factors:

$$\left.\begin{array}{l} x^2 - 2x - 3 = (x - 3)(x + 1) \\ x^2 - 9 \qquad\ = (x - 3)(x + 3) \end{array}\right\} \text{LCD} = (x - 3)(x + 3)(x + 1)$$

Step 2. Change each rational expression to an equivalent expression that has the LCD for a denominator:

$$\frac{5}{x^2 - 2x - 3} = \frac{5}{(x-3)(x+1)} \cdot \frac{(x+3)}{(x+3)} = \frac{5x+15}{(x-3)(x+3)(x+1)}$$

$$\frac{4}{x^2 - 9} = \frac{4}{(x-3)(x+3)} \cdot \frac{(x+1)}{(x+1)} = \frac{4x+4}{(x-3)(x+3)(x+1)}$$

Step 3.   Add numerators of the rational expressions found in step 2:

$$\frac{5x+15}{(x-3)(x+3)(x+1)} + \frac{4x+4}{(x-3)(x+3)(x+1)}$$

$$= \frac{9x+19}{(x-3)(x+3)(x+1)}$$

The numerator and denominator of the resulting expression do not have any factors in common. It is in lowest terms. It is best to leave the denominator in factored form rather than multiply it out.   ▲

▼   **Example 5**   Subtract $\dfrac{2x}{x^2 + 7x + 10} - \dfrac{3}{5x + 10}$.

Solution   Factoring the denominators, we have:

$$\frac{2x}{x^2 + 7x + 10} - \frac{3}{5x + 10} = \frac{2x}{(x+2)(x+5)} + \frac{-3}{5(x+2)}$$

The LCD is $5(x+2)(x+5)$. Completing the problem, we have

$$= \frac{5}{5} \cdot \frac{2x}{(x+2)(x+5)} + \frac{-3}{5(x+2)} \cdot \frac{(x+5)}{(x+5)}$$

$$= \frac{10x}{5(x+2)(x+5)} + \frac{-3x-15}{5(x+2)(x+5)}$$

$$= \frac{7x-15}{5(x+2)(x+5)}$$

Notice that in the first step we wrote subtraction as addition of the opposite. There seems to be less chance for error when this is done.   ▲

▼   **Example 6**   Add $\dfrac{x^2}{x-7} + \dfrac{6x+7}{7-x}$.

Solution Recall Section 4.1 where we were able to reverse the terms in a factor such as $7 - x$ by factoring -1 from each term. In a problem like this, the same result can be obtained by multiplying the numerator and denominator by -1:

$$\frac{x^2}{x - 7} + \frac{6x + 7}{7 - x} \cdot \frac{-1}{-1} = \frac{x^2}{x - 7} + \frac{-6x - 7}{x - 7}$$

$$= \frac{x^2 - 6x - 7}{x - 7} \qquad \text{Add numerators.}$$

$$= \frac{(x - 7)(x + 1)}{(x - 7)} \qquad \text{Factor numerator.}$$

$$= x + 1 \qquad\qquad \text{Divide out } (x - 7).$$

$\blacktriangle$

Combine the following fractions: *Problem Set 4.3*

1. $\frac{3}{4} + \frac{1}{2}$ 2. $\frac{5}{6} + \frac{1}{3}$
3. $\frac{2}{5} - \frac{1}{15}$ 4. $\frac{5}{8} - \frac{1}{4}$
5. $\frac{5}{6} + \frac{7}{8}$ 6. $\frac{3}{4} + \frac{2}{3}$
7. $\frac{9}{48} - \frac{3}{54}$ 8. $\frac{6}{28} - \frac{5}{42}$
9. $\frac{3}{4} - \frac{1}{8} + \frac{2}{3}$ 10. $\frac{1}{3} - \frac{5}{6} + \frac{5}{12}$

Combine the following rational expressions. Reduce all answers to lowest terms.

11. $\dfrac{x}{x + 3} + \dfrac{3}{x + 3}$ 12. $\dfrac{5x}{5x + 2} + \dfrac{2}{5x + 2}$

13. $\dfrac{4}{y - 4} - \dfrac{y}{y - 4}$ 14. $\dfrac{8}{y + 8} + \dfrac{y}{y + 8}$

15. $\dfrac{x}{x^2 - y^2} - \dfrac{y}{x^2 - y^2}$ 16. $\dfrac{x}{x^2 - y^2} + \dfrac{y}{x^2 - y^2}$

17. $\dfrac{5}{3x + 6} + \dfrac{2}{x + 2}$ 18. $\dfrac{7}{5x - 10} - \dfrac{3}{x - 2}$

19. $\dfrac{1}{a} + \dfrac{2}{a^2} + \dfrac{3}{a^3}$ 20. $\dfrac{3}{a} - \dfrac{2}{a^2} + \dfrac{1}{a^3}$

21. $\dfrac{4}{2x^2} - \dfrac{5}{3x}$ 22. $\dfrac{6}{4x^3} - \dfrac{2}{x^2}$

23. $\dfrac{5}{x - 1} + \dfrac{x}{x^2 - 1}$ 24. $\dfrac{1}{x + 3} + \dfrac{x}{x^2 - 9}$

25. $\dfrac{x}{x^2 - 5x + 6} - \dfrac{3}{3 - x}$ 26. $\dfrac{x}{x^2 + 4x + 4} - \dfrac{2}{2 + x}$

27. $\dfrac{3}{a^2 - 5a + 6} - \dfrac{2}{a^2 - a - 2}$ 28. $\dfrac{1}{a^2 + a - 2} + \dfrac{4}{a^2 - a - 6}$

29. $\dfrac{3}{y^2 - 1} - \dfrac{2}{y^2 - y - 12}$

30. $\dfrac{2y}{y^2 + 5y + 4} - \dfrac{3}{y^2 + 3y - 4}$

31. $\dfrac{3a}{2a^2 - 5a - 3} + \dfrac{a}{a^2 - 9}$

32. $\dfrac{1}{a^2 - 2a - 15} + \dfrac{3}{a^2 - 10a + 25}$

33. $x + \dfrac{1}{3 - x} + \dfrac{1}{x^2 - 9}$

34. $x - \dfrac{1}{2 + x} - \dfrac{2}{x^2 - 4}$

35. $\dfrac{3}{x^2 - x - 6} + \dfrac{2}{x^2 - 4} + \dfrac{1}{x^2 - 5x + 6}$

36. $\dfrac{1}{x^2 - 25} + \dfrac{1}{x^2 - 4x - 5} + \dfrac{1}{x^2 + 6x + 5}$

37. $\dfrac{xy}{x^3 - y^3} + \dfrac{1}{x - y}$

38. $\dfrac{1}{x - 2} - \dfrac{2x}{x^3 - 8}$

39. The formula $P = \dfrac{1}{a} + \dfrac{1}{b}$ is used by optometrists to help determine how strong to make the lenses for a pair of eyeglasses. If a is 10 and b is .2, find the corresponding value of p.

4.4
Complex Fractions

The quotient of two fractions or two rational expressions is called a *complex fraction*. This section is concerned with the simplification of complex fractions. There are no new properties. The problems in this section can be worked using the methods already developed in the chapter.

▼   **Example 1**   Simplify $\dfrac{3/4}{5/8}$.

Solution   There are generally two methods that can be used to simplify complex fractions.

Method 1.   Instead of dividing by 5/8 we can multiply by 8/5:

$$\frac{3/4}{5/8} = \frac{3}{4} \times \frac{8}{5} = \frac{24}{20} = \frac{6}{5}$$

Method 2.   We can multiply the numerator and denominator of the complex fraction by the LCD for both of the fractions, which in this case is 8.

$$\frac{3/4}{5/8} = \frac{\frac{3}{4} \cdot 8}{\frac{5}{8} \cdot 8} = \frac{6}{5}$$ ▲

Here are some examples of complex fractions involving rational expressions. Most can be solved using either of the two methods shown in Example 1.

▼   **Example 2**   Simplify $\dfrac{\dfrac{1}{x} + \dfrac{1}{y}}{\dfrac{1}{x} - \dfrac{1}{y}}$.

Solution   This problem is most easily solved using method 2. We begin by multiplying both the numerator and denominator by the quantity xy, which is the LCD for all the fractions:

$$\frac{\dfrac{1}{x} + \dfrac{1}{y}}{\dfrac{1}{x} - \dfrac{1}{y}} = \frac{\left(\dfrac{1}{x} + \dfrac{1}{y}\right) \cdot xy}{\left(\dfrac{1}{x} - \dfrac{1}{y}\right) \cdot xy}$$

$$= \frac{\dfrac{1}{x}(xy) + \dfrac{1}{y}(xy)}{\dfrac{1}{x}(xy) - \dfrac{1}{y}(xy)}$$

Apply the distributive property to distribute xy over both terms in the numerator and denominator.

$$= \frac{y + x}{y - x}$$ ▲

▼   **Example 3**   Simplify $\dfrac{\dfrac{x - 2}{x^2 - 9}}{\dfrac{x^2 - 4}{x + 3}}$.

Solution   Applying method 1, we have

$$\frac{\dfrac{x-2}{x^2-9}}{\dfrac{x^2-4}{x+3}} = \frac{x-2}{x^2-9} \cdot \frac{x+3}{x^2-4}$$

$$= \frac{(x-2)(x+3)}{(x+3)(x-3)(x+2)(x-2)}$$

$$= \frac{1}{(x-3)(x+2)}$$

▲

▼ **Example 4**   Simplify $\dfrac{\dfrac{1}{x}+\dfrac{1}{x-1}}{\dfrac{1}{x-1}-\dfrac{1}{x}}$.

Solution

Method 1.   Simplifying the numerator first, we have

$$\frac{1}{x}+\frac{1}{x-1} = \frac{1}{x}\cdot\frac{(x-1)}{(x-1)}+\frac{1}{x-1}\cdot\frac{(x)}{(x)}$$

$$= \frac{x-1+x}{x(x-1)}$$

$$= \frac{2x-1}{x(x-1)}$$

Simplifying the denominator, we have

$$\frac{1}{x-1}-\frac{1}{x} = \frac{1}{x-1}\cdot\frac{(x)}{(x)}-\frac{1}{x}\cdot\frac{(x-1)}{(x-1)}$$

$$= \frac{x-x+1}{x(x-1)}$$

$$= \frac{1}{x(x-1)}$$

Using the results just obtained, we complete the problem as follows:

$$\frac{\dfrac{1}{x} + \dfrac{1}{x-1}}{\dfrac{1}{x-1} - \dfrac{1}{x}} = \frac{\dfrac{2x-1}{x(x-1)}}{\dfrac{1}{x(x-1)}}$$

$$= \frac{2x-1}{x(x-1)} \cdot \frac{x(x-1)}{1}$$

$$= 2x - 1$$

Method 2.   We multiply the numerator and denominator by $x(x-1)$:

$$\frac{\dfrac{1}{x} + \dfrac{1}{x-1}}{\dfrac{1}{x-1} - \dfrac{1}{x}} = \frac{\left(\dfrac{1}{x} + \dfrac{1}{x-1}\right) \cdot x(x-1)}{\left(\dfrac{1}{x-1} - \dfrac{1}{x}\right) \cdot x(x-1)}$$

$$= \frac{\dfrac{1}{x} \cdot x(x-1) + \dfrac{1}{x-1} \cdot x(x-1)}{\dfrac{1}{x-1} \cdot x(x-1) - \dfrac{1}{x} \cdot x(x-1)}$$

$$= \frac{x - 1 + x}{x - (x-1)}$$

$$= \frac{2x-1}{1}$$

$$= 2x - 1 \qquad\qquad \blacktriangle$$

▼   **Example 5**   Simplify $x - \dfrac{3}{x + \frac{1}{3}}$.

Solution   We begin by simplifying the denominator:

$$x + \frac{1}{3} = \frac{3x}{3} + \frac{1}{3} = \frac{3x+1}{3}$$

Here is the complete solution:

$$x - \frac{3}{x + \frac{1}{3}} = x - \frac{3}{\frac{3x + 1}{3}}$$

$$= x - 3 \cdot \frac{3}{3x + 1}$$

$$= x - \frac{9}{3x + 1}$$

$$= \frac{x}{1} \cdot \frac{3x + 1}{3x + 1} - \frac{9}{3x + 1}$$

$$= \frac{3x^2 + x - 9}{3x + 1}$$

▲

Problem Set 4.4   Simplify each of the following as much as possible:

1. $\dfrac{3/4}{2/3}$

2. $\dfrac{5/9}{7/12}$

3. $\dfrac{\frac{1}{3} - \frac{1}{4}}{\frac{1}{2} + \frac{1}{8}}$

4. $\dfrac{\frac{1}{6} - \frac{1}{3}}{\frac{1}{4} - \frac{1}{8}}$

5. $\dfrac{3 + \frac{2}{5}}{1 - \frac{3}{7}}$

6. $\dfrac{2 + \frac{5}{6}}{1 - \frac{7}{8}}$

7. $\dfrac{\dfrac{1}{x}}{1 + \dfrac{1}{x}}$

8. $\dfrac{1 - \dfrac{1}{x}}{\dfrac{1}{x}}$

9. $\dfrac{1 + \dfrac{1}{a}}{1 - \dfrac{1}{a}}$

10. $\dfrac{1 - \dfrac{2}{a}}{1 - \dfrac{3}{a}}$

11. $\dfrac{\dfrac{1}{x} - \dfrac{1}{y}}{\dfrac{1}{x} + \dfrac{1}{y}}$

12. $\dfrac{\dfrac{1}{x} + \dfrac{2}{y}}{\dfrac{2}{x} + \dfrac{1}{y}}$

13. $\dfrac{\dfrac{x + y}{2x^2}}{\dfrac{x - y}{4x}}$

14. $\dfrac{\dfrac{x - y}{5x^2}}{\dfrac{x - y}{10x}}$

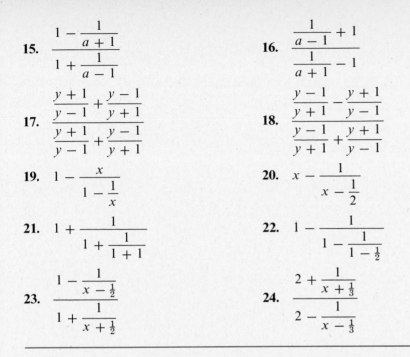

15. $\dfrac{1 - \dfrac{1}{a+1}}{1 + \dfrac{1}{a-1}}$

16. $\dfrac{\dfrac{1}{a-1} + 1}{\dfrac{1}{a+1} - 1}$

17. $\dfrac{\dfrac{y+1}{y-1} + \dfrac{y-1}{y+1}}{\dfrac{y+1}{y-1} + \dfrac{y-1}{y+1}}$

18. $\dfrac{\dfrac{y-1}{y+1} - \dfrac{y+1}{y-1}}{\dfrac{y-1}{y+1} + \dfrac{y+1}{y-1}}$

19. $1 - \dfrac{x}{1 - \dfrac{1}{x}}$

20. $x - \dfrac{1}{x - \dfrac{1}{2}}$

21. $1 + \dfrac{1}{1 + \dfrac{1}{1+1}}$

22. $1 - \dfrac{1}{1 - \dfrac{1}{1 - \frac{1}{2}}}$

23. $\dfrac{1 - \dfrac{1}{x - \frac{1}{2}}}{1 + \dfrac{1}{x + \frac{1}{2}}}$

24. $\dfrac{2 + \dfrac{1}{x + \frac{1}{3}}}{2 - \dfrac{1}{x - \frac{1}{3}}}$

The first step in solving an equation that contains one or more rational expressions is to find the LCD for all denominators in the equation. We then multiply both sides of the equation by the LCD to clear the equation of all fractions. That is, after we have multiplied through by the LCD, each term in the resulting equation will have a denominator of 1.

**4.5
Equations Involving
Rational Expressions**

▼   **Example 1**   Solve $\dfrac{x}{2} - 3 = \dfrac{2}{3}$.

Solution   The LCD for 2 and 3 is 6. Multiplying both sides by 6, we have

$$6\left(\frac{x}{2} - 3\right) = 6\left(\frac{2}{3}\right)$$

$$6\left(\frac{x}{2}\right) - 6(3) = 6\left(\frac{2}{3}\right)$$

$$3x - 18 = 4$$

$$3x = 22$$

$$x = \frac{22}{3} \qquad\qquad ▲$$

Multiplying both sides of an equation by the LCD clears the equation of fractions because the LCD has the property that all the denominators divide it evenly. That is, the LCD contains all factors of each denominator.

▼   **Example 2**   Solve $\dfrac{6}{a-4} = \dfrac{3}{8}$.

Solution   The LCD for $a - 4$ and 8 is $8(a - 4)$. Multiplying both sides by this quantity yields

$$8(a - 4) \cdot \frac{6}{a-4} = 8(a-4) \cdot \frac{3}{8}$$
$$48 = (a - 4) \cdot 3$$
$$48 = 3a - 12$$
$$60 = 3a$$
$$20 = a$$

The solution set is $\{20\}$, which checks in the original equation.   ▲

When we multiply both sides of an equation by an expression containing the variable, we must be sure to check our solutions. The multiplication property of equality does not allow multiplication by 0. If the expression we multiply by contains the variable, then it has the possibility of being 0. In the last example we multiplied both sides by $8(a - 4)$. This gives a restriction $a \neq 4$ for any solution we come up with.

▼   **Example 3**   Solve $\dfrac{x}{x-2} + \dfrac{2}{3} = \dfrac{2}{x-2}$.

Solution   The LCD is $3(x - 2)$. We are assuming $x \neq 2$ when we multiply both sides of the equation by $3(x - 2)$:

$$3(x - 2) \cdot \left[\frac{x}{x-2} + \frac{2}{3} \right] = 3(x-2) \cdot \frac{2}{x-2}$$
$$3x + (x - 2) \cdot 2 = 3 \cdot 2$$
$$3x + 2x - 4 = 6$$
$$5x - 4 = 6$$
$$5x = 10$$
$$x = 2$$

The only possible solution is $x = 2$. Checking this value back in the original equation gives

$$\frac{2}{2 - 2} + \frac{2}{3} = \frac{2}{2 - 2}$$

$$\frac{2}{0} + \frac{2}{3} = \frac{2}{0}$$

The first and last terms are undefined. The proposed solution, $x = 2$, does not check in the original equation. In the process of solving the equation, we multiplied both sides by $3(x - 2)$, solved for x, and got $x = 2$. When $x = 2$, the quantity $3(x - 2)$ is 0, which means we multiplied both sides of our original equation by 0. Multiplying both sides by 0 does not produce an equation equivalent to the original one.

The solution set is $\varnothing$. There is no real number x such that

$$\frac{x}{x - 2} + \frac{2}{3} = \frac{2}{x - 2} \qquad\qquad \blacktriangle$$

When the proposed solution to an equation is not actually a solution, it is called an *extraneous* solution. In the last example, $x = 2$ is an extraneous solution.

▼   **Example 4**   Solve $\dfrac{x}{x^2 - 3x + 2} - \dfrac{1}{x - 2} = \dfrac{1}{3x - 3}$.

Solution   Writing the equation again with the denominators in factored form, we have

$$\frac{5}{(x - 2)(x - 1)} - \frac{1}{x - 2} = \frac{1}{3(x - 1)}$$

The LCD is $3(x - 2)(x - 1)$. Multiplying through by the LCD, we have

$$3(x - 2)(x - 1)\frac{5}{(x - 2)(x - 1)} - 3(x - 2)(x - 1) \cdot \frac{1}{(x - 2)}$$

$$= 3(x - 2)(x - 1) \cdot \frac{1}{3(x - 1)}$$

$$3 \cdot 5 - 3(x - 1) \cdot 1 = (x - 2) \cdot 1$$
$$15 - 3x + 3 \quad\quad = x - 2$$
$$-3x + 18 \quad\quad = x - 2$$
$$-4x + 18 \;\; = -2$$
$$-4x = -20$$
$$x = 5$$

Checking the proposed solution $x = 5$ in the original equation gives

$$\frac{5}{(5 - 2)(5 - 1)} - \frac{1}{5 - 2} = \frac{1}{3(5 - 1)}$$

$$\frac{5}{12} - \frac{1}{3} = \frac{1}{12}$$

$$\frac{5}{12} - \frac{4}{12} = \frac{1}{12}$$

$$\frac{1}{12} = \frac{1}{12}$$

The proposed solution $x = 5$ checks. The solution set is $\{5\}$. ▲

Note We can check the proposed solution in any of the equations obtained before multiplying through by the LCD. We cannot check the proposed solution in an equation obtained after multiplying through by the LCD since, if we have multiplied by 0, the resulting equations will not be equivalent to the original one. Checking solutions is required whenever we have multiplied both sides of the equation by an expression containing the variable.

Problem Set 4.5   Solve each of the following equations:

1. $\dfrac{x}{5} + 4 = \dfrac{5}{3}$

2. $\dfrac{x}{5} = \dfrac{x}{2} - 9$

3. $\dfrac{a}{3} + 2 = \dfrac{4}{5}$

4. $\dfrac{a}{4} + \dfrac{1}{2} = \dfrac{2}{3}$

5. $\dfrac{y}{2} + \dfrac{y}{4} + \dfrac{y}{6} = 3$

6. $\dfrac{y}{3} - \dfrac{y}{6} + \dfrac{y}{2} = 1$

7. $\dfrac{5}{2x} = \dfrac{1}{x} + \dfrac{3}{4}$

8. $\dfrac{1}{2a} = \dfrac{2}{a} - \dfrac{3}{8}$

9. $\dfrac{1}{x} = \dfrac{1}{3} - \dfrac{2}{3x}$

10. $\dfrac{5}{2x} = \dfrac{2}{x} - \dfrac{1}{12}$

11. $\dfrac{2x}{x-3} + 2 = \dfrac{2}{x-3}$

12. $\dfrac{2}{x+5} = \dfrac{2}{5} - \dfrac{x}{x+5}$

13. $\dfrac{x+2}{x+1} = \dfrac{1}{x+1} + 2$

14. $\dfrac{x+6}{x+3} = \dfrac{3}{x+3} + 2$

15. $\dfrac{3}{a-2} = \dfrac{2}{a-3}$

16. $\dfrac{5}{a+1} = \dfrac{4}{a+2}$

17. $\dfrac{1}{x-1} - \dfrac{1}{x+1} = \dfrac{3x}{x^2-1}$

18. $\dfrac{5}{x-1} + \dfrac{2}{x-1} = \dfrac{4}{x+1}$

19. $\dfrac{2}{x-3} + \dfrac{x}{x^2-9} = \dfrac{4}{x+3}$

20. $\dfrac{2}{x+5} + \dfrac{3}{x+4} = \dfrac{2x}{x^2+9x+20}$

21. $\dfrac{3}{2} - \dfrac{1}{x-4} = \dfrac{-2}{2x-8}$

22. $\dfrac{2}{x} - \dfrac{1}{x+1} = \dfrac{-2}{5x+5}$

23. $\dfrac{3}{y-4} - \dfrac{2}{y+1} = \dfrac{5}{y^2-3y-4}$

24. $\dfrac{1}{y+2} - \dfrac{2}{y-3} = \dfrac{-2y}{y^2-y-6}$

25. $\dfrac{2}{1+a} = \dfrac{3}{1-a} + \dfrac{5}{a}$

26. $\dfrac{1}{a+3} - \dfrac{a}{a^2-9} = \dfrac{2}{3-a}$

27. $\dfrac{3}{2x-6} - \dfrac{x+1}{4x-12} = 4$

28. $\dfrac{2x-3}{5x+10} + \dfrac{3x-2}{4x+8} = 1$

29. $\dfrac{4}{2x-6} - \dfrac{12}{4x+12} = \dfrac{12}{x^2-9}$

30. $\dfrac{1}{x+2} + \dfrac{1}{x-2} = \dfrac{4}{x^2-4}$

31. $\dfrac{2}{y^2-7y+12} - \dfrac{1}{y^2-9} = \dfrac{4}{y^2-y-12}$

32. $\dfrac{1}{y^2+5y+4} + \dfrac{3}{y^2-1} = \dfrac{-1}{y^2+3y-4}$

33. The following diagram shows a section of an electronic circuit with a 3-ohm resistor and a 5-ohm resistor connected in parallel.

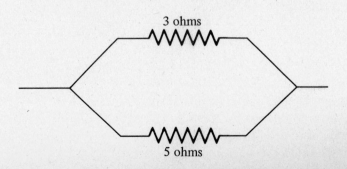

3 ohms

5 ohms

If R is the resistance equivalent to the two resistors connected in parallel, then the relationship between them is given by

$$\frac{1}{R} = \frac{1}{3} + \frac{1}{5}$$

Solve for R.

**4.6
Word Problems**

The procedure used to solve the word problems in this section is the same procedure used in the past to solve word problems. Here, however, translating the problems from words into symbols will result in equations that involve rational expressions.

▼   **Example 1**   One number is twice another. The sum of their reciprocals is 2. Find the numbers.

Solution   Let $x = $ the smaller number. The larger number is $2x$. Their reciprocals are $1/x$ and $1/2x$. The equation that describes the situation is

$$\frac{1}{x} + \frac{1}{2x} = 2$$

Multiplying both sides by the LCD $2x$, we have

$$2x \cdot \frac{1}{x} + 2x \cdot \frac{1}{2x} = 2x(2)$$
$$2 + 1 = 4x$$
$$3 = 4x$$
$$x = \tfrac{3}{4}$$

The smaller number is $\tfrac{3}{4}$. The larger is $2(\tfrac{3}{4}) = \tfrac{6}{4} = \tfrac{3}{2}$. Adding their reciprocals, we have

$$\tfrac{4}{3} + \tfrac{2}{3} = \tfrac{6}{3} = 2$$

The sum of the reciprocals of $\tfrac{3}{4}$ and $\tfrac{2}{3}$ is 2.   ▲

▼   **Example 2**   The speed of a boat in still water is 20 mi/hr. It takes the same amount of time for the boat to travel 3 miles downstream (with the current) as it does to travel 2 miles upstream (against the current). Find the speed of the current.

Solution The following table will be helpful in finding the equation necessary to solve this problem.

	d (distance)	r (rate)	t (time)
Upstream			
Downstream			

If we let $x =$ the speed of the current, the speed (rate) of the boat upstream is $(20 - x)$ since it is traveling against the current. The rate downstream is $(20 + x)$ since the boat is then traveling with the current. The distance traveled upstream is 2 miles, while the distance traveled downstream is 3 miles. Putting the information given here into the table, we have

	d	r	t
Upstream	2	$20 - x$	
Downstream	3	$20 + x$	

To fill in the last two spaces in the table we must use the relationship $d = r \cdot t$. Since we know the spaces to be filled in are in the time column, we solve the equation $d = r \cdot t$ for t and get

$$t = \frac{d}{r}$$

The completed table then is

	d	r	t
Upstream	2	$20 - x$	$\dfrac{2}{20 - x}$
Downstream	3	$20 + x$	$\dfrac{3}{20 + x}$

Reading the problem again, we find that the time moving upstream is equal to the time moving downstream, or

$$\frac{2}{20 - x} = \frac{3}{20 + x}$$

Multiplying both sides by the LCD $(20 - x)(20 + x)$ gives

$$(20 + x) \cdot 2 = 3(20 - x)$$
$$40 + 2x = 60 - 3x$$
$$5x = 20$$
$$x = 4$$

The speed of the current is 4 mi/hr.   ▲

▼   **Example 3**   John can do a certain job in 3 hours, while it takes Bob 5 hours to do the same job. How long will it take them, working together, to get the job done?

Solution   In order to solve a problem like this we must assume that each person works at a constant rate. That is, they do the same amount of work in the first hour as they do in the last hour.

Solving a problem like this seems to be easier if we think in terms of how much work is done by each person in 1 hour.

If it takes John 3 hours to do the whole job, then in 1 hour he must do 1/3 of the job.

If we let x = the amount of time it takes to complete the job working together, then in 1 hour they must do $1/x$ of the job. Here is the equation that describes the situation:

In 1 hour

$$\begin{bmatrix} \text{Amount of work} \\ \text{done by John} \end{bmatrix} + \begin{bmatrix} \text{Amount of work} \\ \text{done by Bob} \end{bmatrix} = \begin{bmatrix} \text{Total amount} \\ \text{of work done} \end{bmatrix}$$
$$\frac{1}{3} \qquad + \qquad \frac{1}{5} \qquad = \qquad \frac{1}{x}$$

Multiplying through by the LCD $15x$, we have

$$15x \cdot \frac{1}{3} + 15x \cdot \frac{1}{5} = 15x \cdot \frac{1}{x}$$
$$5x + 3x = 15$$
$$8x = 15$$
$$x = \frac{15}{8}$$

It takes them $\frac{15}{8}$ hours to do the job when they work together.   ▲

▼ **Example 4** An inlet pipe can fill a pool in 10 hours, while an outlet pipe can empty it in 12 hours. If the pool is empty and both pipes are open, how long will it take to fill the pool?

   Solution This problem is very similar to the problem in Example 3. It is helpful to think in terms of how much work is done by each pipe in 1 hour.
   Let x = the time it takes to fill the pool with both pipes open.
   If the inlet pipe can fill the pool in 10 hours, then in 1 hour it is 1/10 full.
   If the outlet pipe empties the pool in 12 hours, then in 1 hour it is 1/12 empty.
   If the pool can be filled in x hours with both pipes open, then in 1 hour it is 1/x full when both pipes are open.
   Here is the equation:

In 1 hour

$$\begin{bmatrix} \text{Amount full by} \\ \text{inlet pipe} \end{bmatrix} - \begin{bmatrix} \text{Amount empty by} \\ \text{outlet pipe} \end{bmatrix} = \begin{bmatrix} \text{Fraction of pool} \\ \text{filled by both} \end{bmatrix}$$

$$\frac{1}{10} \quad - \quad \frac{1}{12} \quad = \quad \frac{1}{x}$$

Multiplying through by $60x$, we have

$$60x \cdot \frac{1}{10} - 60x \cdot \frac{1}{12} = 60x \cdot \frac{1}{x}$$

$$6x - 5x = 60$$

$$x = 60$$

   It takes 60 hours to fill the pool if both the inlet pipe and the outlet pipe are open.   ▲

Solve each of the following word problems. Be sure to show the equation in each case.   *Problem Set 4.6*

1. One number is 3 times another. The sum of their reciprocals is $\frac{20}{3}$. Find the numbers.

2. One number is 3 times another. The sum of their reciprocals is $\frac{4}{9}$. Find the numbers.

3. If a certain number is added to the numerator and denominator of $\frac{7}{9}$, the result is $\frac{5}{6}$. Find the number.

4. Find the number you would add to both the numerator and denominator of $\frac{8}{11}$ so the result would be $\frac{6}{7}$.

5. The speed of a boat in still water is 5 mi/hr. If the boat travels 3 miles downstream in the same amount of time it takes to travel 1.5 miles upstream, what is the speed of the current?

6. A boat, which moves at 18 mi/hr in still water, travels 14 miles downstream in the same amount of time it takes to travel 10 miles upstream. Find the speed of the current.

7. Train A has a speed 15 mi/hr greater than that of train B. If train A travels 150 miles in the same time train B travels 120 miles, what are the speeds of the two trains?

8. A train travels 30 mi/hr faster than a car. If the train covers 120 miles in the same time the car covers 80 miles, what is the speed of each of them?

9. If Sam can do a certain job in 3 days, while it takes Fred 6 days to do the same job, how long will it take them, working together, to complete the job?

10. Tim can finish a certain job in 10 hours. It takes his wife JoAnn only 8 hours to do the same job. If they work together, how long will it take them to complete the job?

11. Two people working together can complete a job in 6 hours. If one of them works twice as fast as the other, how long would it take the faster person, working alone, to do the job?

12. If two people working together can do a job in 3 hours, how long will it take the slower person to do the same job if one of them is 3 times as fast as the other?

13. A water tank can be filled by an inlet pipe in 8 hours. It takes twice that long for the outlet pipe to empty the tank. How long will it take to fill the tank if both pipes are open?

14. A sink can be filled from the faucet in 5 minutes. It takes only 3 minutes to empty the sink when the drain is open. If the sink is full and both the faucet and the drain are open, how long will it take to empty the sink?

15. It takes 10 hours to fill a pool with the inlet pipe. It can be emptied in 15 hours with the outlet pipe. If the pool is half-full to begin with, how long will it take to fill it from there if both pipes are open?

16. A sink is $\frac{1}{4}$ full when both the faucet and the drain are opened. The faucet alone can fill the sink in 6 minutes, while it takes 8 minutes to empty it with the drain. How long will it take to fill the remaining $\frac{3}{4}$ of the sink?

17. The time (T) in hours it takes a satellite to circle the earth, if it travels at v mi/hr at a height h above the earth, is given by the formula

$$T = \frac{176,000}{7v} + \frac{h}{v}$$

If $h = 600$ miles, and $v = 1200$ mi/hr, find T.

A *rational number* is any number that can be expressed as the ratio of two integers:

$$\text{Rational numbers} = \left\{ \frac{a}{b} \,\middle|\, a \text{ and } b \text{ are integers, } b \neq 0 \right\}$$

A *rational expression* is any quantity that can be expressed as the ratio of two polynomials:

$$\text{Rational expressions} = \left\{ \frac{P}{Q} \,\middle|\, P \text{ and } Q \text{ are polynomials, } Q \neq 0 \right\}$$

If P, Q, and K are polynomials with $Q \neq 0$ and $K \neq 0$, then

$$\frac{P}{Q} = \frac{PK}{QK} \quad \text{and} \quad \frac{P}{Q} = \frac{P/K}{Q/K}$$

which is to say that multiplying or dividing the numerator and denominator of a rational expression by the same nonzero quantity always produces an equivalent rational expression.

To reduce a rational expression to lowest terms we first factor the numerator and denominator and then divide the numerator and denominator by any factors they have in common.

If P, Q, R, and S represent polynomials, then

$$\frac{P}{Q} \cdot \frac{R}{S} = \frac{PR}{QS} \quad (Q \neq 0 \text{ and } S \neq 0)$$

$$\frac{P}{Q} \div \frac{R}{S} = \frac{P}{Q} \cdot \frac{S}{R} = \frac{PS}{QR} \quad (Q \neq 0, \, S \neq 0, \, R \neq 0)$$

The *least common denominator*, LCD, for a set of denominators is the smallest quantity divisible by each of the denominators.

If P, Q, and R represent polynomials, $R \neq 0$, then

$$\frac{P}{R} + \frac{Q}{R} = \frac{P + Q}{R} \quad \text{and} \quad \frac{P}{R} - \frac{Q}{R} = \frac{P - Q}{R}$$

When adding or subtracting rational expressions with different denominators, we must find the LCD for all denominators and change each rational expression to an equivalent expression that has the LCD.

Complex Fractions

A *complex fraction* is the ratio of two rational expressions or an expression that can be written as the ratio of two rational expressions.

Equations Involving Rational Expressions

To solve an equation involving rational expressions we first find the LCD for all denominators appearing on either side of the equation. We then multiply both sides by the LCD to clear the equation of all fractions and solve as usual.

COMMON MISTAKES

1. Attempting to divide the numerator and denominator of a rational expression by a quantity that is not a factor of both. Like this:

$$\frac{\overset{3}{x^2} - \overset{}{9}x - \overset{2}{20}}{\underset{1}{x^2} - \underset{}{3}x - \underset{1}{10}}$$

This makes no sense at all. The numerator and denominator must be factored completely before any factors they have in common can be recognized:

$$\frac{x^2 - 9x + 20}{x^2 - 3x - 10} = \frac{(x - 5)(x - 4)}{(x - 5)(x + 2)}$$

$$= \frac{x - 4}{x + 2}$$

2. Forgetting to check solutions to equations involving rational expressions. When we multiply both sides of an equation by a quantity containing the variable, we must be sure to check for extraneous solutions (see Section 4.5).

**Chapter 4
Test**

Reduce to lowest terms:

1. $\dfrac{x^2 - y^2}{x - y}$

2. $\dfrac{2x^2 - 5x + 3}{2x^2 - x - 3}$

Perform the indicated operations:

3. $\dfrac{a^2 - 16}{5a - 15} \cdot \dfrac{10(a - 3)^2}{a^2 - 7a + 12}$

4. $\dfrac{a^4 - 81}{a^2 + 9} \div \dfrac{a^2 - 8a + 15}{4a - 20}$

5. $\frac{4}{21} + \frac{6}{35}$

6. $\frac{3}{4} - \frac{1}{2} + \frac{5}{8}$

7. $\dfrac{1}{x} + \dfrac{2}{x - 3}$

8. $\dfrac{3}{x^2 - x - 12} - \dfrac{2}{x^2 - 16}$

9. $\dfrac{3 - \dfrac{1}{a + 3}}{3 + \dfrac{1}{a - 3}}$

10. $\dfrac{\dfrac{1}{a + 1} - \dfrac{1}{a - 1}}{\dfrac{1}{a - 1} + \dfrac{1}{a + 1}}$

Solve each of the following equations:

11. $\dfrac{1}{x} + 3 = \dfrac{4}{3}$

12. $\dfrac{x}{x-3} + 3 = \dfrac{3}{x-3}$

13. $\dfrac{y+3}{2y} + \dfrac{5}{y-1} = \dfrac{1}{2}$

Solve the following word problems. Be sure to show the equation in each case.

14. What number must be subtracted from the denominator of $\frac{10}{23}$ to make the result $\frac{1}{3}$?

15. An inlet pipe can fill a pool in 10 hours while an outlet pipe can empty it in 15 hours. If the pool is half-full and both pipes are left open, how long will it take to fill the pool the rest of the way?

5

Rational Exponents and Roots

To the student:

This chapter is concerned with fractional exponents, roots, and complex numbers. As we will see, expressions involving fractional exponents are actually just radical expressions. That is, fractional exponents are used to denote square roots, cube roots, and so forth. We use fractional exponents as an alternative to radical notation (radical notation involves the use of the symbol $\sqrt{}$).

Many of the formulas that describe the characteristics of objects in the universe involve roots. The formula for the length of the diagonal of a square involves a square root. The length of time it takes a pendulum (as on a grandfather clock) to swing through one complete cycle depends on the square root of the length of the pendulum. The formulas that describe the changes in length, mass, and time for objects traveling at velocities close to the speed of light also contain roots.

We begin the chapter with some simple roots and the correlation between fractional exponents and roots. We will then list the properties associated with radicals and use these properties to write some radical expressions in simplified form. Combinations of radical expressions and equations involving radicals are considered next. The chapter concludes with the definition for complex numbers and some applications of this definition. The work we have done previously with exponents and polynomials will be very useful in understanding the concepts developed here. Radical expressions and complex numbers behave like polynomials. As was the case in the preceding chapter, the distributive property is used extensively as justification for many of the properties developed in this chapter.

In Chapter 3 we developed notation (exponents) to give us the square, cube, or any power of a number. For instance, if we wanted the square of 3, we wrote $3^2 = 9$. If we wanted the cube of 3, we wrote $3^3 = 27$. In this section we will develop notation that will take us in the reverse direction, that is, from the square of a number, say 25, back to the original number, 5. There are two kinds of notation that allow us to do this. One is radical notation (square roots, cube roots, etc.), and the other involves fractional exponents. This section is concerned with the definitions associated with the two types of notation.

The number 49 has two square roots, 7 and -7. The positive square root (principal root) of 49 is written $\sqrt{49}$, while the negative square root of 49 is written $-\sqrt{49}$. All positive real numbers have two square roots, one positive and the other negative. The positive square root of 11 is $\sqrt{11}$. The negative square root of 11 is written $-\sqrt{11}$. Here is a definition for square roots:

DEFINITION If x is a positive real number, then the expression $\sqrt{x}$ is called the *positive square root* of x and is such that

$$(\sqrt{x})^2 = x$$

In words: $\sqrt{x}$ is the positive number we square to get x.

The negative square root of x, $-\sqrt{x}$, is defined in a similar manner.

▼ **Example 1** The positive square root of 64 is 8 because 8 is the positive number with the property $8^2 = 64$. The negative square root of 64 is -8 since -8 is the negative number whose square is 64. We can summarize both of these facts by saying

$$\sqrt{64} = 8 \quad \text{and} \quad -\sqrt{64} = -8 \qquad \blacktriangle$$

Note It is a common mistake to assume that an expression like $\sqrt{25}$ indicates both square roots, $+5$ and -5. The expression $\sqrt{25}$ indicates only the positive square root of 25, which is 5. If we want the negative square root, we must use a negative sign: $-\sqrt{25} = -5$.

The higher roots, cube roots, fourth roots, and so on, are defined by definitions similar to that of square roots.

DEFINITION   If x is a real number, and n is a positive integer, then

The positive square root of x, $\sqrt{x}$, is such that $(\sqrt{x})^2 = x$ (x positive)

The cube root of x, $\sqrt[3]{x}$, is such that $(\sqrt[3]{x})^3 = x$

The positive fourth root of x, $\sqrt[4]{x}$, is such that $(\sqrt[4]{x})^4 = x$ (x positive)

The fifth root of x, $\sqrt[5]{x}$, is such that $(\sqrt[5]{x})^5 = x$

$$\vdots \qquad\qquad \vdots \qquad\qquad \vdots$$

The nth root of x, $\sqrt[n]{x}$, is such that $(\sqrt[n]{x})^n = x$ (x positive if n is even)

Note   We have restricted the even roots in this definition to positive numbers. Even roots of negative numbers exist, but are not represented by real numbers. That is, $\sqrt{-4}$ is not a real number since there is no real number whose square is -4. We will have to wait until the last two sections of this chapter to see how to deal with even roots of negative numbers.

Here is a table of the most common roots used in this book. Any of the roots that are unfamiliar should be memorized.

Square roots		Cube roots	Fourth roots
$\sqrt{0} = 0$	$\sqrt{49} = 7$	$\sqrt[3]{0} = 0$	$\sqrt[4]{0} = 0$
$\sqrt{1} = 1$	$\sqrt{64} = 8$	$\sqrt[3]{1} = 1$	$\sqrt[4]{1} = 1$
$\sqrt{4} = 2$	$\sqrt{81} = 9$	$\sqrt[3]{8} = 2$	$\sqrt[4]{16} = 2$
$\sqrt{9} = 3$	$\sqrt{100} = 10$	$\sqrt[3]{27} = 3$	$\sqrt[4]{81} = 3$
$\sqrt{16} = 4$	$\sqrt{121} = 11$	$\sqrt[3]{64} = 4$	
$\sqrt{25} = 5$	$\sqrt{144} = 12$	$\sqrt[3]{125} = 5$	
$\sqrt{36} = 6$	$\sqrt{169} = 13$		

We will now develop a second kind of notation involving exponents that will allow us to designate square roots, cube roots, and so on in another way.

Consider the equation $x = 8^{1/3}$. Although we have not encountered fractional exponents before, let's assume that all the properties of exponents hold in this case. Cubing both sides of the equation, we have

$$x^3 = (8^{1/3})^3$$
$$x^3 = 8^{(1/3)(3)}$$
$$x^3 = 8^1$$
$$x^3 = 8$$

The last line tells us that x is the number whose cube is 8. It must be true, then, that x is the cube root of 8, $x = \sqrt[3]{8}$. Since we started with $x = 8^{1/3}$, it follows that

$$8^{1/3} = \sqrt[3]{8}$$

It seems reasonable, then, to define fractional exponents as indicating roots.

DEFINITION   If x is a real number and n is a positive integer, then

$$x^{1/n} = \sqrt[n]{x} \qquad (x \geq 0 \text{ when } n \text{ is even})$$

In words: the quantity $x^{1/n}$ is the nth root of x.

With this definition we have a way of representing roots with exponents. Here are some examples.

▼   **Example 2**

a. $8^{1/3} = \sqrt[3]{8} = 2$

b. $36^{1/2} = \sqrt{36} = 6$

c. $-25^{1/2} = -\sqrt{25} = -5$

d. $(-25)^{1/2} = \sqrt{-25}$, which is not a real number

e. $(\frac{4}{9})^{1/2} = \sqrt{\frac{4}{9}} = \frac{2}{3}$

f. $16^{1/2} \cdot 27^{1/3} = \sqrt{16}\,\sqrt[3]{27} = 4 \cdot 3 = 12$

g. $125^{1/3} + 81^{1/4} = \sqrt[3]{125} + \sqrt[4]{81} = 5 + 3 = 8$ ▲

Once we become familiar with fractional exponents, the second step in problems like the above will not be necessary.

The properties of exponents developed in Chapter 3 apply to integer exponents only. We will now extend these properties to include rational exponents also. We do so without proof.

PROPERTIES OF EXPONENTS   If a and b are real numbers and r and s are rational numbers, and a and b are positive whenever r or s indicate even roots, then

1. $a^r \cdot a^s = a^{r+s}$

2. $(a^r)^s = a^{rs}$

3. $(ab)^r = a^r b^r$

4. $\left(\dfrac{a}{b}\right)^r = \dfrac{a^r}{b^r}$ $(b \neq 0)$

5. $a^{-r} = \dfrac{1}{a^r}$ $(a \neq 0)$

6. $\dfrac{a^r}{a^s} = a^{r-s}$ $(a \neq 0)$

We can use the properties of exponents to prove the following theorem:

THEOREM 5.1 If a is a positive real number, m is an integer, and n is a positive integer, then

$$a^{m/n} = (a^{1/n})^m = (a^m)^{1/n}$$

Proof

$$
\begin{array}{c|c}
\begin{aligned}
a^{m/n} &= a^{m(1/n)} \\
&= (a^m)^{1/n}
\end{aligned}
&
\begin{aligned}
a^{m/n} &= a^{(1/n)(m)} \\
&= (a^{1/n})^m
\end{aligned}
\end{array}
$$

Here are some examples that use the definition of fractional exponents, the properties of exponents, and Theorem 5.1 to simplify expressions:

▼   **Example 3**   Simplify as much as possible:

a. $\begin{aligned} 8^{2/3} &= (8^{1/3})^2 \\ &= 2^2 \\ &= 4 \end{aligned}$ Theorem 5.1
          Definition of fractional exponents
          The square of 2 is 4.

b. $\begin{aligned} 25^{3/2} &= (25^{1/2})^3 \\ &= 5^3 \\ &= 125 \end{aligned}$ Theorem 5.1
          Definition of fractional exponents
          The cube of 5 is 125.

c. $9^{-3/2} = (9^{1/2})^{-3}$ Theorem 5.1

          $= 3^{-3}$ Definition of fractional exponents

          $= \dfrac{1}{3^3}$ Property 5 for exponents

          $= \dfrac{1}{27}$ The cube of 3 is 27.

d. $\left(\dfrac{27}{8}\right)^{-4/3} = \left[\left(\dfrac{27}{8}\right)^{1/3}\right]^{-4}$ Theorem 5.1

$\qquad\qquad = \left(\dfrac{3}{2}\right)^{-4}$ Definition of fractional exponents

$\qquad\qquad = \left(\dfrac{2}{3}\right)^{4}$ Property 5 for exponents

$\qquad\qquad = \dfrac{16}{81}$ $\left(\dfrac{2}{3}\right)^4 = \dfrac{16}{81}$ ▲

The following examples show the application of the properties of exponents to rational exponents:

▼   Example 4   Assume x, y, and z all represent positive quantities and simplify as much as possible:

a. $x^{1/3} \cdot x^{5/6} = x^{1/3+5/6}$
$\qquad\qquad = x^{2/6+5/6}$
$\qquad\qquad = x^{7/6}$

b. $(y^{2/3})^{3/4} = y^{(2/3)(3/4)}$
$\qquad\qquad = y^{1/2}$

c. $(x^{1/3}y^{2/5}z^{3/7})^4 = x^{4/3}y^{8/5}z^{12/7}$

d. $\dfrac{x^{1/3}}{x^{1/4}} = x^{1/3-1/4}$

$\qquad\qquad = x^{4/12-3/12}$
$\qquad\qquad = x^{1/12}$

e. $\dfrac{x^{4/5}y^{5/6}z^3}{x^{1/5}yz^{2/3}} = x^{3/5}y^{-1/6}z^{7/3}$

$\qquad\qquad = \dfrac{x^{3/5}z^{7/3}}{y^{1/6}}$

f. $\dfrac{(x^3y^{1/2})^{2/3}}{x^{1/6}y^{1/6}} = \dfrac{x^2y^{1/3}}{x^{1/6}y^{1/6}}$

$\qquad\qquad = x^{11/6}y^{1/6}$ ▲

Use the definition of fractional exponents to write each of the following with the appropriate root, then simplify:

1. $36^{1/2}$ 2. $49^{1/2}$
3. $-9^{1/2}$ 4. $-16^{1/2}$

5. $8^{1/3}$ **6.** $-8^{1/3}$
7. $(-8)^{1/3}$ **8.** $-27^{1/3}$
9. $32^{1/5}$ **10.** $81^{1/4}$
11. $\left(\frac{81}{25}\right)^{1/2}$ **12.** $\left(\frac{9}{16}\right)^{1/2}$
13. $\left(\frac{64}{125}\right)^{1/3}$ **14.** $\left(\frac{8}{27}\right)^{1/3}$

Use Theorem 5.1 to simplify each of the following as much as possible:

15. $27^{2/3}$ **16.** $8^{4/3}$
17. $25^{3/2}$ **18.** $9^{3/2}$
19. $16^{3/4}$ **20.** $81^{3/4}$
21. $27^{-1/3}$ **22.** $9^{-1/2}$
23. $81^{-3/4}$ **24.** $4^{-3/2}$
25. $\left(\frac{25}{36}\right)^{-1/2}$ **26.** $\left(\frac{16}{49}\right)^{-1/2}$
27. $\left(\frac{81}{16}\right)^{-3/4}$ **28.** $\left(\frac{27}{8}\right)^{-2/3}$
29. $16^{1/2} + 27^{1/3}$ **30.** $25^{1/2} + 100^{1/2}$
31. $8^{-2/3} + 4^{-1/2}$ **32.** $49^{-1/2} + 25^{-1/2}$
33. $81^{3/4} \cdot 9^{1/2}$ **34.** $64^{2/3} \cdot 16^{3/2}$
35. $9^{-1/2} \cdot 8^{-2/3}$ **36.** $27^{-2/3} \cdot 16^{-1/2}$

Use the properties of exponents to simplify each of the following as much as possible. Assume all bases are positive.

37. $x^{3/5} \cdot x^{1/5}$ **38.** $x^{3/4} \cdot x^{5/4}$
39. $(a^{3/4})^{4/3}$ **40.** $(a^{2/3})^{3/4}$

41. $\dfrac{x^{1/5}}{x^{3/5}}$ **42.** $\dfrac{x^{2/7}}{x^{5/7}}$

43. $(a^{3/4} \cdot b^{1/3})^2$ **44.** $(a^{5/6} \cdot b^{2/5})^3$
45. $x^{2/3} \cdot x^{2/5}$ **46.** $x^{1/3} \cdot x^{3/4}$

47. $\dfrac{x^{5/6}}{x^{2/3}}$ **48.** $\dfrac{x^{7/8}}{x^{8/7}}$

49. $(x^{3/5}y^{5/6}z^{1/3})^{3/5}$ **50.** $(x^{3/4}y^{1/8}z^{5/6})^{4/5}$

51. $\dfrac{x^{3/4}y^{2/3}}{x^{1/4}y^{1/3}}$ **52.** $\dfrac{x^{5/6}y^{3/5}}{x^{1/6}y^{2/5}}$

53. $\dfrac{a^{3/4}b^2}{a^{7/8}b^{1/4}}$ **54.** $\dfrac{a^{1/3}b^4}{a^{3/5}b^{1/3}}$

55. $\dfrac{(y^{2/3})^{3/4}}{(y^{1/3})^{3/5}}$ **56.** $\dfrac{(y^{5/4})^{2/5}}{(y^{1/4})^{4/3}}$

57. $\dfrac{(x^{2/3}y^{-1/3}z^{-2})^3}{(x^{-1/3}y^{4/3}z^4)^2}$ **58.** $\dfrac{(x^{3/5}y^{1/5}z^3)^4}{(x^{4/5}y^{-3/5}z^{-1})^3}$

59. $\dfrac{(a^{1/2}b^{-1/3}c^{1/5})^{3/2}}{(a^{-3/4}b^{5/6}c^{2/5})^{-2/3}}$ **60.** $\dfrac{(a^{3/2}b^{-4/3}c^{5/4})^{1/3}}{(a^{-5/4}b^{4/3}c^{-3/2})^{-1/3}}$

61. The maximum speed (v) that an automobile can travel around a curve of radius r without skidding is given by the equation

$$v = \left(\frac{5r}{2}\right)^{1/2}$$

Where v is in mi/hr and r is measured in feet. What is the maximum speed a car can travel around a curve with a radius of 250 feet without skidding?

Any expression containing a radical is called a *radical expression*. In the expression $\sqrt[3]{8}$, the 3 is called the *index*, the $\sqrt{}$ is the *radical sign*, and 8 is called the *radicand*. The index of a radical must be a positive integer greater than 1. If no index is written, it is assumed to be 2.

There are two properties of radicals that we will prove using the definition of fractional exponents and the properties of exponents. For these two properties we will assume a and b are nonnegative real numbers whenever n is an even number.

**5.2
Simplified Form for
Radicals**

PROPERTY 1

$$\sqrt[n]{ab} = \sqrt[n]{a}\,\sqrt[n]{b}$$

In words: the nth root of a product is the product of the nth roots.

Proof

$$\sqrt[n]{ab} = (ab)^{1/n} \qquad \text{Definition of fractional exponents.}$$
$$= a^{1/n}b^{1/n} \qquad \text{Exponents distribute over products.}$$
$$= \sqrt[n]{a}\,\sqrt[n]{b} \qquad \text{Definition of fractional exponents.}$$

PROPERTY 2

$$\sqrt[n]{\frac{a}{b}} = \frac{\sqrt[n]{a}}{\sqrt[n]{b}} \qquad (b \neq 0)$$

In words: the nth root of a quotient is the quotient of the nth roots.

Proof $\sqrt[n]{\dfrac{a}{b}} = \left(\dfrac{a}{b}\right)^{1/n}$ Definition of fractional exponents.

$= \dfrac{a^{1/n}}{b^{1/n}}$ Exponents distribute over quotients.

$= \dfrac{\sqrt[n]{a}}{\sqrt[n]{b}}$ Definition of fractional exponents.

Note   There is no property for radicals that says the *n*th root of a sum is the sum of the *n*th roots. That is,

$$\sqrt[n]{a + b} \neq \sqrt[n]{a} + \sqrt[n]{b}$$

The two properties of the exponents allow us to change the form and simplify radical expressions without changing their value.

DEFINITION   A radical expression is in *simplified form* if:

1. none of the factors of the radicand (the quantity under the radical sign) can be written as powers greater than or equal to the index—that is, no perfect squares can be factors of the quantity under a square root sign, no perfect cubes as factors of what is under a cube root sign, and so forth;
2. there are no fractions under the radical sign; and
3. there are no radicals in the denominator.

Writing a radical expression in simplified form does not always result in a simpler-looking expression. Simplified form for radicals is a way of writing radicals so they are easiest to work with.

Satisfying the first condition for simplified form actually amounts to taking as much out from under the radical sign as possible. The following examples illustrate the first condition for simplified form:

▼   **Example 1**   Write $\sqrt{50}$ in simplified form.

Solution   The largest perfect square that divides 50 is 25. We write 50 as $25 \cdot 2$ and apply Property 1 for radicals.

$$\sqrt{50} = \sqrt{25 \cdot 2} \qquad 50 = 25 \cdot 2$$
$$= \sqrt{25}\sqrt{2} \qquad \text{Property 1}$$
$$= 5\sqrt{2} \qquad \sqrt{25} = 5$$

We have taken as much as possible out from under the radical sign—in this case, factoring 25 from 50 and then writing $\sqrt{25}$ as 5.

▲

▼ **Example 2** Write in simplified form: $\sqrt{48x^4y^3}$, where $x, y \geq 0$.

3·16 x²y⁷

Solution The largest perfect square that is a factor of the radicand is $16x^4y^2$. Applying Property 1 again, we have

$$\sqrt{48x^4y^3} = \sqrt{16x^4y^2 \cdot 3y}$$
$$= \sqrt{16x^4y^2}\sqrt{3y}$$
$$= 4x^2y\sqrt{3y}$$

▲

▼ **Example 3** Write $\sqrt[3]{40a^5b^4}$ in simplified form.

Solution We now want to factor the largest perfect cube from the radicand. We write $40a^5b^4$ as $8a^3b^3 \cdot 5a^2b$ and proceed as we did in Examples 1 and 2 above.

$$\sqrt[3]{40a^5b^4} = \sqrt[3]{8a^3b^3 \cdot 5a^2b}$$
$$= \sqrt[3]{8a^3b^3}\sqrt[3]{5a^2b}$$
$$= 2ab\sqrt[3]{5a^2b}$$

▲

Here are some further examples concerning the first condition for simplified form:

▼ **Example 4**

a. $\sqrt{12x^7y^6} = \sqrt{4x^6y^6 \cdot 3x}$
$$= \sqrt{4x^6y^6}\sqrt{3x}$$
$$= 2x^3y^3\sqrt{3x}$$

b. $\sqrt[3]{54a^6b^2c^4} = \sqrt[3]{27a^6c^3 \cdot 2b^2c}$
$$= \sqrt[3]{27a^6c^3}\sqrt[3]{2b^2c}$$
$$= 3a^2c\sqrt[3]{2b^2c}$$

c. $\sqrt[5]{32x^{10}y^7} = \sqrt[5]{32x^{10}y^5 \cdot y^2}$
$$= \sqrt[5]{32x^{10}y^5}\sqrt[5]{y^2}$$
$$= 2x^2y\sqrt[5]{y^2}$$

▲

The second property of radicals is used to simplify a radical that contains a fraction.

▼   **Example 5**   Simplify $\sqrt{\dfrac{3}{4}}$.

Solution   Applying Property 2 for radicals, we have

$$\sqrt{\frac{3}{4}} = \frac{\sqrt{3}}{\sqrt{4}} \qquad \text{Property 2}$$

$$= \frac{\sqrt{3}}{2} \qquad \sqrt{4} = 2$$

The last expression is in simplified form because it satisfies all three conditions for simplified form.   ▲

▼   **Example 6**   Write $\sqrt{\dfrac{5}{6}}$ in simplified form.

Solution   Proceeding as in Example 5, above, we have

$$\sqrt{\frac{5}{6}} = \frac{\sqrt{5}}{\sqrt{6}}$$

The resulting expression satisfies the second condition for simplified form since neither radical contains a fraction. It does, however, violate condition 3 since it has a radical in the denominator. Getting rid of the radical in the denominator is called *rationalizing the denominator* and is accomplished, in this case, by multiplying the numerator and denominator by $\sqrt{6}$:

$$\frac{\sqrt{5}}{\sqrt{6}} = \frac{\sqrt{5}}{\sqrt{6}} \cdot \frac{\sqrt{6}}{\sqrt{6}}$$

$$= \frac{\sqrt{30}}{\sqrt{6^2}}$$

$$= \frac{\sqrt{30}}{6} \qquad \qquad ▲$$

The idea behind rationalizing the denominator is to produce a perfect square under the square root sign in the denominator. This is accomplished by multiplying both the numerator and denominator by the appropriate radical.

▼   **Example 7**

a. $\dfrac{4}{\sqrt{3}} = \dfrac{4}{\sqrt{3}} \cdot \dfrac{\sqrt{3}}{\sqrt{3}}$

$\qquad = \dfrac{4\sqrt{3}}{\sqrt{3^2}}$

$\qquad = \dfrac{4\sqrt{3}}{3}$

b. $\dfrac{2\sqrt{3x}}{\sqrt{5y}} = \dfrac{2\sqrt{3x}}{\sqrt{5y}} \cdot \dfrac{\sqrt{5y}}{\sqrt{5y}}$

$\qquad = \dfrac{2\sqrt{15xy}}{\sqrt{(5y)^2}}$

$\qquad = \dfrac{2\sqrt{15xy}}{5y}$ ▲

When the denominator involves a cube root, we must multiply by a radical that will produce a perfect cube under the cube root sign in the denominator.

▼   **Example 8**   Rationalize the denominator in $\dfrac{7}{\sqrt[3]{4}}$.

Solution   Since $4 = 2^2$, we can multiply both numerator and denominator by $\sqrt[3]{2}$ and obtain $\sqrt[3]{2^3}$ in the denominator.

$$\frac{7}{\sqrt[3]{4}} = \frac{7}{\sqrt[3]{2^2}}$$

$$= \frac{7}{\sqrt[3]{2^2}} \cdot \frac{\sqrt[3]{2}}{\sqrt[3]{2}}$$

$$= \frac{7\sqrt[3]{2}}{\sqrt[3]{2^3}}$$

$$= \frac{7\sqrt[3]{2}}{2} \qquad ▲$$

▼   **Example 9**   Rationalize the denominator in $\dfrac{5\sqrt[4]{2}}{\sqrt[4]{3x^3}}$.

Solution We need a perfect fourth power under the radical sign in the denominator. Multiplying numerator and denominator by $\sqrt[4]{3^3 x}$ will give us what we want:

$$\frac{5\sqrt[4]{2}}{\sqrt[4]{3x^3}} = \frac{5\sqrt[4]{2}}{\sqrt[4]{3x^3}} \cdot \frac{\sqrt[4]{3^3 x}}{\sqrt[4]{3^3 x}}$$

$$= \frac{5\sqrt[4]{54x}}{\sqrt[4]{3^4 x^4}}$$

$$= \frac{5\sqrt[4]{54x}}{3x} \qquad \blacktriangle$$

As a last example we consider a radical expression that requires the use of both properties to meet the three conditions for simplified form.

▼ **Example 10** Simplify $\sqrt{\dfrac{12x^5 y^3}{5z}}$.

Solution We use Property 2 to write the numerator and denominator as two separate radicals:

$$\sqrt{\frac{12x^5 y^3}{5z}} = \frac{\sqrt{12x^5 y^3}}{\sqrt{5z}}$$

Simplifying the numerator, we have

$$\frac{\sqrt{12x^5 y^3}}{\sqrt{5z}} = \frac{\sqrt{4x^4 y^2}\sqrt{3xy}}{\sqrt{5z}}$$

$$= \frac{2x^2 y\sqrt{3xy}}{\sqrt{5z}}$$

To rationalize the denominator we multiply the numerator and denominator by $\sqrt{5z}$:

$$\frac{2x^2 y\sqrt{3xy}}{\sqrt{5z}} \cdot \frac{\sqrt{5z}}{\sqrt{5z}} = \frac{2x^2 y\sqrt{15xyz}}{\sqrt{(5z)^2}}$$

$$= \frac{2x^2 y\sqrt{15xyz}}{5z} \qquad \blacktriangle$$

Use Property 1 for radicals to write each of the following expressions in simplified form. (Assume all variables are positive throughout problem set.)

1. $\sqrt{8}$ 2. $\sqrt{32}$

3. $\sqrt{18}$ 4. $\sqrt{98}$

5. $\sqrt{75}$ 6. $\sqrt{12}$

7. $\sqrt{288}$ 8. $\sqrt{128}$

9. $\sqrt{80}$ 10. $\sqrt{200}$

11. $\sqrt{48}$ 12. $\sqrt{27}$

13. $\sqrt{45}$ 14. $\sqrt{20}$

15. $\sqrt[3]{54}$ 16. $\sqrt[3]{24}$

17. $\sqrt[3]{128}$ 18. $\sqrt[3]{162}$

19. $\sqrt[5]{64}$ 20. $\sqrt[4]{48}$

21. $\sqrt{54}$ 22. $\sqrt{63}$

23. $\sqrt[3]{40}$ 24. $\sqrt[3]{48}$

25. $\sqrt{99}$ 26. $\sqrt{44}$

27. $\sqrt{18x^3}$ 28. $\sqrt{27x^5}$

29. $\sqrt{32y^7}$ 30. $\sqrt{20y^3}$

31. $\sqrt[3]{40x^4y^7}$ 32. $\sqrt[3]{128x^6y^2}$

33. $\sqrt{48a^2b^3c^4}$ 34. $\sqrt{72a^4b^3c^2}$

35. $\sqrt[3]{48a^2b^3c^4}$ 36. $\sqrt[3]{72a^4b^3c^2}$

37. $\sqrt[5]{64x^8y^{12}}$ 38. $\sqrt[4]{32x^9y^{10}}$

39. $\sqrt{12x^5y^2z^3}$ 40. $\sqrt{24x^8y^3z^5}$

Rationalize the denominator in each of the following expressions:

41. $\dfrac{2}{\sqrt{3}}$ 42. $\dfrac{3}{\sqrt{2}}$

43. $\dfrac{5}{\sqrt{6}}$ 44. $\dfrac{7}{\sqrt{5}}$

45. $\sqrt{\frac{1}{2}}$ 46. $\sqrt{\frac{1}{3}}$

47. $\sqrt{\frac{1}{5}}$ 48. $\sqrt{\frac{1}{6}}$

49. $\dfrac{4}{\sqrt[3]{2}}$ 50. $\dfrac{5}{\sqrt[3]{3}}$

51. $\dfrac{2}{\sqrt[3]{9}}$ 52. $\dfrac{3}{\sqrt[3]{4}}$

53. $\sqrt{\dfrac{3}{2x}}$ 54. $\sqrt{\dfrac{5}{3x}}$

55. $\sqrt{\dfrac{8}{y}}$ 56. $\sqrt{\dfrac{27}{y}}$

57. $\sqrt[3]{\dfrac{4x}{3y}}$ 58. $\sqrt[3]{\dfrac{7x}{6y}}$

59. $\sqrt[3]{\dfrac{2x}{9y}}$ 60. $\sqrt[3]{\dfrac{5x}{4y}}$

61. $\sqrt[4]{\dfrac{1}{8x}}$ 62. $\sqrt[4]{\dfrac{8}{9x}}$

Write each of the following in simplified form:

63. $\sqrt{\dfrac{27x^3}{5y}}$ 64. $\sqrt{\dfrac{12x^5}{7y}}$

65. $\sqrt{\dfrac{75x^3y^2}{2z}}$ 66. $\sqrt{\dfrac{50x^2y^3}{3z}}$

67. $\sqrt[3]{\dfrac{16a^4b^3}{9c}}$ 68. $\sqrt[3]{\dfrac{54a^5b^4}{25c^2}}$

69. $\sqrt[3]{\dfrac{8x^3y^6}{9z}}$ 70. $\sqrt[3]{\dfrac{27x^6y^3}{2z^2}}$

71. The distance (d) between opposite corners of a rectangular room with length l and width w is given by

$$d = \sqrt{l^2 + w^2}$$

How far is it between opposite corners of a living room that measures 10 by 15 feet? ▲

**5.3
Combinations of
Radical Expressions**

In this section we will combine radical expressions under the four basic operations: addition, subtraction, multiplication, and division. All of the operations involve use of the distributive property, and many of the problems use the two properties of radicals developed in the previous section. We will apply the conditions for simplified form for radicals to many of the expressions in this section.

*Addition and
Subtraction of
Radical Expressions*

In Chapter 3 we found that we could add only similar terms when combining polynomials. The same idea applies to addition and subtraction of radical expressions.

DEFINITION   Two radicals are said to be *similar radicals* if they have the same index and the same radicand.

The expressions $5\sqrt[3]{7}$ and $-8\sqrt[3]{7}$ are similar since the index is ^{3} in both cases and the radicands are 7. The expressions $3\sqrt[4]{5}$ and $7\sqrt[3]{5}$ are not similar since they have different indices, while the expressions $2\sqrt[5]{8}$ and $3\sqrt[5]{9}$ are not similar because the radicands are not the same.

We add and subtract radical expressions in the same way we add and subtract polynomials—by combining similar terms under the distributive property.

▼ **Example 1**   Combine $5\sqrt{3} - 4\sqrt{3} + 6\sqrt{3}$.

Solution   All three radicals are similar. We apply the distributive property to get

$$5\sqrt{3} - 4\sqrt{3} + 6\sqrt{3} = (5 - 4 + 6)\sqrt{3}$$
$$= 7\sqrt{3} \qquad \blacktriangle$$

▼ **Example 2**   Combine $3\sqrt{8} + 5\sqrt{18}$.

Solution   The two radicals do not seem to be similar. We must write each in simplified form before applying the distributive property.

$$3\sqrt{8} + 5\sqrt{18} = 3\sqrt{4 \cdot 2} + 5\sqrt{9 \cdot 2}$$
$$= 3\sqrt{4}\sqrt{2} + 5\sqrt{9}\sqrt{2}$$
$$= 3 \cdot 2\sqrt{2} + 5 \cdot 3\sqrt{2}$$
$$= 6\sqrt{2} + 15\sqrt{2}$$
$$= (6 + 15)\sqrt{2}$$
$$= 21\sqrt{2} \qquad \blacktriangle$$

The result of Example 2 can be generalized to the following rule for sums and differences of radical expressions:

RULE   To add or subtract two radical expressions, put each in simplified form and apply the distributive property if possible. We can add only similar radicals. We must write each expression in simplified form for radicals before we can tell if the radicals are similar.

▼ **Example 3**   Combine $7\sqrt{75xy^3} - 4y\sqrt{12xy}$, where $x, y \geq 0$.

Solution   We write each expression in simplified form and combine similar radicals:

$$7\sqrt{75xy^3} - 4y\sqrt{12xy} = 7\sqrt{25y^2}\sqrt{3xy} - 4y\sqrt{4}\sqrt{3xy}$$
$$= 35y\sqrt{3xy} - 8y\sqrt{3xy}$$
$$= (35y - 8y)\sqrt{3xy}$$
$$= 27y\sqrt{3xy} \qquad \blacktriangle$$

▼ **Example 4** Combine $10\sqrt[3]{8a^4b^2} + 11a\sqrt[3]{27ab^2}$.

Solution Writing each radical in simplified form and combining similar terms, we have

$$10\sqrt[3]{8a^4b^2} + 11a\sqrt[3]{27ab^2} = 10\sqrt[3]{8a^3}\sqrt[3]{ab^2} + 11a\sqrt[3]{27}\sqrt[3]{ab^2}$$
$$= 20a\sqrt[3]{ab^2} + 33a\sqrt[3]{ab^2}$$
$$= 53a\sqrt[3]{ab^2}. \qquad \blacktriangle$$

Multiplication of
Radical Expressions

We use the same process to multiply radical expressions as we have in the past to multiply polynomials.

▼ **Example 5** Multiply $\sqrt{3}(2\sqrt{6} - 5\sqrt{12})$.

Solution Applying the distributive property, we have

$$\sqrt{3}(2\sqrt{6} - 5\sqrt{12}) = \sqrt{3}\cdot 2\sqrt{6} - \sqrt{3}\cdot 5\sqrt{12}$$
$$= 2\sqrt{18} - 5\sqrt{36}$$

Writing each radical in simplified form gives

$$2\sqrt{18} - 5\sqrt{36} = 2\sqrt{9}\sqrt{2} - 5\sqrt{36}$$
$$= 6\sqrt{2} - 30 \qquad \blacktriangle$$

▼ **Example 6** Multiply $(\sqrt{3} + \sqrt{5})(4\sqrt{3} - \sqrt{5})$.

Solution The same principle that applies to multiply two binomials applies to this product. We must multiply each term in the first expression by each term in the second one. Any convenient method can be used.

$$(\sqrt{3} + \sqrt{5})(4\sqrt{3} - \sqrt{5}) = \sqrt{3}\cdot 4\sqrt{3} - \sqrt{3}\sqrt{5} + \sqrt{5}\cdot 4\sqrt{3} - \sqrt{5}\sqrt{5}$$
$$= 4\cdot 3 - \sqrt{15} + 4\sqrt{15} - 5$$
$$= 12 + 3\sqrt{15} - 5$$
$$= 7 + 3\sqrt{15} \qquad \blacktriangle$$

▼ **Example 7** Expand $(3\sqrt{x} - 2\sqrt{y})^2$.

Solution 1 We write this problem as a multiplication problem and proceed as in Example 6:

$$(3\sqrt{x} - 2\sqrt{y})^2 = (3\sqrt{x} - 2\sqrt{y})(3\sqrt{x} - 2\sqrt{y})$$
$$= 3\sqrt{x} \cdot 3\sqrt{x} - 3\sqrt{x} \cdot 2\sqrt{y} - 2\sqrt{y} \cdot 3\sqrt{x} + 2\sqrt{y} \cdot 2\sqrt{y}$$
$$= 9x - 6\sqrt{xy} - 6\sqrt{xy} + 4y$$
$$= 9x - 12\sqrt{xy} + 4y$$

Solution 2 We can apply the formula for the square of a sum, $(a + b)^2 = a^2 + 2ab + b^2$:

$$(3\sqrt{x} - 2\sqrt{y})^2 = (3\sqrt{x})^2 + 2(3\sqrt{x})(-2\sqrt{y}) + (-2\sqrt{y})^2$$
$$= 9x - 12\sqrt{xy} + 4y \qquad \blacktriangle$$

▼ **Example 8** Multiply $(\sqrt{6} + \sqrt{2})(\sqrt{6} - \sqrt{2})$.

Solution We notice the product is of the form $(a + b)(a - b)$, which always gives the difference of two squares, $a^2 - b^2$:

$$(\sqrt{6} + \sqrt{2})(\sqrt{6} - \sqrt{2}) = (\sqrt{6})^2 - (\sqrt{2})^2$$
$$= 6 - 2$$
$$= 4 \qquad \blacktriangle$$

In Example 8 the two expressions $(\sqrt{6} + \sqrt{2})$ and $(\sqrt{6} - \sqrt{2})$ are called *conjugates*. In general, the conjugate of $\sqrt{a} + \sqrt{b}$ is $\sqrt{a} - \sqrt{b}$. Multiplying conjugates of this form always produces a real number.

Division with radical expressions is the same as rationalizing the denominator. In the preceding section we were able to divide $\sqrt{3}$ by $\sqrt{2}$ by rationalizing the denominator:

Division with Radical Expressions

$$\frac{\sqrt{3}}{\sqrt{2}} = \frac{\sqrt{3}}{\sqrt{2}} \cdot \frac{\sqrt{2}}{\sqrt{2}}$$
$$= \frac{\sqrt{6}}{2}$$

We can accomplish the same result with expressions such as

$$\frac{6}{\sqrt{5} - \sqrt{3}}$$

by multiplying the numerator and denominator by the conjugate of the denominator.

▼ **Example 9** Divide $\dfrac{6}{\sqrt{5} - \sqrt{3}}$.

Solution Since the product of two conjugates is a real number, we multiply the numerator and denominator by the conjugate of the denominator:

$$\frac{6}{\sqrt{5} - \sqrt{3}} = \frac{6}{\sqrt{5} - \sqrt{3}} \cdot \frac{(\sqrt{5} + \sqrt{3})}{(\sqrt{5} + \sqrt{3})}$$

$$= \frac{6\sqrt{5} + 6\sqrt{3}}{(\sqrt{5})^2 - (\sqrt{3})^2}$$

$$= \frac{6\sqrt{5} + 6\sqrt{3}}{5 - 3}$$

$$= \frac{6\sqrt{5} + 6\sqrt{3}}{2}$$

$$= \frac{2(3\sqrt{5} + 3\sqrt{3})}{2}$$

$$= 3\sqrt{5} + 3\sqrt{3} \qquad\qquad ▲$$

▼ **Example 10** Rationalize the denominator in $\dfrac{2\sqrt{3} - 1}{5\sqrt{3} + 2}$.

Solution Multiplying the numerator and denominator by the conjugate of the denominator, we have

$$\frac{2\sqrt{3} - 1}{5\sqrt{3} + 2} = \frac{2\sqrt{3} - 1}{5\sqrt{3} + 2} \cdot \frac{(5\sqrt{3} - 2)}{(5\sqrt{3} - 2)}$$

$$= \frac{2\sqrt{3} \cdot 5\sqrt{3} - 2 \cdot 2\sqrt{3} - 5\sqrt{3} + 2}{(5\sqrt{3})^2 - 2^2}$$

$$= \frac{30 - 4\sqrt{3} - 5\sqrt{3} + 2}{75 - 4}$$

$$= \frac{32 - 9\sqrt{3}}{71} \qquad\qquad ▲$$

Combine the following expressions. (Assume any variables under an even root are positive.)   *Problem Set 5.3*

1. $3\sqrt{5} + 4\sqrt{5}$

2. $6\sqrt{3} - 5\sqrt{3}$

3. $8\sqrt{6} - 2\sqrt{6} + 3\sqrt{6}$

4. $7\sqrt{7} - \sqrt{7} + 4\sqrt{7}$

5. $3x\sqrt{2} - 4x\sqrt{2} + x\sqrt{2}$

6. $5x\sqrt{6} - 3x\sqrt{6} - 2x\sqrt{6}$

7. $\sqrt{18} + \sqrt{2}$

8. $\sqrt{12} + \sqrt{3}$

9. $\sqrt{20} - \sqrt{80} + \sqrt{45}$

10. $\sqrt{8} - \sqrt{32} - \sqrt{18}$

11. $4\sqrt{8} - 2\sqrt{50} - 5\sqrt{72}$

12. $\sqrt{48} - 3\sqrt{27} + 2\sqrt{75}$

13. $5x\sqrt{8} + 3\sqrt{32x^2} - 5\sqrt{50x^2}$

14. $2\sqrt{50x^2} - 8x\sqrt{18} - 3\sqrt{72x^2}$

15. $5\sqrt[3]{16} - 4\sqrt[3]{54}$

16. $\sqrt[3]{81} + 3\sqrt[3]{24}$

17. $\sqrt[3]{x^4y^2} + 7x\sqrt[3]{xy^2}$

18. $2\sqrt[3]{x^8y^6} - 3y^2\sqrt[3]{8x^8}$

19. $5a^2\sqrt{27ab^3} - 6b\sqrt{12a^5b}$

20. $9a\sqrt{20a^3b^2} + 7b\sqrt{45a^5}$

21. $b\sqrt[3]{24a^5b} + 3a\sqrt[3]{81a^2b^4}$

22. $7\sqrt[3]{a^4b^3c^2} - 6ab\sqrt[3]{ac^2}$

Find the following products:

23. $\sqrt{3}(\sqrt{2} - 3\sqrt{3})$

24. $\sqrt{2}(5\sqrt{3} + 4\sqrt{2})$

25. $6\sqrt{6}(2\sqrt{2} + 1)$

26. $7\sqrt{5}(3\sqrt{15} - 2)$

27. $(\sqrt{3} + \sqrt{2})(3\sqrt{3} - \sqrt{2})$

28. $(\sqrt{5} - \sqrt{2})(3\sqrt{5} + 2\sqrt{2})$

29. $(3\sqrt{6} + 4\sqrt{2})(\sqrt{6} + 2\sqrt{2})$

30. $(\sqrt{7} - 3\sqrt{3})(2\sqrt{7} - 4\sqrt{3})$

31. $(\sqrt{3} + 4)^2$

32. $(\sqrt{5} - 2)^2$

33. $(2\sqrt{3} + 3\sqrt{2})^2$

34. $(4\sqrt{6} - 3\sqrt{2})^2$

35. $(\sqrt{3} - \sqrt{2})(\sqrt{3} + \sqrt{2})$

36. $(\sqrt{5} - \sqrt{2})(\sqrt{5} + \sqrt{2})$

37. $(2\sqrt{6} - 3)(2\sqrt{6} + 3)$

38. $(3\sqrt{5} - 1)(3\sqrt{5} + 1)$

39. $(4\sqrt{7} - 2\sqrt{5})(4\sqrt{7} + 2\sqrt{5})$

40. $(2\sqrt{11} - 5\sqrt{2})(2\sqrt{11} + 5\sqrt{2})$

Rationalize the denominator in each of the following:

41. $\dfrac{2}{\sqrt{3} + \sqrt{2}}$

42. $\dfrac{3}{\sqrt{5} - \sqrt{2}}$

43. $\dfrac{4\sqrt{2}}{\sqrt{6} - \sqrt{2}}$

44. $\dfrac{3\sqrt{5}}{\sqrt{5} + \sqrt{3}}$

45. $\dfrac{1}{2\sqrt{3} - 1}$

46. $\dfrac{2}{3\sqrt{2} + 1}$

47. $\dfrac{2\sqrt{5}}{2\sqrt{5} - 3}$

48. $\dfrac{3\sqrt{7}}{3\sqrt{7} - 2}$

49. $\dfrac{\sqrt{2} + \sqrt{6}}{\sqrt{2} - \sqrt{6}}$

50. $\dfrac{\sqrt{3} - \sqrt{5}}{\sqrt{3} + \sqrt{5}}$

51. $\dfrac{2\sqrt{3} - \sqrt{7}}{3\sqrt{3} + \sqrt{7}}$

52. $\dfrac{5\sqrt{6} + 2\sqrt{2}}{\sqrt{6} - \sqrt{2}}$

53. $\dfrac{4\sqrt{3} - 2\sqrt{2}}{3\sqrt{3} - 5\sqrt{2}}$

54. $\dfrac{8\sqrt{5} - 2\sqrt{3}}{2\sqrt{5} + 2\sqrt{3}}$

55. If an object is dropped from the top of a 100-ft building, the amount of time (t), in seconds, that it takes for the object to be h feet from the ground is given by the formula

$$t = \frac{\sqrt{100 - h}}{4}$$

How long does it take before the object is 50 feet from the ground? How long does it take to reach the ground? (When it is on the ground, h is 0.)

**5.4
Equations with
Radicals**

This section is concerned with solving equations that involve one or more radicals. The first step in solving an equation that contains a radical is to eliminate the radical from the equation. To do so we need an additional property:

SQUARING PROPERTY OF EQUALITY If both sides of an equation are squared, the solutions to the original equation are solutions to the resulting equation.

We will never lose solutions to our equations by squaring both sides. We may, however, introduce *extraneous solutions*. Extraneous solutions satisfy the equation obtained by squaring both sides of the original equation, but do not satisfy the original equation.

We know that if two real numbers a and b are equal, then so are their squares:

$$\text{If} \quad a = b$$
$$\text{then} \quad a^2 = b^2$$

On the other hand, extraneous solutions are introduced when we square opposites. That is, even though opposites are not equal, their squares are. For example,

$$5 = -5 \qquad \text{A false statement}$$
$$(5)^2 = (-5)^2 \qquad \text{Square both sides}$$
$$25 = 25 \qquad \text{A true statement}$$

It is because of this that extraneous solutions are sometimes introduced when squaring both sides of an equation.

We are free to square both sides of an equation any time it is convenient. We must be aware, however, that doing so may introduce extraneous solutions. We must, therefore, check all our solutions in the original equation if at any time we square both sides of the original equation.

▼ **Example 1** Solve for x: $\sqrt{3x + 4} = 5$.

Solution We square both sides and proceed as usual:

$$\sqrt{3x + 4} = 5$$
$$(\sqrt{3x + 4})^2 = 5^2$$
$$3x + 4 = 25$$
$$3x = 21$$
$$x = 7$$

Checking $x = 7$ in the original equation, we have

$$\sqrt{3(7) + 4} = 5$$
$$\sqrt{21 + 4} = 5$$
$$\sqrt{25} = 5$$
$$5 = 5$$

The solution $x = 7$ satisfies the original equation.   ▲

▼ **Example 2** Solve $\sqrt{4x - 7} = -3$.

Solution Squaring both sides, we have

$$\sqrt{4x - 7} = -3$$
$$(\sqrt{4x - 7})^2 = (-3)^2$$
$$4x - 7 = 9$$
$$4x = 16$$
$$x = 4$$

Checking $x = 4$ in the original equation gives

$$\sqrt{4(4) - 7} = -3$$
$$\sqrt{16 - 7} = -3$$
$$\sqrt{9} = -3$$
$$3 = -3$$ ▲

The solution $x = 4$ produces a false statement when checked in the original equation. Since $x = 4$ was the only possible solution, there is no solution to the original equation. The solution set is $\varnothing$. The possible solution $x = 4$ is an extraneous solution. It satisfies the equation obtained by squaring both sides of the original equation, but does not satisfy the original equation.

▼ **Example 3** Solve $\sqrt{5x - 1} + 3 = 7$.

Solution We must isolate the radical on the left side of the equation. If we attempt to square both sides without doing so, the resulting equation will also contain a radical. Adding -3 to both sides, we have

$$\sqrt{5x - 1} + 3 = 7$$
$$\sqrt{5x - 1} = 4$$

We can now square both sides and proceed as usual:

$$(\sqrt{5x - 1})^2 = (4)^2$$
$$5x - 1 = 16$$
$$5x = 17$$
$$x = \tfrac{17}{5}$$

Checking $x = \tfrac{17}{5}$, we have

$$\sqrt{5(\tfrac{17}{5}) - 1} + 3 = 7$$
$$\sqrt{17 - 1} + 3 = 7$$
$$4 + 3 = 7$$
$$7 = 7$$ ▲

▼ **Example 4** Solve $\sqrt{3y - 4} - \sqrt{y + 6} = 0$.

Solution We add $\sqrt{y + 6}$ to both sides, square both sides, and solve as usual:

$$\sqrt{3y - 4} - \sqrt{y + 6} = 0$$
$$\sqrt{3y - 4} = \sqrt{y + 6}$$
$$(\sqrt{3y - 4})^2 = (\sqrt{y + 6})^2$$
$$3y - 4 = y + 6$$
$$2y = 10$$
$$y = 5$$

Substituting $y = 5$ into the original equation, we have

$$\sqrt{3(5) - 4} - \sqrt{5 + 6} = 0$$
$$\sqrt{15 - 4} - \sqrt{11} = 0$$
$$\sqrt{11} - \sqrt{11} = 0$$
$$0 = 0 \qquad \blacktriangle$$

It is also possible to raise both sides of an equation to powers greater than 2. We only need to check for extraneous solutions when we raise both sides of an equation to an even power. Raising both sides of an equation to an odd power will not produce extraneous solutions.

▼   **Example 5**   Solve $\sqrt[3]{4x + 5} = 3$.

Solution   Cubing both sides we have:

$$(\sqrt[3]{4x + 5})^3 = 3^3$$
$$4x + 5 = 27$$
$$4x = 22$$
$$x = \frac{22}{4}$$
$$x = \frac{11}{2}$$

We do not need to check $x = \frac{11}{2}$ since we raised both sides to an odd power. $\blacktriangle$

Solve each of the following equations: *Problem Set 5.4*

1. $\sqrt{2x + 1} = 3$ 2. $\sqrt{3x + 1} = 4$
3. $\sqrt{4x + 1} = -5$ 4. $\sqrt{6x + 1} = -5$
5. $\sqrt{2y - 1} = 3$ 6. $\sqrt{3y - 1} = 2$
7. $\sqrt{5x - 7} = -1$ 8. $\sqrt{8x + 3} = -6$
9. $\sqrt{2x - 3} - 2 = 4$ 10. $\sqrt{3x + 1} - 4 = 1$
11. $\sqrt{4a + 1} + 3 = 2$ 12. $\sqrt{5a - 3} + 6 = 2$
13. $\sqrt[4]{3x + 1} = 2$ 14. $\sqrt[4]{4x + 1} = 3$
15. $\sqrt[3]{2x - 5} = 1$ 16. $\sqrt[3]{5x + 7} = 2$
17. $\sqrt[3]{3a + 5} = -3$ 18. $\sqrt[3]{2a + 7} = -2$
19. $\sqrt{2x + 4} = \sqrt{x - 1}$ 20. $\sqrt{3x + 4} = -\sqrt{2x + 3}$
21. $\sqrt{4a + 7} = -\sqrt{a + 2}$ 22. $\sqrt{7a - 1} = \sqrt{2a + 4}$
23. $\sqrt[4]{5x - 8} + \sqrt[4]{4x - 1} = 0$ 24. $\sqrt[4]{6x + 7} + \sqrt[4]{x + 5} = 0$

25. $\sqrt{y-8} - \sqrt{8-y} = 0$ **26.** $\sqrt{2y-5} - \sqrt{5y-2} = 0$

27. $\sqrt[3]{3x+5} = \sqrt[3]{5-2x}$ **28.** $\sqrt[3]{4x+9} = \sqrt[3]{3-2x}$

29. $\sqrt[3]{2x+1} + \sqrt[3]{3x+2} = 0$ **30.** $\sqrt[3]{6x-4} + \sqrt[3]{4x-3} = 0$

31. The length of time (T) in seconds it takes the pendulum of a grandfather clock to swing through one complete cycle is given by the formula

$$T = 2\pi\sqrt{\frac{L}{32}}$$

where L is the length, in feet, of the pendulum, and π is approximately $\frac{22}{7}$. How long must the pendulum be if one complete cycle takes 2 seconds?

5.5

Addition and Subtraction of Complex Numbers

The equation $x^2 = -9$ has no real solutions since the square of a real number is always positive. We have been unable to work with square roots of negative numbers like $\sqrt{-25}$ and $\sqrt{-16}$ for the same reason. Complex numbers allow us to expand our work with radicals to include square roots of negative numbers and to solve equations like $x^2 = -9$ and $x^2 = -64$. Our work with complex numbers is based on the following definition:

DEFINITION The number i is such that $i = \sqrt{-1}$. (Which is the same as saying $i^2 = -1$.)

The number i is not a real number. The number i can be used to eliminate the negative sign under a square root.

▼   **Example 1**   Write (a) $\sqrt{-25}$, (b) $-\sqrt{-49}$, (c) $\sqrt{-12}$, and (d) $-\sqrt{-17}$ in terms of i.

a. $\sqrt{-25} = \sqrt{25(-1)} = \sqrt{25}\sqrt{-1} = 5i$

b. $-\sqrt{-49} = -\sqrt{49(-1)} = -\sqrt{49}\sqrt{-1} = -7i$

c. $\sqrt{-12} = \sqrt{12(-1)} = \sqrt{12}\sqrt{-1} = 2\sqrt{3}i = 2i\sqrt{3}$

d. $-\sqrt{-17} = -\sqrt{17(-1)} = -\sqrt{17}\sqrt{-1} = -\sqrt{17}i = -i\sqrt{17}$ ▲

If we assume all the properties of exponents hold when the base is i, we can write any power of i as either i, -1, $-i$, or 1. Using the fact that $i^2 = -1$, we have

$$i^1 = i$$
$$i^2 = -1$$
$$i^3 = i^2 \cdot i = -1(i) = -i$$
$$i^4 = i^2 \cdot i^2 = -1(-1) = 1$$

Since $i^4 = 1$, i^5 will simplify to i, and we will begin repeating the sequence i, -1, $-i$, 1 as we simplify higher powers of i:

$$i^5 = i^4 \cdot i \ = 1(i) = i$$
$$i^6 = i^4 \cdot i^2 = 1(-1) = -1$$
$$i^7 = i^4 \cdot i^3 = 1(-i) = -i$$
$$i^8 = i^4 \cdot i^4 = 1(1) = 1$$
$$\vdots$$

These results can be summarized by

$$i = i^1 = i^5 = i^9 \cdots$$
$$-1 = i^2 = i^6 = i^{10} \cdots$$
$$-i = i^3 = i^7 = i^{11} \cdots$$
$$1 = i^4 = i^8 = i^{12} \cdots$$

▼   **Example 2**   Simplify as much as possible:

a. $i^{30} = (i^4)^7 \cdot i^2 = 1(-1) = -1$
b. $i^{21} = (i^4)^5 \cdot i \ = 1(i) \quad = i$
c. $i^{40} = (i^4)^{10} \quad = 1$ ▲

Complex numbers are defined in terms of the number i.

DEFINITION   A *complex number* is any number that can be put in the form

$$a + bi$$

where a and b are real numbers and $i = \sqrt{-1}$. The form $a + bi$ is called *standard form* for complex numbers. The number a is called the *real part* of the complex number. The number b is called the *imaginary part* of the complex number.

▼   **Example 3**

a.   The number $3 + 4i$ is a complex number since it has the correct form. The number 3 is the real part and 4 is the imaginary part.
b.   The number $-6i$ is a complex number since it can be written as $0 + (-6i)$. The real part is 0. The imaginary part is -6.
c.   The number 8 is a complex number since it can be written as $8 + 0i$. The real part is 8. The imaginary part is 0.   ▲

From part c in Example 3 it is apparent that all real numbers can be considered complex numbers. The real numbers are a subset of the complex numbers.

Equality for Complex Numbers

Two complex numbers are equal if and only if their real parts are equal and their imaginary parts are equal. That is, for real numbers a, b, c, and d,

$$a + bi = c + di \quad \text{if and only if} \quad a = c \quad \text{and} \quad b = d$$

▼ **Example 4** Find x and y if $3x + 4i = 12 - 8yi$.

Solution Since the two complex numbers are equal, their real parts are equal and their imaginary parts are equal:

$$3x = 12 \quad \text{and} \quad 4 = -8y$$
$$x = 4 \qquad\qquad y = -\tfrac{1}{2}$$ ▲

▼ **Example 5** Find x and y if $(4x - 3) + 7i = 5 + (2y - 1)i$.

Solution The real parts are $4x - 3$ and 5. The imaginary parts are 7 and $2y - 1$.

$$4x - 3 = 5 \quad \text{and} \quad 7 = 2y - 1$$
$$4x = 8 \qquad\qquad 8 = 2y$$
$$x = 2 \qquad\qquad y = 4$$ ▲

Addition and Subtraction of Complex Numbers

To add two complex numbers, add their real parts and add their imaginary parts. That is, if a, b, c, and d are real numbers, then

$$(a + bi) + (c + di) = (a + c) + (b + d)i$$

If we assume that the commutative, associative, and distributive properties hold for the number i, then the definition of addition is simply an extension of these properties.

We define subtraction in a similar manner. If a, b, c, and d are real numbers, then

$$(a + bi) - (c + di) = (a - c) + (b - d)i$$

▼ **Example 6**

a. $(3 + 4i) + (7 - 6i) = (3 + 7) + (4 - 6)i$
$$= 10 - 2i$$
b. $(8 - 3i) + (6 - 2i) = (8 + 6) + (-3 - 2)i$
$$= 14 - 5i$$

c. $(7 + 3i) - (5 + 6i) = (7 - 5) + (3 - 6)i$
$$= 2 - 3i$$
d. $(5 - 2i) - (9 - 4i) = (5 - 9) + (-2 + 4)i$
$$= -4 + 2i$$
e. $(3 - 5i) - (4 + 2i) + (6 - 8i) = (3 - 4 + 6) + (-5 - 2 - 8)i$
$$= 5 - 15i \qquad \blacktriangle$$

Write the following in terms of i and simplify as much as possible.   *Problem Set 5.5*

1. $\sqrt{-36}$ 2. $\sqrt{-49}$
3. $-\sqrt{-25}$ 4. $-\sqrt{-81}$
5. $\sqrt{-72}$ 6. $\sqrt{-48}$
7. $-\sqrt{-12}$ 8. $-\sqrt{-75}$
9. $\sqrt{-162}$ 10. $\sqrt{-400}$

Write each of the following as i, -1, $-i$, or 1:

11. i^{17} 12. i^{18}
13. i^{19} 14. i^{20}
15. i^{28} 16. i^{31}
17. i^{26} 18. i^{37}
19. i^{75} 20. i^{42}

Find x and y so each of the following equations is true:

21. $2x + 3yi = 6 - 3i$ 22. $4x - 2yi = 4 + 8i$
23. $2 - 5i = -x + 10yi$ 24. $4 + 7i = 6x - 14yi$
25. $2x + 10i = -16 - 2yi$ 26. $4x - 5i = -2 + 3yi$
27. $4y - 6i = 3 + 5xi$ 28. $7y - 3i = 4 + 9xi$
29. $(2x - 4) - 3i = 10 - 6yi$ 30. $(4x - 3) - 2i = 8 + yi$
31. $(7x - 1) + 4i = 2 + (5y + 2)i$
32. $(5x + 2) - 7i = 4 + (2y + 1)i$

Combine the following complex numbers:

33. $(2 + 3i) + (3 + 6i)$ 34. $(4 + i) + (3 + 2i)$
35. $(3 - 5i) + (2 + 4i)$ 36. $(7 + 2i) + (3 - 4i)$
37. $(5 + 2i) - (3 + 6i)$ 38. $(6 + 7i) - (4 + i)$
39. $(3 - 5i) - (2 + i)$ 40. $(7 - 3i) - (4 + 10i)$
41. $(11 - 6i) - (2 - 4i)$ 42. $(10 - 12i) - (6 - i)$
43. $(-3 + 2i) + (6 - 5i)$ 44. $(-8 + 3i) + (4 - 8i)$
45. $[(3 + 2i) - (6 + i)] + (5 + i)$

46. $[(4 - 5i) - (2 + i)] + (2 + 5i)$

47. $[(7 - i) - (2 + 4i)] - (6 + 2i)$

48. $[(3 - i) - (4 + 7i)] - (3 - 4i)$

49. $(3 + 2i) - [(3 - 4i) - (6 + 2i)]$

50. $(7 - 4i) - [(-2 + i) - (3 + 7i)]$

51. $(4 - 9i) + [(2 - 7i) - (4 + 8i)]$

52. $(10 - 2i) - [(2 + i) - (3 - i)]$

**5.6
Multiplication and
Division of Complex
Numbers**

Since complex numbers have the same form as binomials, we find the product of two complex numbers the same way we find the product of two binomials.

▼ **Example 1** Multiply $(3 - 4i)(2 + 5i)$.

Solution Multiplying each term in the second complex number by each term in the first, we have

$$(3 - 4i)(2 + 5i) = 3 \cdot 2 + 3 \cdot 5i - 2 \cdot 4i - 5i(4i)$$
$$= 6 + 15i - 8i - 20i^2$$

Combining similar terms and using the fact that $i^2 = -1$, we can simplify as follows:

$$6 + 15i - 8i - 20i^2 = 6 + 7i - 20(-1)$$
$$= 6 + 7i + 20$$
$$= 26 + 7i$$

The product of the complex numbers $3 - 4i$ and $2 + 5i$ is the complex number $26 + 7i$. ▲

We can generalize multiplication for complex numbers by finding the product of $a + bi$ and $c + di$.

If a, b, c, and d are real numbers, then

$$(a + bi)(c + di) = ac + adi + bci + bdi^2$$
$$= ac + adi + bci - bd$$
$$= (ac - bd) + (ad + bc)i$$

This gives a formula for the product of two complex numbers:

$$(a + bi)(c + di) = (ac - bd) + (ad + bc)i$$

Although we can use the formula for the product of any two complex numbers, it is probably just as easy to multiply them in the same way we multiply binomials.

▼   **Example 2**   Multiply $2i(4 - 6i)$.

Solution   Applying the distributive property gives us

$$2i(4 - 6i) = 2i \cdot 4 - 2i(6i)$$
$$= 8i - 12i^2$$
$$= 12 + 8i$$

Notice that we could have written $2i$ as $0 + 2i$ and applied the formula for the product of two complex numbers.   ▲

▼   **Example 3**   Expand $(3 + 5i)^2$.

Solution   We treat this like the square of a binomial. Remember: $(a + b)^2 = a^2 + 2ab + b^2$.

$$(3 + 5i)^2 = 3^2 + 2(3)(5i) + (5i)^2$$
$$= 9 + 30i + 25i^2$$
$$= 9 + 30i - 25$$
$$= -16 + 30i$$ ▲

▼   **Example 4**   Multiply $(2 - 3i)(2 + 3i)$.

Solution   This product has the form $(a - b)(a + b)$, which we know results in the difference of two squares, $a^2 - b^2$:

$$(2 - 3i)(2 + 3i) = 2^2 - (3i)^2$$
$$= 4 - 9i^2$$
$$= 4 + 9$$
$$= 13$$ ▲

The product of the two complex numbers $2 - 3i$ and $2 + 3i$ is the real number 13. The two complex numbers $2 - 3i$ and $2 + 3i$ are called complex conjugates. The fact that their product is a real number is very useful.

DEFINITION   The complex numbers $a + bi$ and $a - bi$ are called *complex conjugates*. One important property they have is that their product is the real number $a^2 + b^2$. Here's why:

$$(a + bi)(a - bi) = a^2 - (bi)^2$$
$$= a^2 - b^2 i^2$$
$$= a^2 - b^2(-1)$$
$$= a^2 + b^2$$

The fact that the product of two complex conjugates is a real number is the key to division with complex numbers.

▼   **Example 5**   Divide $\dfrac{2 + i}{3 - 2i}$.

Solution   We want a complex number in standard form that is equivalent to the quotient $(2 + i)/(3 - 2i)$. We need to eliminate i from the denominator. Multiplying the numerator and denominator by $3 + 2i$ will give us what we want:

$$\frac{2 + i}{3 - 2i} = \frac{2 + i}{3 - 2i} \cdot \frac{(3 + 2i)}{(3 + 2i)}$$
$$= \frac{6 + 4i + 3i + 2i^2}{9 - 4i^2}$$
$$= \frac{6 + 7i - 2}{9 + 4}$$
$$= \frac{4 + 7i}{13}$$
$$= \tfrac{4}{13} + \tfrac{7}{13}i$$

Dividing the complex number $2 + i$ by $3 - 2i$ gives the complex number $\tfrac{4}{13} + \tfrac{7}{13}i$. The second step in Example 5 is shown for clarity. It takes a while to get used to the idea that $i^2 = -1$.   ▲

Here are some further examples illustrating division with complex numbers.

▼   **Example 6**   Divide $\dfrac{7 - 4i}{i}$.

Solution   The conjugate of the denominator is $-i$. Multiplying numerator and denominator by this amount, we have

$$\frac{7 - 4i}{i} = \frac{7 - 4i}{i} \cdot \frac{-i}{-i}$$

$$= \frac{-7i + 4i^2}{-i^2}$$

$$= \frac{-7i + 4(-1)}{-(-1)}$$

$$= -4 - 7i$$ ▲

▼   **Example 7**   Divide $\dfrac{6}{3 - 5i}$.

Solution   The conjugate of $3 - 5i$ is $3 + 5i$.

$$\frac{6}{3 - 5i} = \frac{6}{3 - 5i} \cdot \frac{(3 + 5i)}{(3 + 5i)}$$

$$= \frac{18 + 30i}{9 + 25}$$

$$= \frac{18 + 30i}{34}$$

$$= \tfrac{18}{34} + \tfrac{30}{34}i$$

$$= \tfrac{9}{17} + \tfrac{15}{17}i$$ ▲

Find the following products: *Problem Set 5.6*

1. $3i(4 + 5i)$
2. $2i(3 + 4i)$
3. $-7i(1 + i)$
4. $-6i(3 - 8i)$
5. $6i(4 - 3i)$
6. $11i(2 - i)$
7. $(3 + 2i)(4 + i)$
8. $(2 - 4i)(3 + i)$
9. $(4 + 9i)(3 - i)$
10. $(5 - 2i)(1 + i)$
11. $(-3 - 4i)(2 - 5i)$
12. $(-6 - 2i)(3 - 4i)$
13. $(2 + 5i)^2$
14. $(3 + 2i)^2$
15. $(1 - i)^2$
16. $(1 + i)^2$
17. $(3 - 4i)^2$
18. $(6 - 5i)^2$
19. $(2 + i)(2 - i)$
20. $(3 + i)(3 - i)$
21. $(6 - 2i)(6 + 2i)$
22. $(5 + 4i)(5 - 4i)$
23. $(2 + 3i)(2 - 3i)$
24. $(2 - 7i)(2 + 7i)$
25. $(10 + 8i)(10 - 8i)$
26. $(11 - 7i)(11 + 7i)$

Find the following quotients. Write all answers in standard form for complex numbers.

27. $\dfrac{2 - 3i}{i}$ 28. $\dfrac{3 + 4i}{i}$

29. $\dfrac{5 + 2i}{-i}$ 30. $\dfrac{4 - 3i}{-i}$

31. $\dfrac{4}{2 - 3i}$ 32. $\dfrac{3}{4 - 5i}$

33. $\dfrac{6}{-3 + 2i}$ 34. $\dfrac{-1}{-2 - 5i}$

35. $\dfrac{2 + 3i}{2 - 3i}$ 36. $\dfrac{4 - 7i}{4 + 7i}$

37. $\dfrac{5 + 4i}{3 + 6i}$ 38. $\dfrac{2 + i}{5 - 6i}$

39. $\dfrac{3 - 7i}{9 - 5i}$ 40. $\dfrac{4 + 10i}{3 + 6i}$

Chapter 5
Summary and Review
Square Roots

Every positive real number x has two square roots. The *positive square root* of x is written $\sqrt{x}$, while the *negative square root* of x is written $-\sqrt{x}$. Both the positive and the negative square roots of x are numbers we square to get x. That is,

$$\left.\begin{array}{c} (\sqrt{x})^2 = x \\ \text{and} \quad (-\sqrt{x})^2 = x \end{array}\right\} \quad \text{for } x \geq 0$$

Higher Roots

In the expression $\sqrt[n]{a}$, n is the index, a is the *radicand,* and $\sqrt{}$ is the *radical sign.* The expression $\sqrt[n]{a}$ is such that

$$(\sqrt[n]{a})^n = a \qquad a \geq 0 \text{ when } n \text{ is even}$$

Fractional Exponents

Fractional exponents are used to indicate roots. Using fractional exponents is an alternative to radicals. The relationship between fractional exponents and roots is given by

$$a^{1/n} = \sqrt[n]{a} \qquad a \geq 0 \text{ when } n \text{ is even}$$
$$\text{and}$$
$$a^{m/n} = (a^{1/n})^m = (a^m)^{1/n} \qquad a \geq 0 \text{ when } n \text{ is even}$$

All the properties of exponents developed previously apply to rational exponents.

If a and b are nonnegative real numbers whenever n is even, then

Properties of Radicals

1. $\sqrt[n]{ab} = \sqrt[n]{a}\sqrt[n]{b}$

2. $\sqrt[n]{\dfrac{a}{b}} = \dfrac{\sqrt[n]{a}}{\sqrt[n]{b}}$ $(b \neq 0)$

A radical expression is said to be in *simplified form*

Simplified Form for Radicals

1. if there is no factor of the radicand that can be written as a power greater than or equal to the index;
2. if there are no fractions under the *radical sign*; and
3. if there are no radicals in the denominator.

The process of combining radical expressions under the four basic operations—addition, subtraction, multiplication, and division—is similar to the process of combining polynomials.

Operations with Radical Expressions

   Two radical expressions are *similar* if they have the same index and the same radicand.

We may square both sides of an equation any time it is convenient to do so, as long as we check all resulting solutions in the original equation.

Squaring Property of Equality

A *complex number* is any number that can be put in the form

Complex Numbers

$$a + bi$$

where a and b are real numbers and $i = \sqrt{-1}$. The *real part* of the complex number is a, and b is the *imaginary part*.

   If a, b, c, and d are real numbers, then we have the following definitions associated with complex numbers:

1. Equality

$$a + bi = c + di \quad \text{if and only if} \quad a = c \text{ and } b = d$$

2. Addition

$$(a + bi) + (c + di) = (a + c) + (b + d)i$$

3. Subtraction

$$(a + bi) - (c + di) = (a - c) + (b - d)i$$

4. Multiplication

$$(a + bi)(c + di) = (ac - bd) + (ad + bc)i$$

5.   Conjugates

$$a + bi \quad \text{and} \quad a - bi \quad \text{are conjugates}$$

COMMON MISTAKES

1.   The most common mistake when working with radicals is to assume that the square root of a sum is the sum of the square roots—or:

$$\sqrt{x + y} = \sqrt{x} + \sqrt{y}$$

The problem with this is it just isn't true. If we try it with 16 and 9, the mistake becomes obvious:

$$\sqrt{16 + 9} \stackrel{?}{=} \sqrt{16} + \sqrt{9}$$
$$\sqrt{25} \stackrel{?}{=} 4 + 3$$
$$5 \neq 7$$

2.   A common mistake when working with complex numbers is to mistake i for -1. The letter i is not -1, it is the square root of -1. That is, $i = \sqrt{-1}$.

**Chapter 5
Test**

Write with radical notation:

1. $8^{3/5}$

2. $17^{4/3}$

Write with rational exponents:

3. $\sqrt[4]{3x^3}$

4. $\sqrt[3]{7x^2}$

Simplify each of the following:

5. $27^{-2/3}$

6. $\left(\frac{25}{49}\right)^{-1/2}$

7. $a^{3/4} \cdot a^{-1/3}$

8. $\dfrac{(x^{2/3}y^{-3})^{1/2}}{(x^{3/4}y^{1/2})^{-1}}$

Write in simplified form:

9. $\sqrt{125x^3y^5}$

10. $\sqrt[3]{40x^7y^8}$

11. $\sqrt{\frac{2}{3}}$

12. $\sqrt{\dfrac{12a^4b^3}{5c}}$

Combine:

13. $3\sqrt{12} - 4\sqrt{27}$

14. $2\sqrt[3]{24a^3b^3} - 5a\sqrt[3]{3b^3}$

Multiply:

15. $(\sqrt{7} - \sqrt{2})(3\sqrt{7} + 2\sqrt{2})$

16. $(3\sqrt{2} - \sqrt{3})^2$

Rationalize the denominator:

17. $\dfrac{5}{\sqrt{3} - 1}$

18. $\dfrac{\sqrt{5} - \sqrt{2}}{\sqrt{5} + \sqrt{2}}$

Solve for x:

19. $\sqrt{5x - 1} = 7$

20. $\sqrt{3x + 2} + 4 = 0$

21. $\sqrt[3]{2x + 7} = -1$

22. $\sqrt[4]{3x + 1} = -2$

Solve for x and y so that each of the following equations is true:

23. $3x - 4i = 2 - 8yi$

24. $(2x + 5) - 4i = 6 - (y - 3)i$

Perform the indicated operations:

25. $(3 - 7i) - (4 + 2i)$

26. $(3 + 2i) - [(7 - i) - (4 + 3i)]$

27. $(2 - 3i)(4 + 3i)$

28. $(5 - 4i)^2$

29. $\dfrac{3 + 2i}{i}$

30. $\dfrac{5}{2 - 3i}$

31. $\dfrac{2 - 3i}{2 + 3i}$

32. Show that i^{38} can be written as -1.

6 Quadratic Equations

To the student:

If an object is thrown straight up into the air with an initial velocity of 32 feet/second, and we neglect the friction of the air on the object, then its height h above the ground, t seconds later, can be found by using the equation

$$h = 32t - 16t^2$$

Notice that the height depends only on t. The height of the object does not depend on the size or weight of the object. If we neglect the resistance of air on the object, then any object, whether it is a golf ball or a bowling ball, that is thrown into the air with an initial velocity of 32 feet/second, will reach the same height. If we want to find how long it takes the object to hit the ground, we let $h = 0$ (it is 0 feet above the ground when it hits the ground) and solve for t in

$$0 = 32t - 16t^2$$

This last equation is called a quadratic equation because it contains a polynomial of degree 2. Until now we have solved only first-degree equations. This chapter is about solving second-degree equations in one variable: quadratic equations. The chapter begins with three basic methods of solving quadratic equations: factoring, completing the square, and using the quadratic formula. We will also include applications of quadratic equations and some additional topics on the properties of the solutions to quadratic equations. To be successful in this chapter you should have a working knowledge of factoring, binomial squares, square roots, and complex numbers.

In the last two sections of Chapter 3 we spent some time learning to factor polynomials. In Chapter 4 we used our knowledge of factoring to deal with rational expressions. This first section of Chapter 6 will give us another application of factoring.

We are going to combine our ability to solve first-degree equations with our knowledge of factoring to solve quadratic equations. Before we do so, however, we must do two things. First of all we must define what we mean by a quadratic equation. Secondly, we must prove a theorem about the number 0 that we will use in the process of solving quadratic equations by factoring.

DEFINITION   Any equation that can be written in the form

$$ax^2 + bx + c = 0$$

where a, b, and c are constants and a is not 0 ($a \neq 0$), is called a *quadratic equation*. The form $ax^2 + bx + c = 0$ is called *standard form* for quadratic equations.

▼   **Example 1**   Each of the following is a quadratic equation:

$$2x^2 = 5x + 3 \qquad 5x^2 = 75 \qquad 4x^2 - 3x + 2 = 0$$

The third equation is clearly a quadratic equation because it is written in the form $ax^2 + bx + c = 0$. (Notice in this case that a is 4, b is -3, and c is 2.) The first two are also quadratic equations because both could be written in the form $ax^2 + bx + c = 0$ by using the addition property of equality to rearrange terms.   ▲

NOTATION   For a quadratic equation written in standard form, the first term, ax^2, is called the *quadratic term;* the second term, bx, is the *linear term;* and the last term, c, is called the *constant term.*

In the past we have noticed that the number 0 is a special number. That is, 0 has some unique properties. In some situations it does not behave like other numbers. For example, division by 0 does not make sense, whereas division by all other real numbers does. Note also that 0 is the only number without a reciprocal. There is another property of 0 that is the key to solving quadratic equations. It is called the *zero-factor property* and we state it as a theorem.

THEOREM 6.1 (ZERO-FACTOR PROPERTY)   For all real numbers r and s,

$$r \cdot s = 0 \quad \text{if and only if} \quad r = 0 \quad \text{or} \quad s = 0 \quad \text{(or both)}$$

The statement "or both" is there to remind us that the word *or* in mathematics is an inclusive *or*—that it denotes one *or* the other *or* both.

What Theorem 6.1 says in words is that every time we multiply two numbers and the product (answer) is 0, then one or both of the original numbers must have been 0.

Proof We begin with

$$r \cdot s = 0$$

Now suppose $r \neq 0$. (If $r = 0$, we are finished, since there would be nothing left to prove.) We can multiply both sides by the reciprocal of r:

$$\frac{1}{r}(r \cdot s) = 0\left(\frac{1}{r}\right) \qquad \text{Multiply both sides by } \frac{1}{r}$$

$$1 \cdot s = 0\left(\frac{1}{r}\right) \qquad \text{Reciprocals multiply to 1}$$

$$1 \cdot s = 0 \qquad \text{Multiplication by 0}$$

$$s = 0 \qquad \text{Multiplication by 1}$$

So we see that if $r \cdot s = 0$ and r is not 0, then s must be 0. We could reverse the beginning of the proof and assume $s \neq 0$. We would then find that $r = 0$. In either case, if the product is 0, then one of the two original numbers must also be 0.

Suppose we want to solve the equation

$$x^2 - 2x - 24 = 0$$

We begin by factoring the left side as $(x - 6)(x + 4)$ and get

$$(x - 6)(x + 4) = 0$$

Now both $(x - 6)$ and $(x + 4)$ represent real numbers. We notice that their product is 0. By the zero-factor property, one or both of them must be 0:

$$x - 6 = 0 \quad \text{or} \quad x + 4 = 0$$

We have used factoring and the zero-factor property to rewrite our original second-degree equation as two first-degree equations connected by the word *or*. Completing the solution, we solve the two first-degree equations:

$$x - 6 = 0 \quad \text{or} \quad x + 4 = 0$$
$$x = 6 \quad \text{or} \quad x = -4$$

Our solution set is $\{6, -4\}$. Both numbers satisfy the original quadratic equation $x^2 - 2x - 24 = 0$. We can check this as follows:

When $x = 6$, the equation $x^2 - 2x - 24 = 0$ becomes

$$6^2 - 2(6) - 24 = 0$$
$$36 - 12 - 24 = 0$$
$$0 = 0$$

When $x = -4$, the equation $x^2 - 2x - 24 = 0$ becomes

$$(-4)^2 - 2(-4) - 24 = 0$$
$$16 + 8 - 24 = 0$$
$$0 = 0$$

In both cases the result is a true statement, which means that both 6 and -4 are solutions to the original equation.

▼   **Example 2**   Solve $2x^2 = 5x + 3$.

Solution   We begin by adding $-5x$ and -3 to both sides in order to rewrite the equation in standard form:

$$2x^2 - 5x - 3 = 0$$

We then factor the left side and use the zero-factor property to set each factor to zero:

$$(2x + 1)(x - 3) = 0 \qquad \text{Factor}$$
$$2x + 1 = 0 \quad \text{or} \quad x - 3 = 0 \qquad \text{Zero-factor property}$$

Solving each of the resulting first-degree equations, we have

$$x = -\tfrac{1}{2} \quad \text{or} \quad x = 3$$

The solution set is $\{-\tfrac{1}{2}, 3\}$.   ▲

To generalize the above example, here are the steps used in solving a quadratic equation by factoring:

Step 1.   Write the equation in standard form.
Step 2.   Factor the left side.
Step 3.   Use the zero-factor property to set each factor equal to 0.
Step 4.   Solve the resulting first-degree equations.

▼ **Example 3** Solve $x^2 = 3x$.

Solution We begin by writing the equation in standard form and factoring:

$$x^2 = 3x$$
$$x^2 - 3x = 0 \qquad \text{Standard form}$$
$$x(x - 3) = 0 \qquad \text{Factor}$$

Using the zero-factor property to set each factor to 0, we have

$$x = 0 \quad \text{or} \quad x - 3 = 0$$
$$x = 3$$

The solution set is $\{0, 3\}$.   ▲

▼ **Example 4** Solve $(x - 2)(x + 1) = 4$.

Solution We begin by multiplying the two factors on the left side. (Notice that it would be incorrect to set each of the factors on the left side equal to 4. The fact that the product is 4 does not imply that either of the factors must be 4.)

$$(x - 2)(x + 1) = 4$$
$$x^2 - x - 2 = 4 \qquad \text{Multiply the left side}$$
$$x^2 - x - 6 = 0 \qquad \text{Standard form}$$
$$(x - 3)(x + 2) = 0 \qquad \text{Factor}$$
$$x - 3 = 0 \quad \text{or} \quad x + 2 = 0 \qquad \text{Zero-factor property}$$
$$x = 3 \quad \text{or} \quad x = -2$$

The solution set is $\{3, -2\}$.   ▲

▼ **Example 5** Solve $3 + \dfrac{1}{x} = \dfrac{10}{x^2}$.

Solution To clear the equation of denominators we multiply both sides by x^2:

$$3(x^2) + \left(\frac{1}{x}\right)(x^2) = \left(\frac{10}{x^2}\right)(x^2)$$
$$3x^2 + x = 10$$

Rewrite in standard form and solve:

$$3x^2 + x - 10 = 0$$
$$(3x - 5)(x + 2) = 0$$
$$3x - 5 = 0 \quad \text{or} \quad x + 2 = 0$$
$$x = \tfrac{5}{3} \quad \text{or} \quad x = -2$$

The solution set is $\{-2, \tfrac{5}{3}\}$.   ▲

▼ **Example 6**   Solve $\sqrt{6x^2 + 5x} = 5$.

Solution   To eliminate the radical we begin by squaring both sides of the equation. We then write the resulting equation in standard form and solve as before:

$$(\sqrt{6x^2 + 5x})^2 = 5^2 \qquad \text{Square both sides}$$
$$6x^2 + 5x = 25 \qquad \text{Simplify}$$
$$6x^2 + 5x - 25 = 0 \qquad \text{Standard form}$$
$$(3x - 5)(2x + 5) = 0 \qquad \text{Factor}$$
$$3x - 5 = 0 \quad \text{or} \quad 2x + 5 = 0 \qquad \text{Zero-factor property}$$
$$x = \tfrac{5}{3} \quad \text{or} \quad x = -\tfrac{5}{2}$$

Since we raised both sides of the original equation to an even power, we have the possibility that one or both of the solutions are extraneous. We must check each solution in the original equation.

When $x = \tfrac{5}{3}$,   When $x = -\tfrac{5}{2}$,

the equation $\sqrt{6x^2 + 5x} = 5$ the equation $\sqrt{6x^2 + 5x} = 5$

becomes $\sqrt{6(\tfrac{25}{9}) + 5(\tfrac{5}{3})} = 5$ becomes $\sqrt{6(\tfrac{25}{4}) + 5(-\tfrac{5}{2})} = 5$

or $\sqrt{\tfrac{50}{3} + \tfrac{25}{3}} = 5$ or $\sqrt{\tfrac{75}{2} - \tfrac{25}{2}} = 5$

$\sqrt{\tfrac{75}{3}} = 5$ $\sqrt{\tfrac{50}{2}} = 5$

$\sqrt{25} = 5$ $\sqrt{25} = 5$

$5 = 5$ $5 = 5$

Since both checks result in a true statement, neither of the two possible solutions is extraneous.
The solution set is $\{\tfrac{5}{3}, -\tfrac{5}{2}\}$.   ▲

It is clear from the examples in this section that solving a quadratic equation by factoring is fairly straightforward once the equation is in standard form. Sometimes the most difficult part is writing the original equation in standard form.

1-29

The following quadratic equations are in standard form. Factor the left side and solve:

1. $x^2 - 5x - 6 = 0$ 2. $x^2 + 5x - 6 = 0$
3. $x^2 - 5x + 6 = 0$ 4. $x^2 + 5x + 6 = 0$
5. $3y^2 + 11y - 4 = 0$ 6. $3y^2 - y - 4 = 0$
7. $6x^2 - 13x + 6 = 0$ 8. $9x^2 + 6x - 8 = 0$
9. $t^2 - 25 = 0$ 10. $4t^2 - 49 = 0$

Write each of the following in standard form and solve for the indicated variable:

11. $x^2 = 4x + 21$ 12. $x^2 = -4x + 21$
13. $2y^2 - 20 = -3y$ 14. $3y^2 + 10 = 17y$
15. $9x^2 - 12x = 0$ 16. $4x^2 + 4x = 0$
17. $2r + 1 = 15r^2$ 18. $2r - 1 = -8r^2$
19. $9a^2 = 16$ 20. $16a^2 = 25$
21. $-10x = x^2$ 22. $8x = x^2$
23. $(x + 6)(x - 2) = -7$ 24. $(x - 7)(x + 5) = -20$
25. $(y - 4)(y + 1) = -6$ 26. $(y - 6)(y + 1) = -12$
27. $(x + 1)^2 = 3x + 7$ 28. $(x + 2)^2 = 9x$
29. $(2r + 3)(2r - 1) = -(3r + 1)$ 30. $(3r + 2)(r - 1) = -(7r - 7)$

Multiply each of the following by its least common denominator and solve for the indicated variable:

31. $1 - \dfrac{1}{x} = \dfrac{12}{x^2}$ 32. $2 + \dfrac{5}{x} = \dfrac{3}{x^2}$

33. $x - \dfrac{4}{3x} = -\dfrac{1}{3}$ 34. $\dfrac{x}{2} - \dfrac{4}{x} = -\dfrac{7}{2}$

35. $1 = \dfrac{9}{y^2}$ 36. $1 = \dfrac{4}{x^2}$

37. $\dfrac{6}{(r - 1)(r + 1)} - \dfrac{1}{2} = \dfrac{1}{(r + 1)}$ 38. $2 + \dfrac{5}{x - 1} = \dfrac{12}{(x - 1)^2}$

Square both sides of the following equations and solve for the indicated variable:

39. $\sqrt{6x^2 + 5x} = 2$ 40. $\sqrt{2x^2 - 5x} = 5$
41. $\sqrt{y + 2} = y - 4$ 42. $\sqrt{11 - 5x} = x - 3$
43. $\sqrt{5y + 1} = 11 - 5y$ 44. $4\sqrt{x + 2} = x + 5$
45. $\sqrt{3x^2 + 4x} - 2 = 0$ 46. $\sqrt{2x^2 + 3x} - 3 = 0$

47. In the introduction to this chapter we said the height h of an object thrown straight up into the air with an initial velocity of 32 feet/second could be found by using the equation $h = 32t - 16t^2$. When does the object hit the ground?

In the preceding section we developed a method of solving quadratic equations by factoring them. The problem with that particular method is that it only works when the left side of the standard-form quadratic equation is factorable. The method fails to give us solutions when the left side of the equation cannot be factored.

In this section we will produce another method of solving quadratic equations. The method is called *completing the square*. Completing the square on a quadratic equation always allows us to obtain solutions, regardless of whether or not the equation can be factored. More importantly, the method of completing the square will give us the tools necessary to derive the *quadratic formula*, which we will do in the next section.

Consider the equation

$$x^2 = 16$$

We could solve it by writing it in standard form, factoring the left side, and proceeding as we did in the last section. However, we can shorten our work considerably if we simply notice that x must be either the positive square root of 16 or the negative square root of 16. That is,

$$
\begin{aligned}
&\text{If}\quad x^2 = 16, \\
&\text{then}\quad x = \sqrt{16}\quad \text{or}\quad x = -\sqrt{16} \\
&\qquad\quad x = 4\qquad \text{or}\quad x = -4
\end{aligned}
$$

We can generalize this result into a theorem as follows:

THEOREM 6.2   If $a^2 = b$ where b is a real number, then $a = \sqrt{b}$ or $a = -\sqrt{b}$.

Proof   We prove the theorem by considering the expression $a^2 = b$ to be a quadratic equation in a:

$$
\begin{aligned}
a^2 &= b \\
a^2 - b &= 0
\end{aligned}
$$

To get the left side in a form we can factor, all we have to do is notice that $b = (\sqrt{b})^2$:

$$a^2 - (\sqrt{b})^2 = 0$$

The left side now factors as the difference of two squares:

$$(a - \sqrt{b})(a + \sqrt{b}) = 0$$

$$a - \sqrt{b} = 0 \quad \text{or} \quad a + \sqrt{b} = 0 \qquad \text{Zero-factor property}$$

$$a = \sqrt{b} \quad \text{or} \quad a = -\sqrt{b}$$

NOTATION   The expression $a = \sqrt{b}$ or $a = -\sqrt{b}$ can be written in shorthand form as $a = \pm\sqrt{b}$. The symbol $\pm$ is read "plus or minus."
   We can apply Theorem 1 to some fairly complicated quadratic equations.

▼   **Example 1**   Solve $(2x - 3)^2 = 25$.

   Solution
$$(2x - 3)^2 = 25$$
$$2x - 3 = \pm\sqrt{25} \qquad \text{Theorem 6.2}$$
$$2x - 3 = \pm 5 \qquad \sqrt{25} = 5$$
$$2x = 3 \pm 5 \qquad \text{Add 3 to both sides}$$
$$x = \frac{3 \pm 5}{2} \qquad \text{Divide both sides by 2}$$

The last equation can be written as two separate statements:

$$x = \frac{3 + 5}{2} \quad \text{or} \quad x = \frac{3 - 5}{2}$$
$$x = 4 \qquad \text{or} \quad x = -1$$

The solution set is $\{4, -1\}$.   ▲

▼   **Example 2**   Solve for x: $(3x - 1)^2 = -12$.

Solution
$$(3x - 1)^2 = -12$$
$$3x - 1 = \pm\sqrt{-12} \qquad \text{Theorem 6.2}$$
$$3x - 1 = \pm 2\sqrt{3}i \qquad \sqrt{-12} = \sqrt{12}i = \sqrt{4}\sqrt{3}i = 2\sqrt{3}i$$
$$3x = 1 \pm 2\sqrt{3}i \qquad \text{Add 1 to both sides}$$
$$x = \frac{1 \pm 2\sqrt{3}i}{3} \qquad \text{Divide both sides by 3}$$

The solution set is

$$\left\{ \frac{1 + 2\sqrt{3}i}{3}, \quad \frac{1 - 2\sqrt{3}i}{3} \right\}$$

Both solutions are complex. Although the arithmetic is somewhat more complicated, we check each solution in the usual manner. Here is a check of the first solution:

$$\text{When} \quad x = \frac{1 + 2\sqrt{3}i}{3},$$

$$\text{the equation} \quad (3x - 1)^2 = -12$$

$$\text{becomes} \quad \left(3 \cdot \frac{1 + 2\sqrt{3}i}{3} - 1\right)^2 = -12$$

$$\text{or} \quad (1 + 2\sqrt{3}i - 1)^2 = -12$$
$$(2\sqrt{3}i)^2 = -12$$
$$4 \cdot 3 \cdot i^2 = -12$$
$$12(-1) = -12$$
$$-12 = -12 \qquad \blacktriangle$$

The method of completing the square is simply a way of transforming any quadratic equation into an equation of the form found in the preceding two examples.

The key to understanding the method of completing the square lies in recognizing the relationship between the last two terms of any perfect square trinomial whose leading coefficient is 1.

Consider the following list of perfect square trinomials and their corresponding binomial squares:

$$x^2 - 6x + 9 = (x - 3)^2$$
$$x^2 + 8x + 16 = (x + 4)^2$$
$$x^2 - 10x + 25 = (x - 5)^2$$
$$x^2 + 12x + 36 = (x + 6)^2$$

In each case the leading coefficient is 1. A more important observation comes from noticing the relationship between the linear and constant terms (middle and last terms) in each trinomial. Observe that the constant term in each case is the square of half the coefficient of x in the middle term. For example, in the last expression, the constant term, 36, is the square of half of 12, where 12 is the coefficient of x in the middle term. (Notice also that the second terms in all the binomials on the right side are half the coefficients of the middle terms of the trinomials on the left side.) We can use these observations to build our own perfect square trinomials, and in doing so, solve some quadratic equations. Consider the following equation:

$$x^2 + 6x = 3$$

We can think of the left side as having the first two terms of a perfect square trinomial. We need only add the correct constant term. If we take half the coefficient of x, we get 3. If we then square this quantity, we have 9. Adding the 9 to both sides, the equation becomes

$$x^2 + 6x + 9 = 3 + 9$$

The left side is the perfect square $(x + 3)^2$; the right side is 12:

$$(x + 3)^2 = 12$$

The equation is now in the correct form. We can apply Theorem 6.2 and finish the solution:

$$
\begin{aligned}
(x + 3)^2 &= 12 \\
x + 3 &= \pm\sqrt{12} \qquad \text{Theorem 6.2} \\
x + 3 &= \pm 2\sqrt{3} \\
x &= -3 \pm 2\sqrt{3}
\end{aligned}
$$

The solution set is $\{-3 + 2\sqrt{3}, -3 - 2\sqrt{3}\}$. The method just used is called *completing the square,* since we complete the square on the left side of the original equation by adding the appropriate constant term.

▼   **Example 3**   Solve by completing the square: $x^2 + 5x - 2 = 0$.

Solution   We must begin by adding 2 to both sides. (The left side of the equation, as it is, is not a perfect square because it does not have the correct constant term. We will simply "move" that term to the other side and use our own constant term.)

$$x^2 + 5x = 2$$

We complete the square by adding the square of half the coefficient of the linear term to both sides:

$$x^2 + 5x + \frac{25}{4} = 2 + \frac{25}{4} \qquad \text{Half of 5 is } \frac{5}{2}$$

$$\left(x + \frac{5}{2}\right)^2 = \frac{33}{4}$$

$$x + \frac{5}{2} = \pm\sqrt{\frac{33}{4}}$$

$$x + \frac{5}{2} = \pm\frac{\sqrt{33}}{2}$$

$$x = -\frac{5}{2} \pm \frac{\sqrt{33}}{2}$$

$$x = \frac{-5 \pm \sqrt{33}}{2}$$

The solution set is

$$\left\{ \frac{-5 + \sqrt{33}}{2}, \frac{-5 - \sqrt{33}}{2} \right\}$$

▲

▼ Example 4 Solve for x: $3x^2 - 8x + 7 = 0$.

Solution

$$3x^2 - 8x + 7 = 0$$
$$3x^2 - 8x = -7 \qquad \text{Add } -7 \text{ to both sides}$$

We cannot complete the square on the left side because the leading coefficient is not 1. We take an extra step and divide both sides by 3:

$$\frac{3x^2}{3} - \frac{8x}{3} = -\frac{7}{3}$$

$$x^2 - \frac{8}{3}x = -\frac{7}{3}$$

Half of $\frac{8}{3}$ is $\frac{4}{3}$, the square of which is $\frac{16}{9}$.

$$x^2 - \frac{8}{3}x + \frac{16}{9} = -\frac{7}{3} + \frac{16}{9}$$

$$\left(x - \frac{4}{3} \right)^2 = -\frac{5}{9}$$

$$x - \frac{4}{3} = \pm \frac{\sqrt{5}i}{3}$$

$$x = \frac{4}{3} \pm \frac{\sqrt{5}i}{3}$$

$$x = \frac{4 \pm \sqrt{5}i}{3}$$

The solution set is

$$\left\{ \frac{4 + \sqrt{5}i}{3}, \frac{4 - \sqrt{5}i}{3} \right\}$$

▲

To Solve a Quadratic Equation by Completing the Square

To summarize the method used in the preceding two examples, we list the following steps:

Step 1. Write the equation in the form $ax^2 + bx = c$.

Step 2. If the leading coefficient is not 1, divide both sides by the coefficient so that the resulting equation has a leading coefficient of 1. That is, if $a \neq 1$, then divide both sides by a.

Step 3. Add the square of half the coefficient of the linear term to both sides of the equation.

Step 4. Write the left side of the equation as the square of a binomial and simplify the right side if possible.

Step 5. Apply Theorem 6.2 and solve as usual.

Problem Set 6.2

Solve the following by applying Theorem 6.2:

1. $(x - 5)^2 = 9$

2. $(x + 2)^2 = 16$

3. $(y + 3)^2 = 18$

4. $(y - 1)^2 = 24$

5. $(2x - 1)^2 = 25$

6. $(3x + 7)^2 = 1$

7. $(2a + 3)^2 = -9$

8. $(3a - 5)^2 = -49$

9. $(5x + 2)^2 = -8$

10. $(6x - 7)^2 = -75$

Solve each of the following quadratic equations by completing the square:

11. $x^2 + 4x = 12$

12. $x^2 - 2x = 8$

13. $x^2 + 12x = -27$

14. $x^2 - 6x = 16$

15. $a^2 - 2a + 5 = 0$

16. $a^2 + 10a + 22 = 0$

17. $y^2 - 8y + 1 = 0$

18. $y^2 + 6y - 1 = 0$

19. $x^2 - 5x - 3 = 0$

20. $x^2 - 5x - 2 = 0$

21. $2x^2 - 4x - 8 = 0$

22. $3x^2 - 9x - 12 = 0$

23. $3t^2 - 8t + 1 = 0$

24. $5t^2 + 12t - 1 = 0$

25. $4x^2 - 3x + 5 = 0$

26. $7x^2 - 5x + 2 = 0$

For each of the following equations, multiply both sides by the least common denominator. Solve the resulting equation.

27. $3 + \dfrac{2}{x} = \dfrac{4}{x^2}$

28. $5 - \dfrac{1}{x} = \dfrac{3}{x^2}$

29. $\dfrac{5}{x} + 3 = 2x$

30. $\dfrac{2}{x} - 1 = 3x$

31. $\dfrac{1}{y - 1} + \dfrac{1}{y + 1} = 1$

32. $\dfrac{2}{y + 2} - \dfrac{3}{y - 2} = 4$

33. $\dfrac{10}{x^2} = \dfrac{-3}{x}$

34. $\dfrac{5}{x} = \dfrac{4}{x^2}$

Clear each equation of radicals. Solve the resulting equation.

35. $\sqrt{x^2 - 6x} = 1$ **36.** $\sqrt{x^2 + 4x} = 3$
37. $\sqrt{4y - 1} = y$ **38.** $\sqrt{10y + 3} = y$
39. $\sqrt{2a + 1} = a + 3$ **40.** $\sqrt{2a + 5} = a - 4$
41. $\sqrt{2x + 3} = x - 4$ **42.** $\sqrt{3x - 2} = x + 3$

In this section we will use the method of completing the square from the preceding section to derive the quadratic formula. The quadratic formula is a very useful tool in mathematics. It allows us to solve all types of quadratic equations.

6.3
The Quadratic
Formula

Kevd

THEOREM 6.3 (THE QUADRATIC THEOREM) For any quadratic equation in the form $ax^2 + bx + c = 0$, where $a \neq 0$, the two solutions are

$$x = \frac{-b + \sqrt{b^2 - 4ac}}{2a} \quad \text{and} \quad x = \frac{-b - \sqrt{b^2 - 4ac}}{2a}$$

Proof We will prove the quadratic theorem by completing the square on $ax^2 + bx + c = 0$.

$$ax^2 + bx + c = 0$$
$$ax^2 + bx \quad\;\; = -c \qquad \text{Add } -c \text{ to both sides}$$
$$x^2 + \frac{b}{a}x \quad\;\; = -\frac{c}{a} \qquad \text{Divide both sides by } a$$

To complete the square on the left side we add the square of 1/2 of b/a to both sides. (1/2 of b/a is $b/2a$.)

$$x^2 + \frac{b}{a}x + \left(\frac{b}{2a}\right)^2 = -\frac{c}{a} + \left(\frac{b}{2a}\right)^2$$

We now simplify the right side as a separate step. We square the second term and combine the two terms by writing each with the least common denominator $4a^2$:

$$-\frac{c}{a} + \left(\frac{b}{2a}\right)^2 = -\frac{c}{a} + \frac{b^2}{4a^2} = \frac{4a}{4a}\left(\frac{-c}{a}\right) + \frac{b^2}{4a^2} = \frac{-4ac + b^2}{4a^2}$$

It is convenient to write this last expression as

$$\frac{b^2 - 4ac}{4a^2}$$

Continuing with the proof, we have

$$x^2 + \frac{b}{a}x + \left(\frac{b}{2a}\right)^2 = \frac{b^2 - 4ac}{4a^2}$$

$$\left(x + \frac{b}{2a}\right)^2 = \frac{b^2 - 4ac}{4a^2}$$

$$x + \frac{b}{2a} = \pm\frac{\sqrt{b^2 - 4ac}}{2a}$$

$$x = -\frac{b}{2a} \pm \frac{\sqrt{b^2 - 4ac}}{2a}$$

$$x = \frac{-b \pm \sqrt{b^2 - 4ac}}{2a}$$

which is equivalent to

$$x = \frac{-b + \sqrt{b^2 - 4ac}}{2a} \quad \text{or} \quad x = \frac{-b - \sqrt{b^2 - 4ac}}{2a}$$

Our proof is now complete. What we have is this: If our equation is in the form $ax^2 + bx + c = 0$ (standard form), where $a \neq 0$, the two solutions are always given by the formula

$$x = \frac{-b \pm \sqrt{b^2 - 4ac}}{2a}$$

This formula is known as the *quadratic formula*. If we substitute the coefficients a, b, and c of any quadratic equation in standard form in the formula, we need only perform some basic arithmetic to arrive at the solution set.

▼   **Example 1**   Use the quadratic formula to solve $6x^2 + 7x - 5 = 0$.

Solution   Using the coefficients $a = 6$, $b = 7$, and $c = -5$ in the formula

$$x = \frac{-b \pm \sqrt{b^2 - 4ac}}{2a}$$

we have

$$x = \frac{-7 \pm \sqrt{49 - 4(6)(-5)}}{2(6)}$$

or
$$x = \frac{-7 \pm \sqrt{49 + 120}}{12}$$

$$= \frac{-7 \pm \sqrt{169}}{12}$$

$$= \frac{-7 \pm 13}{12}$$

We separate the last equation into the two statements

$$x = \frac{-7 + 13}{12} \quad \text{or} \quad x = \frac{-7 - 13}{12}$$

$$x = \frac{1}{2} \quad \text{or} \quad x = -\frac{5}{3}$$

The solution set is $\{\frac{1}{2}, -\frac{5}{3}\}$.   ▲

▼ **Example 2**   Solve $\dfrac{x^2}{3} - x = -\dfrac{1}{2}$.

Solution   Multiplying through by 6 and writing the result in standard form, we have

$$2x^2 - 6x + 3 = 0$$

In this case $a = 2$, $b = -6$, and $c = 3$. The two solutions are given by

$$x = \frac{-(-6) \pm \sqrt{36 - 4(2)(3)}}{2(2)}$$

$$= \frac{6 \pm \sqrt{12}}{4}$$

$$= \frac{6 \pm 2\sqrt{3}}{4}$$

We can reduce this last expression to lowest terms by dividing the numerator and denominator by 2:

$$x = \frac{3 \pm \sqrt{3}}{2}$$

The solution set is

$$\left\{ \frac{3 + \sqrt{3}}{2}, \frac{3 - \sqrt{3}}{2} \right\}$$ ▲

▼   **Example 3**   Solve $x^4 + 3x^2 = 10$.

Solution   This equation is quadratic in x^2. We can make it more reasonable by using the substitution $y = x^2$. (Note that the choice of the letter y is arbitrary. We could just as easily use the substitution $m = x^2$.)

If $y = x^2$, the original equation becomes

$$y^2 + 3y = 10$$

$$\text{or }\;\; y^2 + 3y - 10 = 0$$

Using $a = 1$, $b = 3$, and $c = -10$ in the quadratic formula, we have

$$y = \frac{-3 \pm \sqrt{9 - 4(1)(-10)}}{2(1)}$$

$$= \frac{-3 \pm \sqrt{49}}{2}$$

$$= \frac{-3 \pm 7}{2}$$

$$y = 2 \;\; \text{or} \;\; y = -5$$

This last statement gives the solutions in terms of y. The original equation is in terms of x, however, so we want our solutions in terms of x also. This is easily taken care of by the original substitution $y = x^2$:

The equations

$$y = 2 \;\; \text{and} \;\; y = -5$$

are equivalent to

$$x^2 = 2 \;\; \text{and} \;\; x^2 = -5$$

Using Theorem 6.2, we have

$$x = \pm\sqrt{2} \;\; \text{and} \;\; x = \pm\sqrt{-5} = \pm\sqrt{5}i$$

The solution set is $\{ \sqrt{2}, -\sqrt{2}, \sqrt{5}i, -\sqrt{5}i \}$.   ▲

▼ **Example 4** Solve $\sqrt{x+1} + \sqrt{2x} = 1$.

Solution This equation has two separate terms involving radical signs. In this situation it is usually best to separate the radical terms on opposite sides of the equal sign. [*Note:* If we were to square both sides of the equation in its present form, we would not get $(x+1) - (2x)$ for the left side. The square of the left side is $(\sqrt{x+1} + \sqrt{2x})^2 = (x+1) + 2\sqrt{x+1}\sqrt{2x} + 2x.$]

Adding $-\sqrt{2x}$ to both sides, we have

$$\sqrt{x+1} = 1 - \sqrt{2x}$$

Squaring both sides gives

$$x + 1 = 1 - 2\sqrt{2x} + 2x \qquad \text{Recall: } (a+b)^2 = a^2 + 2ab + b^2$$
$$-x = -2\sqrt{2x} \qquad \text{Add } -2x \text{ and } -1 \text{ to both sides}$$
$$x^2 = 4(2x) \qquad \text{Square both sides}$$
$$x^2 - 8x = 0 \qquad \text{Standard form}$$

The coefficients are $a = 1$, $b = -8$, and $c = 0$.

$$x = \frac{-(-8) \pm \sqrt{64 - 4(1)(0)}}{2(1)}$$

$$x = \frac{8 \pm 8}{2}$$

$$x = 8 \quad \text{or} \quad x = 0$$

Since we squared both sides, we have the possibility that one or both of the solutions are extraneous. We must check each one in the original equation:

When $x = 8$,
we have $\sqrt{8+1} + \sqrt{2 \cdot 8} = 1$
$\sqrt{9} + \sqrt{16} = 1$
$3 + 4 = 1$
$7 = 1$
which implies $x = 8$
is extraneous

When $x = 0$,
we have $\sqrt{0+1} + \sqrt{2(0)} = 1$
$\sqrt{1} + \sqrt{0} = 1$
$1 + 0 = 1$
$1 = 1$
which implies $x = 0$
is a solution ▲

Some of the examples in this section could have been solved by simply factoring. The same will be true of some of the problems in the problem set. It is a good idea to use the quadratic formula to solve these problems, even though factoring may be faster. The reason is that we need to be able to understand and use the quadratic formula. The more problems we work, the better we will understand the formula and the longer we will remember it.

Problem Set 6.3

Use the quadratic formula to solve each of the following:

1. $x^2 + 5x + 6 = 0$ 2. $x^2 + 5x - 6 = 0$
3. $x^2 - 3x + 2 = 0$ 4. $x^2 + x - 2 = 0$
5. $y^2 - 5y = 0$ 6. $2y^2 + 10y = 0$
7. $x^2 + 6x - 8 = 0$ 8. $2x^2 - 3x + 5 = 0$
9. $2r^2 + 5r + 3 = 0$ 10. $3r^2 - 4r - 5 = 0$
11. $x^2 - 2x + 1 = 0$ 12. $x^2 - 6x + 9 = 0$
13. $2x^2 - 5 = 2x$ 14. $7x^2 - 8 = 3x$
15. $3r^2 = r - 4$ 16. $5r^2 = 8r + 2$
17. $\dfrac{3}{x} - 1 = \dfrac{4}{x^2}$ 18. $\dfrac{5}{x} + 2 = \dfrac{1}{x^2}$

19. $\dfrac{1}{y+2} + \dfrac{1}{y-3} = 2$ 20. $\dfrac{3}{y-5} + \dfrac{4}{y+4} = 1$

Make an appropriate substitution and solve:

21. $x^4 - x^2 = 20$ 22. $x^4 + 2x^2 = 63$
23. $x^4 + 20 = 9x^2$ 24. $x^4 + 30 = 11x^2$
25. $6r^4 = r^2 + 2$ 26. $3r^4 = 20 - 7r^2$
27. $9a^4 = 49$ 28. $36a^4 = 9$
29. $6x^4 = -x^2 + 5$ 30. $3x^4 = 8 - 2x^2$

Solve the following equations by first clearing each of the radicals:

31. $\sqrt{x+5} = \sqrt{x} + 1$ 32. $\sqrt{x-2} = 2 - \sqrt{x}$
33. $\sqrt{x+4} = 2 - \sqrt{2x}$ 34. $\sqrt{5y+1} = 1 + \sqrt{3y}$
35. $\sqrt{y+21} + \sqrt{y} = 7$ 36. $\sqrt{y-3} - \sqrt{y} = -1$
37. $\sqrt{2(x+2)} = \sqrt{x+3} + 1$ 38. $\sqrt{2x-1} = \sqrt{x-4} + 2$
39. $\sqrt{y+9} - \sqrt{y-6} = 3$ 40. $\sqrt{y+7} - \sqrt{y+2} = 1$

41. A manufacturer can produce x items at a total cost of $C = -x^2 + 40x - 100$. He sells each item for $11.00, so his total revenue for selling x items is $R = 11x$. The manufacturer will break even when his total revenue is equal to his total cost—that is, when $R = C$. How many items must he sell to break even?

The quadratic formula

$$x = \frac{-b \pm \sqrt{b^2 - 4ac}}{2a}$$

gives the solutions to any quadratic equation in standard form. There are times, when working with quadratic equations, when it is only important to know what kind of solutions the equation has.

DEFINITION   The expression under the radical in the quadratic equation is called the *discriminant:*

$$\text{Discriminant} = D = b^2 - 4ac$$

The discriminant gives the number and type of solutions to a quadratic equation, when the original equation has integer coefficients. For example, when the discriminant is negative, the quadratic formula will contain the square root of a negative number. Hence, the equation will have imaginary solutions. If the discriminant were 0, the formula would be

$$x = \frac{-b \pm 0}{2a} = \frac{-b}{2a}$$

and the equation would have one rational solution: the number $-b/2a$.
   The following table gives the relationship between the discriminant and the type of solutions to the equation.
   For the equation $ax^2 + bx + c = 0$ where a, b, and c are integers and $a \neq 0$:

If the discriminant $b^2 - 4ac$ is	Then the equation will have
Negative	Two imaginary solutions
Zero	One rational solution
A positive number that is also a perfect square	Two rational solutions
A positive number that is not a perfect square	Two irrational solutions

In the second and third cases, when the discriminant is 0 or a positive perfect square, the solutions are rational numbers. The quadratic equations in these two cases are the ones that can be factored.

▼ **Example 1** For each equation give the number and kind of solutions:

$$x^2 - 3x - 40 = 0$$

Solution Using $a = 1$, $b = -3$, and $c = -40$ in $b^2 - 4ac$, we have $9 - 4(1)(-40) = 9 + 160 = 169$.

The discriminant is a perfect square. Therefore, the equation has two rational solutions (*Note:* Solving this equation by factoring would be the fastest method.)

b. $2x^2 - 3x + 4 = 0$

Solution Using $a = 2$, $b = -3$, and $c = 4$, we have $b^2 - 4ac = 9 - 4(2)(4) = 9 - 32 = -23$.

The discriminant is negative, implying the equation has two imaginary solutions.   ▲

▼ **Example 2** Find an appropriate k so that the equation $4x^2 - kx = -9$ has exactly one rational solution.

Solution We begin by writing the equation in standard form:

$$4x^2 - kx + 9 = 0$$

Using $a = 4$, $b = -k$, and $c = 9$, we have

$$b^2 - 4ac = (-k)^2 - 4(4)(9)$$
$$= k^2 - 144$$

An equation has exactly one rational solution when the discriminant is 0. We set the discriminant equal to 0 and solve:

$$k^2 - 144 = 0$$
$$k^2 = 144$$
$$k = \pm 12$$

Choosing k to be 12 or -12 will result in an equation with one rational solution.   ▲

The Sum and Product of Solutions

There is an interesting relationship between the coefficients of a quadratic equation in standard form and the numbers obtained by adding and multiplying the two solutions to the equation.

The quadratic theorem gives the two solutions to $ax^2 + bx + c = 0$, $a \neq 0$, as

$$x_1 = \frac{-b + \sqrt{b^2 - 4ac}}{2a} \quad \text{and} \quad x_2 = \frac{-b - \sqrt{b^2 - 4ac}}{2a}$$

NOTATION x_1 is read "x sub 1." The 1 and 2 as used above are called *subscripts*. We denote the solutions as x_1 and x_2 so we can tell them apart. The symbols x_1 and x_2 are different variable names.
  We add the solutions x_1 and x_2 to find their sum:

$$
\begin{aligned}
x_1 + x_2 &= \frac{-b + \sqrt{b^2 - 4ac}}{2a} + \frac{-b - \sqrt{b^2 - 4ac}}{2a} \\
&= \frac{-b + \sqrt{b^2 - 4ac} - b - \sqrt{b^2 - 4ac}}{2a} \\
&= \frac{-2b}{2a} \\
&= \frac{-b}{a} \qquad \text{Sum of the solutions } x_1 \text{ and } x_2
\end{aligned}
$$

Their product is given by

$$
\begin{aligned}
x_1 \cdot x_2 &= \left[\frac{-b + \sqrt{b^2 - 4ac}}{2a}\right]\left[\frac{-b - \sqrt{b^2 - 4ac}}{2a}\right] \\
&= \frac{(-b)^2 - (\sqrt{b^2 - 4ac})^2}{(2a)^2} \qquad \text{Difference of two squares} \\
&= \frac{b^2 - (b^2 - 4ac)}{4a^2} \\
&= \frac{4ac}{4a^2} \\
&= \frac{c}{a} \qquad \text{Product of the solutions}
\end{aligned}
$$

  If x_1 and x_2 are solutions to $ax^2 + bx + c = 0$, then their sum is $-b/a$ and their product is c/a. The converse of this statement is also true, although we will not prove it here.
  We can summarize these results in the form of a theorem:

THEOREM 6.4 x_1 and x_2 are solutions to the equation $ax^2 + bx + c = 0$, $a \neq 0$, *if and only if*

$$x_1 + x_2 = -\frac{b}{a} \quad \text{and} \quad x_1 \cdot x_2 = \frac{c}{a}$$

▼   **Example 3**   Find the sum and product of the solutions to $5x^2 - 2ix - 4 = 0$.

Solution   Since $a = 5$, $b = -2i$, and $c = -4$, we have

$$-\frac{b}{a} = -\left(\frac{-2i}{5}\right) = \frac{2i}{5} \quad \text{and} \quad \frac{c}{a} = -\frac{4}{5}$$

The sum of the solutions is $2i/5$ and the product of the solutions is $-\frac{4}{5}$.   ▲

Theorem 6.4 also gives us a way of checking solutions to quadratic equations other than simply checking them in the original equation.

▼   **Example 4**   Is the solution set to $2x^2 - 5x + 4 = 0$ the set $\left\{\dfrac{5 + \sqrt{7}i}{4}, \dfrac{5 - \sqrt{7}i}{4}\right\}$?

Solution   From the equation $2x^2 - 5x + 4 = 0$, we see that

$$-\frac{b}{a} = -\left(\frac{-5}{2}\right) = \frac{5}{2} \quad \text{and} \quad \frac{c}{a} = \frac{4}{2} = 2$$

The sum and product of the proposed solutions are

$$\text{Sum} = \frac{5 + \sqrt{7}i}{4} + \frac{5 - \sqrt{7}i}{4} = \frac{10}{4} = \frac{5}{2}$$

$$\text{Product} = \left(\frac{5 + \sqrt{7}i}{4}\right)\left(\frac{5 - \sqrt{7}i}{4}\right) = \frac{25 + 7}{16} = \frac{32}{16} = 2$$

The sum of the solutions is $-b/a$ and the product matches c/a. The solution set must be correct.   ▲

Problem Set 6.4

Use the discriminant to find the number and kind of solution for each of the following equations:

1. $x^2 - 6x + 5 = 0$ 2. $x^2 - x - 12 = 0$
3. $4x^2 - 4x = -1$ 4. $9x^2 + 12x = -4$
5. $x^2 + x - 1 = 0$ 6. $x^2 - 2x + 3 = 0$
7. $2y^2 = 3y + 1$ 8. $3y^2 = 4y - 2$
9. $x^2 - 9 = 0$ 10. $4x^2 - 81 = 0$
11. $5a^2 - 4a = 5$ 12. $3a = 4a^2 - 5$

Determine k so that each of the following has exactly one real solution:

13. $x^2 - kx + 25 = 0$ **14.** $x^2 + kx + 25 = 0$
15. $x^2 = kx - 36$ **16.** $x^2 = kx - 49$
17. $4x^2 - 12x + k = 0$ **18.** $9x^2 + 30x + k = 0$
19. $kx^2 - 40x = 25$ **20.** $kx^2 - 2x = -1$
21. $3x^2 - kx + 2 = 0$ **22.** $5x^2 + kx + 1 = 0$

Use Theorem 6.4 to find the sum and product of the solutions for each equation. Identify any equations that would best be solved by factoring:

23. $x^2 + 3x + 2 = 0$ **24.** $x^2 - 11x + 30 = 0$
25. $2y^2 = -18 + 15y$ **26.** $3y^2 = 20 - 7y$
27. $4x^2 - 3x + 2 = 0$ **28.** $2x^2 - 5x + 7 = 0$
29. $t^2 - 6t = -9$ **30.** $4t^2 + 4t = -1$
31. $2x^2 + 4x - 9 = 0$ **32.** $3x^2 - 3x + 5 = 0$
33. $y^2 - 49 = 0$ **34.** $9y^2 - 1 = 0$

Use Theorem 6.4 to check the two proposed solutions to each equation:

35. $3x^2 + 2x - 5 = 0$; $\{-\frac{5}{3}, 1\}$
36. $x^2 + 3x - 28 = 0$; $\{-7, 4\}$
37. $9x^2 = 9x - 2$; $\{-\frac{1}{3}, \frac{2}{3}\}$
38. $5x^2 = -4x + 12$; $\{-2, \frac{6}{5}\}$
39. $4x^2 - 8x + 1 = 0$; $\{(2 \pm \sqrt{3})/2\}$
40. $x^2 - 3x + 1 = 0$; $\{(3 \pm \sqrt{5})/2\}$
41. $2x^2 - 6x + 5 = 0$; $\{(3 \pm i)/2\}$
42. $2x^2 - 2x + 1 = 0$; $\{(1 \pm i)/2\}$

We will use the same four steps in solving word problems in this section that we have used in the past. The most important part of solving word problems is finding an equation that describes the situation. In this section, the equation will be second-degree. We can solve the equations by any convenient method.

6.5
Word Problems

▼ **Example 1** The sum of the squares of two consecutive integers is 25. Find the two integers.

Solution Let $x = $ the first integer; then $x + 1 = $ the next consecutive integer. The sum of the squares of x and $x + 1$ is 25.

$$x^2 + (x + 1)^2 = 25$$
$$x^2 + x^2 + 2x + 1 = 25$$
$$2x^2 + 2x - 24 = 0$$
$$x^2 + x - 12 = 0 \qquad \text{Divide both sides by 2}$$
$$(x + 4)(x - 3) = 0$$
$$x = -4 \quad \text{or} \quad x = 3$$

These are the possible values for the first integer:

$$\begin{array}{ll} \text{If} \quad x = -4 & \text{If} \quad x = 3 \\ \text{then} \quad x + 1 = -3 & \text{then} \quad x + 1 = 4 \end{array}$$

There are two pairs of consecutive integers, the sum of whose squares is 25. They are $\{-4, -3\}$ and $\{3, 4\}$. ▲

Many word problems dealing with area can best be described by quadratic equations.

▼ **Example 2** A vegetable garden, 20 feet by 30 feet, has a walkway of uniform width around it. If the area of the garden and walkway together is 704 square feet, what is the width of the walkway?

Solution Let x = the width of the walkway. Figure 6-1 shows a picture of the garden and walkway.

From the diagram the dimensions on the outside of the walkway are $(20 + 2x)$ and $(30 + 2x)$. The enclosed area is given as 704 square feet. Since the area is length times width, we have

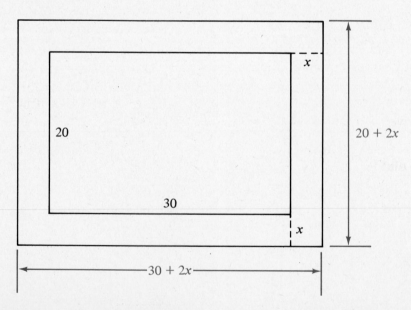

Figure 6-1

$$(30 + 2x)(20 + 2x) = 704$$
$$600 + 100x + 4x^2 = 704$$
$$4x^2 + 100x - 104 = 0$$
$$x^2 + 25x \quad - 26 \quad = 0 \qquad \text{Divide both sides by 4}$$
$$(x + 26)(x - 1) = 0$$
$$x = -26 \quad \text{or} \quad x = 1$$

Our two possible solutions are 1 foot and -26 feet. Since -26 feet is impossible, the width of the walkway must be 1 foot. ▲

Another application of quadratic equations involves the Pythagorean theorem, an important theorem from geometry. The theorem gives the relationship between the sides of any right triangle (a triangle with a $90°$ angle). We state it here without proof.

PYTHAGOREAN THEOREM In any right triangle, the square of the longest side (hypotenuse) is equal to the sum of the squares of the other two sides (legs).

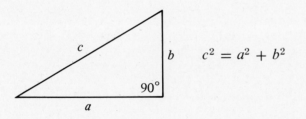

$$c^2 = a^2 + b^2$$

▼ **Example 3** The lengths of the three sides of a right triangle are given by three consecutive integers. Find the lengths of the three sides.

Solution Let x = first integer (shortest side)

Then $x + 1$ = next consecutive integer
$x + 2$ = last consecutive integer (longest side)

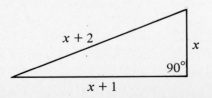

By the Pythagorean theorem, we have

$$(x + 2)^2 = (x + 1)^2 + x^2$$
$$x^2 + 4x + 4 = x^2 + 2x + 1 + x^2$$
$$x^2 - 2x - 3 = 0$$
$$(x - 3)(x + 1) = 0$$
$$x = 3 \quad \text{or} \quad x = -1$$

The shortest side is 3. The other two sides are 4 and 5.   ▲

▼   **Example 4**   If an object is thrown downward with an initial velocity of 10 feet/second, the distance (s) it travels in an amount of time t is given by the equation $s = 20t + 16t^2$. (This is because of the acceleration of the object due to gravity. The equation does not take into account the force of friction on the object due to its falling through air.) How long does it take the object to fall 40 feet?

Solution   In this example, the equation that describes the situation is given. We simply let $s = 40$ and solve for t:

$$\text{When} \quad s = 40,$$
$$\text{the equation} \quad s = 20t + 16t^2$$
$$\text{becomes} \quad 40 = 20t + 16t^2$$
$$\text{or} \quad 16t^2 + 20t - 40 = 0$$
$$4t^2 + 5t - 10 = 0 \qquad \text{Divide by 4}$$

Using the quadratic formula, we have

$$t = \frac{-5 \pm \sqrt{25 - 4(4)(-10)}}{2(4)}$$

$$= \frac{-5 \pm \sqrt{185}}{8}$$

$$t = \frac{-5 + \sqrt{185}}{8} \quad \text{or} \quad t = \frac{-5 - \sqrt{185}}{8}$$

The second solution is impossible since it is a negative number and t must be positive.

It takes

$$t = \frac{-5 + \sqrt{185}}{8}$$

or approximately

$$\frac{-5 + 13.60}{8} = 1.08 \text{ seconds}$$

for the object to fall 40 feet.   ▲

▼   **Example 5**   The current of a river is 3 miles/hour. It takes a motorboat a total of 3 hours to travel 12 miles upstream and return 12 miles downstream. What is the speed of the boat in still water?

Solution   Let $x =$ the speed of the boat in still water. The basic equation that gives the relationship between distance, rate, and time is $d = rt$.   ▲

It is usually helpful to summarize rate problems like this by using a table:

	Distance	Rate	Time
Upstream			
Downstream			

We fill in as much of the table as possible using the information given in the problem. For instance, since we let $x =$ the speed of the boat in still water, the rate upstream (against the current) must be $x - 3$. The rate downstream (with the current) is $x + 3$.

	d	r	t
Upstream	12	$x - 3$	
Downstream	12	$x + 3$	

The last two boxes can be filled in using the relationship $d = r \cdot t$. Since the boxes correspond to t, we solve $d = r \cdot t$ for t and get

$$t = \frac{d}{r}$$

Completing the table, we have

	d	r	t
Upstream	12	x − 3	12/(x − 3)
Downstream	12	x + 3	12/(x + 3)

The total time for the trip up and back is 3 hours:

$$\text{Time upstream} + \text{Time downstream} = \text{Total time}$$

$$\frac{12}{x-3} + \frac{12}{x+3} = 3$$

Multiplying both sides by $(x − 3)(x + 3)$, we have

$$12(x + 3) + 12(x - 3) = 3(x^2 - 9)$$
$$12x + 36 + 12x - 36 = 3x^2 - 27$$
$$3x^2 - 24x - 27 = 0$$
$$x^2 - 8x - 9 = 0 \qquad \text{Divide both sides by 3}$$
$$(x - 9)(x + 1) = 0$$
$$x = 9 \quad \text{or} \quad x = -1$$

The speed of the motorboat in still water is 9 miles/hour.

Problem Set 6.5

1. The sum of the squares of two consecutive odd integers is 34. Find the two integers.
2. The sum of the squares of two consecutive even integers is 100. Find the two integers.
3. The square of the sum of two consecutive integers is 81. Find the two integers.
4. Find two consecutive even integers whose sum squared is 100.
5. One integer is 3 more than twice another. Their product is 65. Find the two integers.
6. The product of two integers is 150. One integer is 5 less than twice the other. Find the integers.
7. The sum of a number and its reciprocal is $\frac{10}{3}$. Find the number.
8. The sum of a number and twice its reciprocal is $\frac{27}{5}$. Find the number.
9. The sum of the reciprocals of two consecutive integers is $\frac{7}{12}$. Find the two integers.

10. Find two consecutive even integers, the sum of whose reciprocals is $\frac{3}{4}$.

11. A rectangular pool, 15 feet by 30 feet, has a strip of grass of uniform width around it. If the total area of the grass and the pool is 646 square feet, find the width of the strip of grass.

12. A rectangular garden is surrounded by a path of uniform width. The area of the path is 168 square feet. If the garden measures 10 feet by 12 feet, find the width of the path.

13. An 8-inch-by-10-inch picture is surrounded by a frame of uniform width. The area of the picture and frame together is 120 square inches. What is the width of the frame?

14. The longest side of a right triangle is twice the shortest side. The third side measures 6 inches. Find the length of the shortest side.

15. One leg of a right triangle is 3 times the other leg. The hypotenuse is $2\sqrt{10}$ centimeters. What are the lengths of the legs?

16. The longest side of a right triangle is 2 less than twice the shortest side. The third side is 2 more than the shortest side. Find the length of all three sides.

17. An object is thrown downward with an initial velocity of 5 feet/second. The relationship between the distance (s) it travels and time (t) is given by $s = 5t + 16t^2$. How long does it take the object to fall 74 feet?

18. The distance an object falls from rest is given by the equation $s = 16t^2$, where s = distance and t = time. How long does it take an object dropped from a 100-foot cliff to hit the ground?

19. An object is thrown upward with an initial velocity of 20 feet/second. The equation that gives the height (h) of the object at any time (t) is $h = 20t - 16t^2$. At what times will the object be 4 feet off the ground?

20. An object is propelled upward with an initial velocity of 32 feet/second from a height of 16 feet above the ground. The equation giving the object's height (h) at any time (t) is $h = 16 + 32t - 16t^2$. Does the object ever reach a height of 32 feet?

21. The current of a river is 2 miles/hour. A boat travels to a point 8 miles upstream and back again in 3 hours. What is the speed of the boat in still water?

22. A boat travels 15 miles/hour in still water. It takes twice as long for the boat to go 20 miles upstream as it does to go downstream. Find the speed of the current.

23. A man can ride his bicycle 3 times as fast as he can walk. He rode his bike for 15 miles, then walked an additional 2.5 miles. The total time for the trip was $1\frac{1}{2}$ hours. How fast does he ride his bicycle?

24. Train A can travel the 50 miles between the farm and the city 10 miles/hour faster than train B. If a person takes train A to the city and train B back to the farm, the total trip takes $2\frac{1}{4}$ hours. Find the speed of each train.

Theorem 6.1 (Zero-Factor Property)

For all real numbers r and s, if $r \cdot s = 0$, then $r = 0$ or $s = 0$ (or both).

To Solve a Quadratic Equation by Factoring

Step 1. Write the equation in standard form:
$$ax^2 + bx + c = 0, \quad a \neq 0$$
Step 2. Factor the left side.
Step 3. Use the zero-factor property to set each factor equal to zero.
Step 4. Solve the resulting first-degree equations.

Theorem 6.2

If $a^2 = b$, where b is a real number, then
$$a = \sqrt{b} \quad \text{or} \quad a = -\sqrt{b} \quad \text{or} \quad a = \pm\sqrt{b}$$

To Solve a Quadratic Equation by Completing the Square

Step 1. Write the equation in the form $ax^2 + bx = c$.
Step 2. If $a \neq 1$, divide through by the constant a so the coefficient of x^2 is 1.
Step 2. Complete the square on the left side by adding the square of $\frac{1}{2}$ the coefficient of x to both sides.
Step 4. Write the left side of the equation as the square of a binomial. Simplify the right side if possible.
Step 5. Apply Theorem 6.2 and solve as usual.

Theorem 6.3 (The Quadratic Theorem)

For any quadratic equation in the form $ax^2 + bx + c = 0$, $a \neq 0$, the two solutions are
$$x = \frac{-b \pm \sqrt{b^2 - 4ac}}{2a}$$

This last expression is known as the *quadratic formula*.

The Discriminant

The expression $b^2 - 4ac$ which appears under the radical sign in the quadratic formula is known as *the discriminant*.
We can classify the solutions to $ax^2 + bx + c = 0$:

The solutions are	When the discriminant is
Two imaginary numbers	Negative
One rational number	Zero
Two rational numbers	A positive perfect square
Two irrational numbers	A positive number, but not a perfect square

If the equation $ax^2 + bx + c = 0$, $a \neq 0$, has solutions x_1 and x_2, then

Sum and Product of Solutions

$$x_1 + x_2 = -\frac{b}{a}$$

$$\text{and} \quad x_1 \cdot x_2 = \frac{c}{a}$$

---

1. Attempting to apply the zero-factor property to numbers other than 0. For example, consider the equation

COMMON MISTAKES

$$(x + 2)(x - 3) = 7$$

The mistake takes place when we try to solve it by setting each factor equal to 7.

$$x + 2 = 7 \quad \text{or} \quad x - 3 = 7$$
$$x = 5 \quad \text{or} \quad x = 10$$

Neither of these two numbers is a solution to the original equation. The mistake arises when we assume that since the product of $(x + 2)$ and $(x - 3)$ is 7, one of the two factors must also be 7.

2. When both sides of an equation are squared in the process of solving the equation, a common mistake occurs when the resulting solutions are not checked in the original equation. Remember, every time we square both sides of an equation, there is the possibility we have introduced an extraneous root.

3. When squaring a quantity that has two terms involving radicals, it is a common mistake to omit the middle term in the result. For example,

$$(\sqrt{x + 3} + \sqrt{2x})^2 = (x + 3) + (2x)$$

is a common mistake. It should look like this:

$$(\sqrt{x + 3} + \sqrt{2x})^2 = (x + 3) + 2\sqrt{2x}\sqrt{x + 3} + (2x)$$

Remember: $(a + b)^2 = a^2 + 2ab + b^2$.

Solve the following by factoring:

Chapter 6
Test

1. $x^2 - 4x = 21$ **2.** $25x^2 - 81 = 0$
3. $4x^2 = 6 - 5x$

Solve by completing the square:

4. $x^2 - 4x = -2$ **5.** $2x^2 - 8x = 5$
6. $3y^2 + 8y + 5 = 0$

Solve by using the quadratic formula:

7. $2x^2 - x = 3$ 8. $x^2 - 7x - 7 = 0$
9. $4x^4 - 7x^2 - 2 = 0$

Solve by any method:

10. $\sqrt{2x + 5} = x + 1$ 11. $\sqrt{2x + 1} = 1 + \sqrt{x}$
12. $\dfrac{1}{x - 1} = \dfrac{5}{4} - \dfrac{1}{x - 4}$

13. Determine k so that $kx^2 - 12x + 4$ has one rational solution.
14. Use the discriminant to identify the number and type of solutions to $2x^2 - 5x = 7$.
15. Give the sum and product of the solutions to $4x^2 - \sqrt{3}x + 5 = 0$.
16. Are $x = (3 + i)/5$ and $x = (3 - i)/5$ solutions to $5x^2 - 6x + 2 = 0$?
17. One number is 1 less than twice another. The sum of their squares is 34. Find the two numbers.
18. A motorboat travels at 4 miles/hour in still water. It goes 12 miles upstream and 12 miles back again in a total of 8 hours. Find the speed of the current of the river.

Linear Equations and Inequalities

To the student:

As we have said before, mathematics is a language that can describe certain aspects of our world better than English. One important aspect of the world is the idea of a path, track, orbit, or course. Some simple paths we are familiar with are circles, cloverleaf interchanges on a highway, the sloping straight line of a sewer pipe, and the elliptical route taken by a satellite. Mathematics can be used to describe these paths very accurately. The simplest of paths—a straight line—can be described by an equation such as $3x - 2y = 5$, which we call a linear equation in two variables. In this chapter we will concern ourselves with linear equations (and inequalities) in two variables. We will begin the chapter by graphing some straight lines. We will then define what we mean by the slope of a line and use this definition to find two special forms of the equation of a straight line. The remainder of the chapter is concerned with linear inequalities and variation problems.

The main concepts you will want to master are (1) how to find the graph of a straight line given the equation of that line, and (2) how to find the equation of a line given the coordinates of two points on the line or given one point and the slope of the line.

To make a successful attempt at Chapter 7 you should be familiar with the concepts developed in Chapter 2. The main concept is how to solve a linear equation in one variable.

Up to this point we have always considered equations with one variable. Solution sets for these equations are sets of single numbers.

Let us now consider the equation $2x - y = 5$. The equation contains

7.1
Graphing in Two Dimensions

213

two variables. A solution, therefore, must be in the form of a pair of numbers, one for x and one for y, that make the equation a true statement. One pair of numbers that works is $x = 3$ and $y = 1$, because when we substitute them for x and y in the equation we get a true statement. That is,

$$2(3) - 1 = 5$$
$$5 = 5 \qquad \leftarrow \text{A true statement}$$

The pair of numbers $x = 3$ and $y = 1$ can be written as $(3, 1)$. This is called an *ordered pair,* because it is a pair of numbers written in a specific order. The first number in the ordered pair is always associated with the variable x, the second number with the variable y. The first number is called the *x-coordinate* (or x-component) of the ordered pair and the second number is called the *y-coordinate* (or y-component) of the ordered pair.

With equations that contain *one* variable, solution sets are graphed on the real number line. Solution sets for equations in two variables consist of ordered pairs of numbers which we will graph on a rectangular (cartesian) coordinate system.

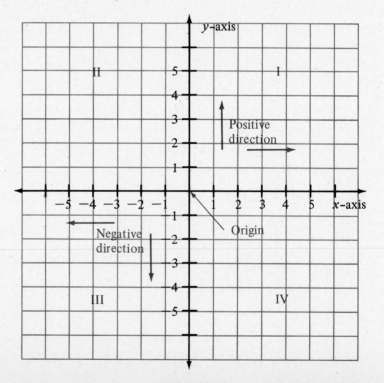

Figure 7-1

A *rectangular coordinate system* is made by drawing two real number lines at right angles to each other. The two number lines, called *axes,* cross each other at 0. This point is called the *origin.* Positive directions are to the right and up. Negative directions are down and to the left. The rectangular coordinate system is shown in Figure 7-1.

The horizontal number line is called the *x-axis* and the vertical number line is called the *y-axis.* The two number lines divide the coordinate system into four quadrants which we number I through IV in a counterclockwise direction. Points on the axes are not considered as being in any quadrant.

To graph the ordered pair (*a, b*) on a rectangular coordinate system, we start at the origin and move *a* units right or left (right if *a* is positive, left if *a* is negative). Then we move *b* units up or down (up if *b* is positive and down if *b* is negative). The point where we end is the graph of the ordered pair (*a, b*).

Graphing Ordered Pairs

▼ **Example 1** Plot (graph) the ordered pairs (2, 5), (−2, 5), (−2, −5), and (2, −5).

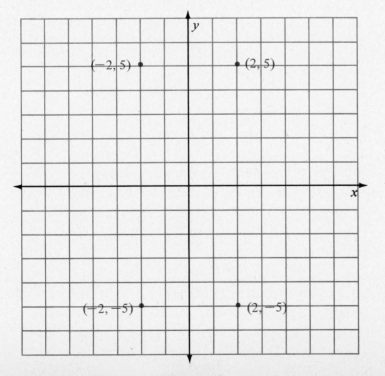

Figure 7-2

Solution   To graph the ordered pair (2, 5), we start at the origin and move 2 units to the right, then 5 units up. We are now at the point whose coordinates are (2, 5). We graph the other three ordered pairs in a similar manner (see Figure 7-2).   ▲

From Example 1 we see that any point in quadrant I has both its x- and y- coordinates positive $(+, +)$. Points in quadrant II have negative x-coordinates and positive y-coordinates $(-, +)$. In quadrant III both coordinates are negative $(-, -)$. In quadrant IV the form is $(+, -)$.

Every point has a unique set of coordinates and every ordered pair (a, b) has as its graph a point in the coordinate plane. That is, for every point there corresponds an ordered pair (a, b), and for every ordered pair (a, b) there corresponds a point. We say there is a *one-to-one correspondence* between points on the coordinate system and ordered pairs (a, b) where a and b are real numbers.

▼   **Example 2**   Graph the ordered pairs $(1, -3)$, $(\frac{1}{2}, 2)$, $(3, 0)$, $(0, -2)$, $(-1, 0)$ and $(0, 5)$.

Solution

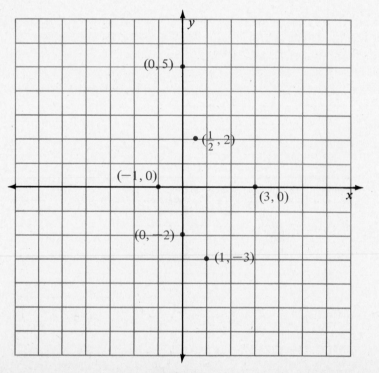

Figure 7-3   ▲

From Example 2 we see that any point on the x-axis has a y-coordinate of 0 (it has no vertical displacement), and any point on the y-axis has an x-coordinate of 0 (no horizontal displacement).

We will now turn our attention to graphing some straight lines.

DEFINITION Any equation that can be put in the form $ax + by = c$, where a, b, and c are real numbers and a and b are not both 0, is called a *linear equation* in two variables. The graph of any equation of this form is a straight line (that is why these equations are called "linear"). The form $ax + by = c$ is called *standard form*.

To graph a linear equation in two variables, we simply graph its solution set. That is, we will draw a line through all the points whose coordinates satisfy the equation.

▼ **Example 3** Graph $y = 2x - 3$.

Solution Since $y = 2x - 3$ can be put in the form $ax + by = c$, it is a linear equation in two variables. Hence, the graph of its solution set is a straight line. We can find some specific solutions by substituting numbers for x and then solving for the corresponding values of y. We are free to choose any convenient numbers for x, so let's use the numbers -2, 0, and 2:

$$\begin{aligned}
\text{When} \quad & x = -2, \\
\text{the equation} \quad & y = 2x - 3 \\
\text{becomes} \quad & y = 2(-2) - 3 \\
& y = -7
\end{aligned}$$

The ordered pair $(-2, -7)$ is a solution.

$$\begin{aligned}
\text{When} \quad & x = 0, \\
\text{we have} \quad & y = 2(0) - 3 \\
& y = -3
\end{aligned}$$

The ordered pair $(0, -3)$ is also a solution.

$$\begin{aligned}
\text{Using} \quad & x = 2, \\
\text{we have} \quad & y = 2(2) - 3 \\
& y = 1
\end{aligned}$$

The ordered pair (2, 1) is another solution.

Graphing these three ordered pairs and drawing a line through them, we have the graph of $y = 2x - 3$:

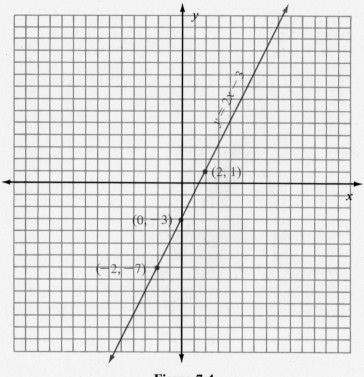

Figure 7-4

There is a one-to-one correspondence between points on the line and solutions to the equation $y = 2x - 3$.   ▲

It actually takes only two points to determine a straight line. We have included a third point in the example above for insurance. If all three points do not line up in a straight line, we have made a mistake.

Two important points on the graph of a straight line, if they exist, are the points where the graph crosses the axes.

DEFINITION   The *x-intercept* of the graph of an equation is the *x*-coordinate of the point where the graph crosses the *x*-axis. The *y-intercept* is defined similarly.

Since any point on the x-axis has a y-coordinate of 0, we can find the x-intercept by letting $y = 0$ and solving the equation for x. We find the y-intercept by letting $x = 0$ and solving for y.

▼   **Example 4**   Find the x- and y-intercepts for $2x + 3y = 6$; then graph the solution set.

Solution   To find the y-intercept we let $x = 0$.

$$\text{When} \qquad x = 0,$$
$$\text{we have} \quad 2(0) + 3y = 6$$
$$3y = 6$$
$$y = 2$$

The y-intercept is 2, and the graph crosses the y-axis at the point $(0, 2)$.

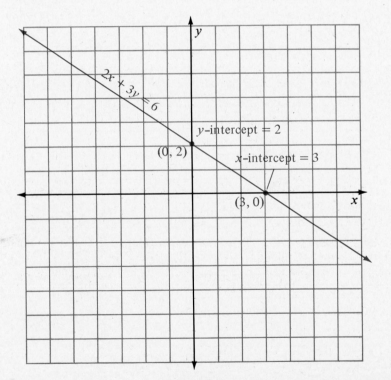

Figure 7-5

$$\text{When} \qquad y = 0$$
$$\text{we have} \quad 2x + 3(0) = 6$$
$$x = 3$$

The x-intercept is 3, so the graph crosses the x-axis at the point $(3, 0)$. We use these results to graph the solution set for $2x + 3y = 6$ (see Figure 7-5). ▲

Graphing straight lines by finding the intercepts works best when the coefficients of x and y are factors of the constant term, as was the case in Example 4.

▼   **Example 5**   Graph the line $x = 3$ and the line $y = -2$.

Solution   The line $x = 3$ is the set of all points whose x-coordinate is 3. The variable y does not appear in the equation, so the y-coordinates can be any number.

The line $y = -2$ is the set of all points whose y-coordinate is -2. The x-coordinate can be any number.

Here are the graphs:

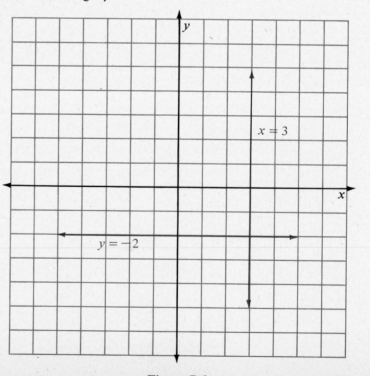

Figure 7-6

Graph each of the following ordered pairs on a rectangular coordinate system: *Problem Set 7.1*

1-41

1. (1, 2)   2. (−1, 2)
3. (−1, −2)   4. (1, −2)
5. (3, 4)   6. (−3, −4)
7. (5, 0)   8. (0, −3)
9. (0, 2)   10. (4, 0)
11. (−7, −9)   12. (−4, −1)
13. ($\frac{1}{2}$, 2)   14. (3, $\frac{1}{4}$)
15. (5, −2)

Give the coordinates of each of the following points:

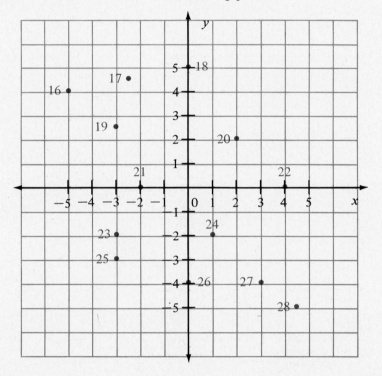

Graph each of the following linear equations by first finding the intercepts:

29. $2x - 3y = 6$ 30. $3x - 2y = 6$
31. $y + 2x = 4$ 32. $y - 2x = 4$
33. $4x - 5y = 20$ 34. $4x + 5y = 20$
35. $2x - y = 8$ 36. $x - 2y = 8$
37. $7x - 2y = 14$ 38. $2x + 7y = 14$
39. $3x + 5y = 15$ 40. $5x - 3y = 15$
41. $2x + 5y = 10$ 42. $2x - 5y = 10$

Graph each of the following straight lines:

43. $y = 2x + 3$ **44.** $y = 3x - 2$
45. $2y = 4x - 8$ **46.** $2y = 4x + 8$
47. $x - y = 5$ **48.** $x + y = 6$
49. $2x - y = 3$ **50.** $3x - y = 2$
51. $y = 5$ **52.** $y = -3$
53. $x = -7$ **54.** $x = 2$
55. $y = 4x - 1$ **56.** $y = 3x + 3$
57. $y = 5x + 1$ **58.** $y = 4x + 3$

59. In New York City a taxi ride costs 75¢ for the first $\frac{1}{7}$ of a mile and 10¢ for every additional $\frac{1}{7}$ of a mile. If x is the number of additional sevenths of a mile, then the total cost y of a taxi ride is

$$y = 10x + 75$$

a. Graph the equation.
b. How much does it cost to ride a taxi for 10 miles in New York City?

7.2
The Slope of a Line

In defining the slope of a straight line, we are looking for a number to associate with a straight line that does two things. First of all, we want the slope of a line to measure the "steepness" of the line. That is, in comparing two lines, the slope of the steeper line should have the larger numerical value. Secondly, we want a line that *rises* going from left to right to have a *positive* slope. We want a line that *falls* going from left to right to have a *negative* slope. (A line that neither rises nor falls going from left to right must, therefore, have 0 slope.)

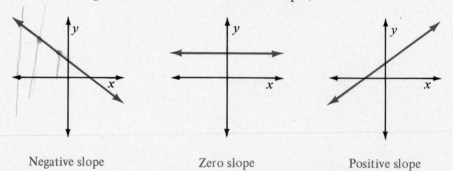

Negative slope   Zero slope   Positive slope

Geometrically, we can define the *slope* of a line as the ratio of the vertical change to the horizontal change encountered when moving from one point to another on the line. The vertical change is sometimes called the *rise*. The horizontal change is called the *run*.

▼   **Example 1**   Find the slope of the line $y = 2x - 3$.

Solution   In order to use our geometric definition, we first graph $y = 2x - 3$. We then pick any two convenient points and find the ratio of rise to run.

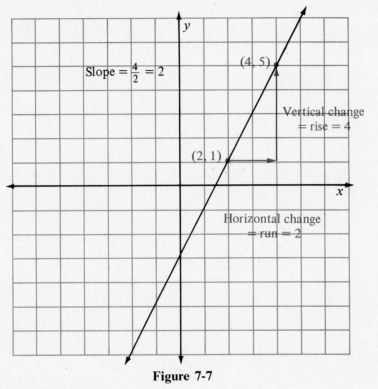

Slope $= \frac{4}{2} = 2$

(4, 5)

Vertical change $= \text{rise} = 4$

(2, 1)

Horizontal change $= \text{run} = 2$

Figure 7-7

Our line has a slope of 2.   ▲

   Notice that we can measure the vertical change by subtracting the y-coordinates of the two points shown: $5 - 1 = 4$. The horizontal change is the difference of the x-coordinates: $4 - 2 = 2$. This gives us a second way of defining the slope of a line. Algebraically, we say the slope of a line between two points whose coordinates are given is the ratio of the difference in the y-coordinates to the difference in the x-coordinates. We can summarize the above discussion by formalizing our definition for slope.

DEFINITION   The slope of the line between two points (x_1, y_1) and (x_2, y_2) is given by

$$\text{Slope} = m = \frac{\text{rise}}{\text{run}} = \frac{y_2 - y_1}{x_2 - x_1}$$

The letter m is usually used to designate slope. Our definition includes both the geometric form (rise/run) and the algebraic form $(y_2 - y_1)/(x_2 - x_1)$.

▼   **Example 2**   Find the slope of the line through $(-2, -3)$ and $(-5, 1)$.

Solution

$$m = \frac{y_2 - y_1}{x_2 - x_1} = \frac{1 - (-3)}{-5 - (-2)} = \frac{4}{-3} = -\frac{4}{3}$$

Looking at the graph of the line between the two points, we can see our geometric approach does not conflict with our algebraic approach:

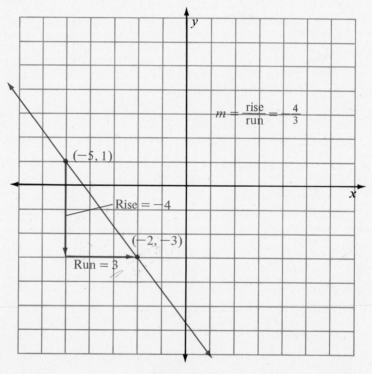

Figure 7-8

We should note here that it does not matter which ordered pair we call (x_1, y_1) and which we call (x_2, y_2). If we were to reverse the order of subtraction of both the *x*- and *y*-coordinates in the above example, we would have

$$m = \frac{-3 - 1}{-2 - (-5)} = \frac{-4}{3} = -\frac{4}{3}$$

which is the same as our previous result.   ▲

▼ **Example 3** Find the slope of the line containing $(3, -1)$ and $(3, 4)$.

Solution   Using the definition for slope, we have

$$m = \frac{-1 - 4}{3 - 3} = \frac{-5}{0}$$

The expression $-5/0$ is undefined. That is, there is no real number to associate with it. In this case, we say the line has *no slope*. The graph of our line is as follows:

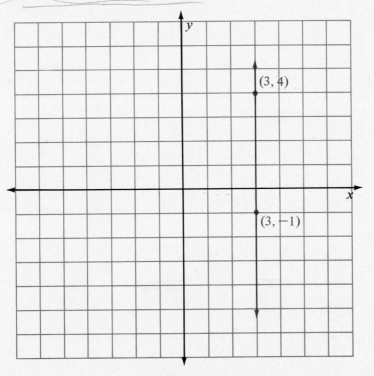

Figure 7-9

Our line with no slope is a vertical line. All vertical lines have no slope. (And all horizontal lines, as was mentioned earlier, have 0 slope.)   ▲

Slope of Parallel and
Perpendicular Lines

In geometry we call lines in the same plane that never intersect parallel. In order for two lines to be nonintersecting, they must rise or fall at the same rate. That is, the ratio of vertical change to horizontal change must be the same for each line. In other words, two lines are *parallel* if and only if they have the *same slope*.

Although it is not as obvious, it is also true that two nonvertical lines are *perpendicular* if and only if the *product of their slopes is* -1. This is the same as saying their slopes are negative reciprocals.

We can state these facts with symbols as follows.

If line l_1 has slope m_1, and line l_2 has slope m_2, then

$$l_1 \text{ and } l_2 \text{ are parallel} \Leftrightarrow m_1 = m_2$$

and

$$l_1 \text{ and } l_2 \text{ are perpendicular} \Leftrightarrow m_1 \cdot m_2 = -1$$

$$\left(\text{or } m_1 = \frac{-1}{m_2}\right)$$

To clarify this, if a line has a slope of $\frac{2}{3}$, then any line parallel to it has a slope of $\frac{2}{3}$. Any line perpendicular to it has a slope of $-\frac{3}{2}$ (the negative reciprocal of $\frac{2}{3}$).

▼   **Example 4**   Find a if the line through $(3, a)$ and $(-2, -8)$ is perpendicular to a line with slope $-\frac{4}{5}$.

     Solution   The slope of the line through the two points is

$$m = \frac{a - (-8)}{3 - (-2)} = \frac{a + 8}{5}$$

Since the line through the two points is perpendicular to a line with slope $-\frac{4}{5}$, we can also write its slope as $\frac{5}{4}$:

$$\frac{a + 8}{5} = \frac{5}{4}$$

Multiplying both sides by 20, we have

$$4(a + 8) = 5 \cdot 5$$
$$4a + 32 = 25$$
$$4a = -7$$
$$a = -\frac{7}{4}$$

                                                                   ▲

Find the slope of each of the following lines from the given graph: *Problem Set 7.2*

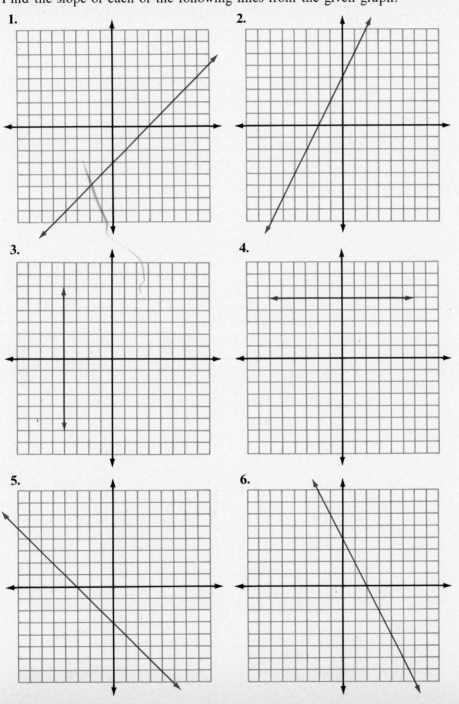

Find the slope of the line through the following pairs of points:

7. $(4, 1), (2, 5)$ **8.** $(7, 3), (2, -2)$
9. $(5, -1), (4, 3)$ **10.** $(2, -3), (5, 6)$
11. $(1, -2), (3, -5)$ **12.** $(4, -1), (5, -3)$
13. $(1, 7), (1, 2)$ **14.** $(-3, -4), (-3, -1)$
15. $(5, 4), (-1, 4)$ **16.** $(2, -1), (6, -1)$
17. $(\frac{1}{2}, 3), (\frac{2}{3}, -\frac{1}{4})$ **18.** $(\frac{3}{4}, \frac{5}{6}), (-\frac{1}{2}, \frac{1}{3})$

Solve for the indicated variable if the line through the two given points has the given slope:

19. $(5, a), (4, 2)$; $m = 3$ **20.** $(3, a), (1, 5)$; $m = -4$
21. $(2, 6), (3, y)$; $m = -7$ **22.** $(-4, 9), (-5, y)$; $m = 3$
23. $(-2, -3), (x, 5)$; $m = -\frac{8}{3}$ **24.** $(4, 9), (x, -2)$; $m = -\frac{7}{3}$
25. $(x, 1), (8, 9)$; $m = \frac{1}{2}$ **26.** $(x, 4), (1, -3)$; $m = \frac{2}{3}$

27. Find the slope of any line parallel to the line through $(\sqrt{2}, 3)$ and $(-\sqrt{8}, 1)$.

28. Find the slope of any line parallel to the line through $(2, \sqrt{27})$ and $(5, -\sqrt{3})$.

29. Line l contains the points $(5\sqrt{2}, -6)$ and $(\sqrt{50}, 2)$. Give the slope of any line perpendicular to l.

30. Line l contains points $(3\sqrt{6}, 4)$ and $(-3\sqrt{24}, 1)$. Give the slope of any line perpendicular to l.

31. Line l has a slope of $\frac{2}{3}$. A horizontal change of 12 will always be accompanied by how much of a vertical change?

32. For any line with slope $\frac{4}{5}$, a vertical change of 8 is always accompanied by how much of a horizontal change?

33. The line through $(2, y^2)$ and $(1, y)$ is perpendicular to a line with slope $-\frac{1}{6}$. What are the possible values for y?

34. The line through $(7, y^2)$ and $(3, 6y)$ is parallel to a line with slope -2. What are the possible values for y?

35. A pile of sand at a construction site is in the shape of a cone. If the slope of the side of the pile is $\frac{2}{3}$ and the pile is 8 feet high, how wide is the diameter of the base of the pile?

7.3

The Equation of a Straight Line

In the first section of this chapter we defined the y-intercept of a line to be the y-coordinate of the point where the graph crosses the y-axis. We can use this definition, along with the definition of slope from the preceding section, to derive the slope-intercept form of the equation of a straight line.

Suppose line l has slope m and y-intercept b. What is the equation of l?

Since the y-intercept is b, we know that the point $(0, b)$ is on the line.

If (x, y) is any other point on l, then using the definition for slope, we have

$$\frac{y - b}{x - 0} = m \qquad \text{Definition of slope}$$

$$y - b = mx \qquad \text{Multiply both sides by } x$$
$$y = mx + b \qquad \text{Add } b \text{ to both sides}$$

This last equation is known as the *slope-intercept form* of the equation of a straight line.   *Slope-Intercept Form*

The equation of any line with slope m and y-intercept b is given by

$$y = \underset{\text{Slope}}{\nearrow} mx + \underset{\uparrow}{b}$$
$$\text{Slope} \qquad\qquad y\text{-intercept}$$

This form of the equation of a straight line is very useful. When the equation is in this form, the *slope* of the line is always the *coefficient of x*, and the *y-intercept* is always the *constant term*. Both slope and y-intercept are easy to identify.

▼   **Example 1**   Write the equation of the line with slope -2 and y-intercept 3.

Solution   Since $m = -2$ and $b = 3$, we have

$$y = mx + b$$
$$\text{or} \quad y = -2x + 3$$

It is just as easy as it seems. The equation of the line with slope -2 and y-intercept 3 is $y = -2x + 3$.   ▲

▼   **Example 2**   Give the slope and y-intercept for the line $2x - 3y = 5$.

Solution   To use the slope-intercept form we must solve the equation for y in terms of x:

$$2x - 3y = 5$$
$$-3y = -2x + 5 \qquad \text{Add } -2x \text{ to both sides}$$
$$y = \tfrac{2}{3}x - \tfrac{5}{3} \qquad \text{Divide by } -3$$

The last equation has the form $y = mx + b$. The slope must be $m = \tfrac{2}{3}$, and the y-intercept is $b = -\tfrac{5}{3}$.   ▲

▼ **Example 3**   Graph the equation $2x + 3y = 6$.

Solution   Although we could graph this equation using the methods developed in Section 6.1 (by finding ordered pairs that are solutions to the equation and drawing a line through their graphs), it is sometimes easier to graph a line using the slope-intercept form of the equation.

Solving the equation for y, we have

$$2x + 3y = 6$$
$$3y = -2x + 6 \qquad \text{Add } -2x \text{ to both sides.}$$
$$y = -\tfrac{2}{3}x + 2 \qquad \text{Divide by 3.}$$

The slope is $m = -\tfrac{2}{3}$ and the y-intercept is $b = 2$. Therefore, the point $(0, 2)$ is on the graph and the ratio rise/run going from $(0, 2)$ to any other point on the line is $-\tfrac{2}{3}$. If we start at $(0, 2)$ and move 2 units up (that's a rise of 2) and 3 units to the left (a run of -3), we will be at another point on the graph. (We could also go down 2 units and right 3 units and also be assured of ending up at another point on the line, since $2/-3$ is the same as $-2/3$.)

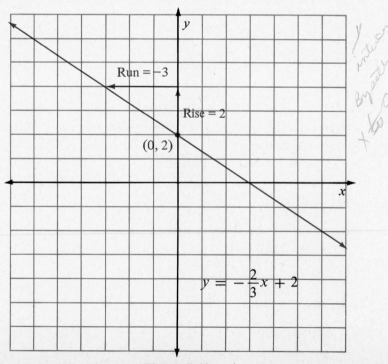

Figure 7-10

A second useful form of the equation of a straight line is the point-slope form.

Let line l contain the point (x_1, y_1) and have slope m. If (x, y) is any other point on l, then by the definition of slope we have

$$\frac{y - y_1}{x - x_1} = m$$

Multiplying both sides by $(x - x_1)$ gives us

$$(x - x_1) \cdot \frac{y - y_1}{x - x_1} = m(x - x_1)$$
$$y - y_1 = m(x - x_1)$$

This last equation is known as the *point-slope form* of the equation of a straight line.

Point-Slope Form

The equation of the line through (x_1, y_1) with slope m is given by

$$y - y_1 = m(x - x_1)$$

This form of the equation of a straight line is also very easy to work with.

▼ **Example 4**   Find the equation of the line with slope 5 that contains the point $(3, -2)$.

Solution   We have

$$(x_1, y_1) = (3, -2) \quad \text{and} \quad m = 5$$
$$y - y_1 = m(x - x_1)$$
$$y + 2 = 5(x - 3)$$
$$y + 2 = 5x - 15$$
$$y = 5x - 17$$

▲

▼ **Example 5**   Find the equation of the line through $(-3, 1)$ and $(2, 5)$.

Solution We begin by using the two given points to find the slope of the line:

$$m = \frac{5-1}{2-(-3)} = \frac{4}{5}$$

We can use either of the two given points in the point-slope form. Let's choose $(x_1, y_1) = (2, 5)$:

$$y - y_1 = m(x - x_1)$$
$$y - 5 = \tfrac{4}{5}(x - 2)$$
$$5y - 25 = 4x - 8$$
$$5y = 4x + 17$$
$$y = \tfrac{4}{5}x + \tfrac{17}{5}$$ ▲

▼ **Example 6** Give the equation of the line through $(-1, 4)$ whose graph is perpendicular to the graph of $2x - y = -3$.

Solution To find the slope of $2x - y = -3$, we solve for y:

$$2x - y = -3$$
$$y = 2x + 3$$

The slope of this line is 2. The line we are interested in is perpendicular to the line with slope 2 and must, therefore, have a slope of $-\tfrac{1}{2}$.

Using $(x_1, y_1) = (-1, 4)$ and $m = -\tfrac{1}{2}$, we have

$$y - y_1 = m(x - x_1)$$
$$y - 4 = -\tfrac{1}{2}(x + 1)$$
$$2y - 8 = -x - 1$$
$$2y = -x + 7$$
$$y = -\tfrac{1}{2}x + \tfrac{7}{2}$$

Our answer is in slope-intercept form. If we were to write it in standard form, we would have

$$x + 2y = 7$$ ▲

As a final note, we should mention again that all horizontal lines have equations of the form $y = b$ and slopes of 0. Vertical lines have no slope and have equations of the form $x = a$. These two special cases do not lend themselves to either the slope-intercept form or the point-slope form (although the equation $y = b$ could be written in slope-intercept form as $y = 0x + b$).

Give the equation of the line with the following slope and y-intercept: *Problem Set 7.3*

1. $m = 2, b = 3$
2. $m = -4, b = 2$
3. $m = 1, b = -5$
4. $m = -5, b = -3$
5. $m = \frac{1}{2}, b = \frac{3}{2}$
6. $m = \frac{2}{3}, b = \frac{5}{6}$
7. $m = 0, b = 4$
8. $m = 0, b = -2$
9. $m = -\sqrt{2}, b = 3\sqrt{2}$
10. $m = 2\sqrt{5}, b = -\sqrt{5}$

Give the slope and y-intercept for each of the following equations. Sketch the graph using the slope and y-intercept. Give the slope of any line perpendicular to the given line.

11. $y = 3x - 2$
12. $y = 2x + 3$
13. $2x - y = 4$
14. $3x + y = -2$
15. $2x - 3y = 12$
16. $3x - 2y = 12$
17. $4x + 5y = 20$
18. $5x - 4y = 20$
19. $3x - 5y = 10$
20. $4x - 3y = -9$

Find the equation of the line that contains the given point and has the given slope:

21. $(1, 3); m = 2$
22. $(3, -2); m = 4$
23. $(5, -1); m = -3$
24. $(6, -4); m = -2$
25. $(-2, 3); m = -\frac{1}{2}$
26. $(-1, -1); m = \frac{1}{4}$
27. $(\frac{1}{2}, -\frac{3}{2}); m = \frac{2}{3}$
28. $(\frac{2}{3}, -\frac{3}{4}); m = -\frac{3}{2}$

Find the equation of the line that contains the given pair of points:

29. $(2, 3), (1, 5)$
30. $(4, 6), (2, -2)$
31. $(-5, -8), (0, 2)$
32. $(2, -7), (-6, -3)$
33. $(3, -2), (1, 5)$
34. $(-4, 1), (-2, 4)$
35. $(0, 5), (-3, 0)$
36. $(0, -7), (4, 0)$
37. $(3, 5), (-2, 5)$
38. $(-8, 2), (4, 2)$
39. $(5, -1), (5, 4)$
40. $(-2, 1), (-2, 3)$

41. Give the slope and y-intercept, and sketch the graph, of $y = -2$.
42. Give the slope and y-intercept, and sketch the graph, of $y = 3\sqrt{2}$.
43. For the line $x = \sqrt{5}$ sketch the graph, give the slope, and name any intercepts.
44. For the line $x = -3$ sketch the graph, give the slope, and name any intercepts.
45. Find the equation of the line parallel to the graph of $3x - y = 5$ that contains the point $(-1, 4)$.
46. Find the equation of the line parallel to the graph of $2x - 4y = 5$ that contains the point $(0, 3)$.
47. Line l is perpendicular to the graph of $2x - 5y = 10$ and contains the point $(-4, -3)$. Find the equation for l.

48. Line l is perpendicular to the graph of $-3x - 5y = 2$ and contains the point $(2, -6)$. Find the equation for l.

49. Give the equation of the line perpendicular to $y = -4x + 2$ that has an x-intercept of -1.

50. Write the equation of the line parallel to the graph of $7x - 2y = 14$ that has an x-intercept of 5.

51. Give the equation of the line with x-intercept 3 and y-intercept -2.

52. Give the equation of the line with x-intercept $\frac{1}{2}$ and y-intercept $-\frac{1}{4}$.

53. Sound travels at approximately 1100 feet/second. It is for this reason that people who observe lightning strike the earth do not hear the sound (thunder) associated with the strike until after they have seen the flash. If we let t represent time (in seconds) and d represent distance (in feet), then we can write a linear equation in d and t that gives the relationship between the distance between us and the lightning strike and the time that elapses before we hear the thunder. If $d = 1100$ when $t = 1$, and $d = 2200$ when $t = 2$, find

    a. The equation that describes the relationship between d and t.

    b. How far away a lightning strike is if it takes 4 seconds before thunder is heard.

    c. How long it will be before we hear a lightning strike that is 1 mile (5280 feet) away.

**7.4
Linear Inequalities in
Two Variables**

A linear inequality in two variables is any expression that can be put in the form

$$ax + by < c$$

where a, b, and c are real numbers (a and b not both 0). The inequality symbol can be any one of the following four: $<, \leq, >, \geq$.

Some examples of linear inequalities are

$$2x + 3y < 6 \qquad y \geq 2x + 1 \qquad x - y \leq 0$$

Although not all of the above have the form $ax + by < c$, each one can be put in that form.

The solution set for a linear inequality is a section of the coordinate plane. The boundary for the section is found by replacing the inequality symbol with an equal sign and graphing the resulting equation. The boundary is included in the solution set (and represented with a solid line) if the inequality symbol used originally is $\leq$ or $\geq$. The boundary is not included (and is represented with a dotted line) if the original symbol is $<$ or $>$.

Let's look at some examples.

▼ **Example 1** Graph the solution set for $x + y \leq 4$.

Solution The boundary for the graph is the graph of $x + y = 4$; the x- and y-intercepts are both 4. (Remember, the x-intercept is found by letting $y = 0$, and the y-intercept by letting $x = 0$.) The boundary is included in the solution set because the inequality symbol is $\leq$.

Here is the graph of the boundary:

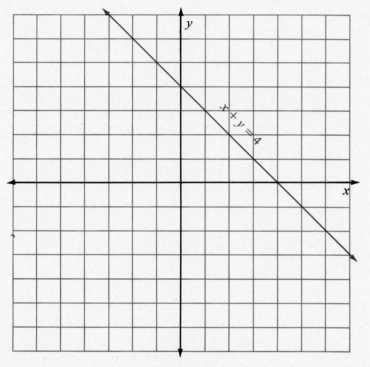

Figure 7-11

The boundary separates the coordinate plane into two sections or regions—the region above the boundary and the region below the boundary. The solution set for $x + y \leq 4$ is one of these two regions along with the boundary. To find the correct region, we simply choose any convenient point that is *not* on the boundary. We then substitute the coordinates of the point into the original inequality $x + y \leq 4$. If the point we choose satisfies the inequality, then it is a member of the solution set, and we can assume that all points on the same side of the boundary as the chosen point are also in the solution set. If the coordinates of our point do not satisfy the original inequality, then the solution set lies on the other side of the boundary.

In this example, a convenient point off the boundary is the origin.

$$\text{Substituting} \quad (0, 0)$$
$$\text{into} \quad x + y \leq 4$$
$$\text{gives us} \quad 0 + 0 \leq 4$$
$$0 \leq 4 \qquad \leftarrow \text{A true statement}$$

Since the origin is a solution to the inequality $x + y \leq 4$, and the origin is below the boundary, all other points below the boundary are also solutions.

Here is the graph of $x + y \leq 4$:

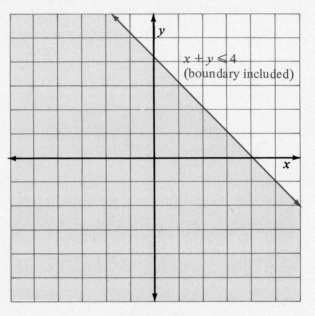

Figure 7-12

The region above the boundary is described by the inequality $x + y > 4$.   ▲

▼   Example 2   Graph all linear inequalities with boundary $x + y = 4$.

Solution   Extending the results from Example 1, we have four different graphs with boundary $x + y = 4$. We have the region above the boundary and the region below the boundary. The boundary can be included in the solution set or not included in the solution set (see Figures 7-13 through 7-16).

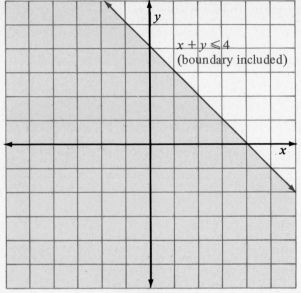

Figure 7-13

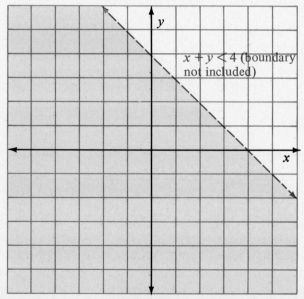

Figure 7-14

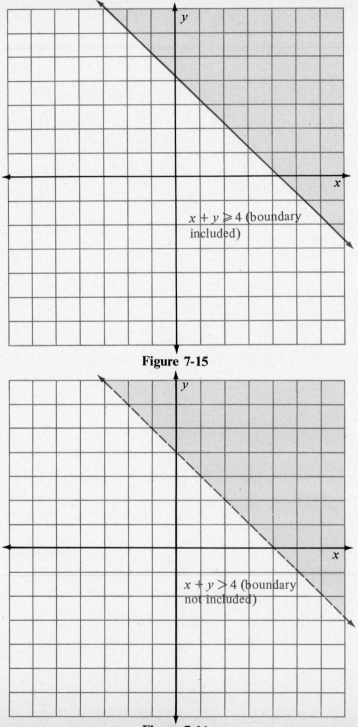

$x + y \geqslant 4$ (boundary included)

Figure 7-15

$x + y > 4$ (boundary not included)

Figure 7-16

Here is a list of steps to follow when graphing the solution set for linear inequalities in two variables:

Step 1. Replace the inequality symbol with an equal sign. The resulting equation represents the boundary for the solution set.

Step 2. Graph the boundary found in step 1 using a *solid line* if the boundary is included in the solution set (that is, if the original inequality symbol was either $\leq$ or $\geq$). Use a *broken line* to graph the boundary if it is *not* included in the solution set. (It is not included if the original inequality was either $<$ or $>$.)

Step 3. Choose any convenient point off the boundary and substitute the coordinates into the *original* inequality. If the resulting statement is *true,* the graph lies on the *same* side of the boundary as the chosen point. If the resulting statement is *false,* the solution set lies on the *opposite* side of the boundary.

▼   **Example 3**   Graph the solution set for $y < 2x - 3$.

Solution   The boundary is the graph of $y = 2x - 3$: a line with slope 2 and y-intercept -3. The boundary is not included since the original inequality symbol is $<$. Therefore, we use a broken line to represent the boundary (see Figure 7-17).

A convenient test point is again the origin:

$$\begin{aligned} \text{Using} \quad & (0, 0) \\ \text{in} \quad & y < 2x - 3, \\ \text{we have} \quad & 0 < 2(0) - 3 \\ & 0 < -3 \qquad \leftarrow \text{A false statement} \end{aligned}$$

Since our test point gives us a false statement and it lies above the boundary, the solution set must lie on the other side of the boundary.

The complete graph is shown in Figure 7-18.   ▲

▼   **Example 4**   Graph the solution set for $x \leq 5$.

Solution   The boundary is $x = 5$, which is a vertical line. All points to the left have x-coordinates less than 5 and all points to the right have x-coordinates greater than 5 (see Figure 7-19).   ▲

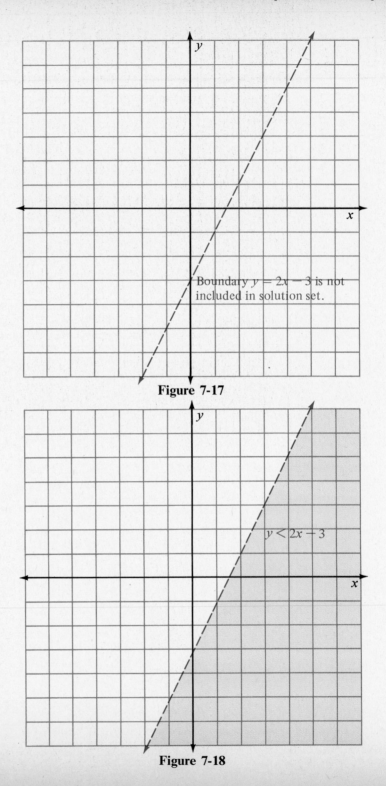

Figure 7-17

Figure 7-18

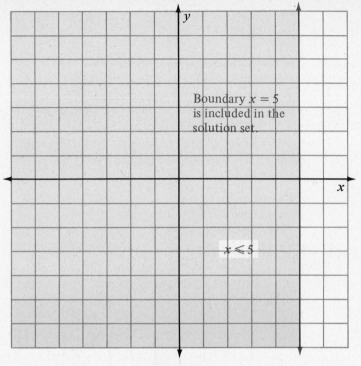

Boundary $x = 5$ is included in the solution set.

$x \leqslant 5$

Figure 7-19

Graph the solution set for each of the following: *Problem Set 7.4*

1. $x + y < 5$ 2. $x + y \leq 5$
3. $x - y \geq -3$ 4. $x - y > -3$
5. $2x + 3y \leq 6$ 6. $2x - 3y > -6$
7. $x - 2y < 4$ 8. $x + 2y > -4$
9. $2x + y < 5$ 10. $2x + y < -5$
11. $3x + 5y \geq 7$ 12. $3x - 5y > 7$
13. $y < 2x - 1$ 14. $y \geq 2x - 1$
15. $y \geq -3x - 4$ 16. $y < -3x + 4$
17. $x > 3$ 18. $x \leq -2$
19. $y \leq 4$ 20. $y > -5$

Find all graphs with the following boundaries. For each graph give the corresponding inequality. (There are four graphs for each boundary.)

21. $x + y = 2$ 22. $x - y = 3$
23. $2x + y = 6$ 24. $3x + y = 6$
25. $4x + 5y = 20$ 26. $3x + 4y = 12$
27. $y = x + 1$ 28. $y = -x + 1$
29. $y = 2x + 5$ 30. $y = 3x - 2$

31. $y = -3$ **32.** $y = 5$
33. $x = 2$ **34.** $x = -2$

35. Give the inequality whose graph lies above a boundary with slope -2 and y-intercept 4, if the boundary is included.
36. Give the inequality whose graph lies above a boundary with slope $+3$ and y-intercept -1, if the boundary is not included.
37. Give the inequality whose graph lies below a boundary with x-intercept 2 and y-intercept -3, if the boundary is not included.
38. Give the inequality whose graph lies below a boundary with x-intercept -3 and y-intercept 2, if the boundary is included.

**7.5
Variation**

There are two main types of variation—direct variation and inverse variation. Variation problems are most common in the sciences, particularly in chemistry and physics. They are also found in nonscience subjects such as business and economics. For the most part, variation problems are just a matter of translating specific English phrases into algebraic equations.

Direct Variation

We say the variable y varies *directly* with the variable x if y increases as x increases (which is the same as saying y decreases as x decreases). A change in one variable produces a corresponding and similar change in the other variable. Algebraically, the relationship is written $y = Kx$ where K is a nonzero constant called the *constant of variation* (proportionality constant).

Another way of saying y varies directly with x is to say y is *directly proportional* to x.

Study the following list. It gives the mathematical equivalent of some direct-variation statements.

English phrase	Algebraic equation
y varies directly with x	$y = Kx$
s varies directly with the square of t	$s = Kt^2$
y is directly proportional to the cube of z	$y = Kz^3$
u is directly proportional to the square root of v	$u = K\sqrt{v}$

Here are some sample problems involving direct-variation statements.

▼   **Example 1** y varies directly with x. If y is 15 when x is 5, find y when x is 7.

Solution The first sentence gives us the general relationship between x and y. The equation equivalent to the statement "y varies directly with x" is

$$y = Kx$$

The first part of the second sentence in our example gives us the information necessary to evaluate the constant K:

$$
\begin{array}{rl}
\text{When} & y = 15 \\
\text{and} & x = 5, \\
\text{the equation} & y = Kx \\
\text{becomes} & 15 = K \cdot 5 \\
\text{or} & K = 3
\end{array}
$$

The equation can now be written specifically as

$$y = 3x$$

Letting $x = 7$, we have

$$
\begin{array}{l}
y = 3 \cdot 7 \\
y = 21
\end{array}
$$ ▲

Almost every problem involving variation can be solved by the above procedure. That is, begin by writing a general variation equation that gives the relationship between the variables. Next, use the information in the problem to evaluate the constant of variation. This results in a specific variation equation. Finally, use the specific equation and the rest of the information in the problem to find the quantity asked for.

▼ **Example 2** The distance a body falls from rest toward the earth is directly proportional to the square of the time it has been falling. If a body falls 64 feet in 2 seconds, how far will it fall in 3.5 seconds?

Solution We will let $d =$ distance and $t =$ time. Since distance is directly proportional to the square of time, we have

$$d = Kt^2$$

Next we evaluate the constant K:

$$
\begin{array}{rl}
\text{When} & t = 2 \\
\text{and} & d = 64, \\
\text{the equation} & d = Kt
\end{array}
$$

$$\text{becomes} \quad 64 = K(2)^2$$
$$\text{or} \quad 64 = 4K$$
$$\text{and} \quad K = 16$$

Specifically, then, the relationship between d and t is

$$d = 16t^2$$

Finally, we find d when $t = 3.5$:

$$d = 16(3.5)^2$$
$$d = 16(12.25)$$
$$d = 196$$

After 3.5 seconds the body will have fallen 196 feet.   ▲

We should note in the above example that the equation $d = 16t^2$ is a specific example of the equation used in physics to find the distance a body falls from rest after a time t. In physics, the equation is $d = \frac{1}{2}gt^2$, where g is the acceleration of gravity. On earth the acceleration of gravity is $g = 32$ ft/sec^2. The equation, then, is $d = \frac{1}{2} \cdot 32t^2$ or $d = 16t^2$.

Inverse Variation

If two variables are related so that an *increase* in one produces a proportional *decrease* in the other, then the variables are said to *vary inversely*. If y varies inversely with x, then $y = K\dfrac{1}{x}$ or $y = \dfrac{K}{x}$. We can also say y is inversely proportional to x. The constant K is again called the *constant of variation or proportionality constant*.

The following list gives the relationship between some inverse-variation statements and their corresponding equations:

English phrase	Algebraic equation
y is inversely proportional to x	$y = \dfrac{K}{x}$
s varies inversely with the square of t	$s = \dfrac{K}{t^2}$
y is inversely proportional to x^4	$y = \dfrac{K}{x^4}$
z varies inversely with the cube root of t	$z = \dfrac{K}{\sqrt[3]{t}}$

The procedure used to solve inverse-variation problems is the same as that used to solve direct-variation problems.

▼   **Example 3** y varies inversely with the square of x. If y is 4 when x is 5, find y when x is 10.

Solution Since y is inversely proportional to the square of x, we can write

$$y = \frac{K}{x^2}$$

Evaluating K using the information given, we have

$$\text{When} \quad x = 5$$
$$\text{and} \quad y = 4,$$

$$\text{the equation} \quad y = \frac{K}{x^2}$$

$$\text{becomes} \quad 4 = \frac{K}{5^2}$$

$$\text{or} \quad 4 = \frac{K}{25}$$

$$\text{and} \quad K = 100$$

Now we write the equation again as

$$y = \frac{100}{x^2}$$

We finish by substituting $x = 10$ into the last equation:

$$y = \frac{100}{10^2}$$

$$y = \frac{100}{100}$$

$$y = 1 \qquad\qquad\qquad ▲$$

▼   **Example 4** The volume of a gas is inversely proportional to the pressure of the gas on its container. If a pressure of 48 pounds per square inch corresponds to a volume of 50 cubic feet, what pressure is needed to produce a volume of 100 cubic feet?

Solution We can represent volume with V and pressure by P:

$$V = \frac{K}{P}$$

Using $P = 48$ and $V = 50$, we have

$$50 = \frac{K}{48}$$

$$K = 50(48)$$

$$K = 2400$$

The equation that describes the relationship between P and V is

$$V = \frac{2400}{P}$$

Substituting $V = 100$ into this last equation, we get

$$100 = \frac{2400}{P}$$

$$100P = 2400$$

$$P = \frac{2400}{100}$$

$$P = 24$$

A volume of 100 cubic feet is produced by a pressure of 24 pounds per square inch.   ▲

The relationship between pressure and volume as given in the example above is known in chemistry as Boyle's law. It was Robert Boyle who, in 1662, published the results of some of his experiments which showed, among other things, that the volume of a gas decreases as the pressure increases. This is known as inverse variation.

Joint Variation and Other Variation Combinations

Many times relationships among different quantities are described in terms of more than two variables. If the variable y varies directly with *two* other variables, say x and z, then we say y varies *jointly* with x and z. In addition to joint variation, there are many other combinations of direct and inverse variation involving more than two variables. The following table is a list of some variation statements and their equivalent mathematical form:

English phrase	Algebraic equation
y varies jointly with x and z	$y = Kxz$
z varies jointly with r and the square of s	$z = Krs^2$

English phrase	Algebraic equation
V is directly proportional to T and inversely proportional to P	$V = \dfrac{KT}{P}$
F varies jointly with m_1 and m_2 and inversely with the square of r	$F = \dfrac{Km_1 \cdot m_2}{r^2}$

▼   **Example 5** y varies jointly with x and the square of z. When x is 5 and z is 3, y is 180. Find y when x is 2 and z is 4.

Solution   The general equation is given by

$$y = Kxz^2$$

Substituting $x = 5$, $z = 3$, and $y = 180$, we have

$$180 = K(5)(3)^2$$
$$180 = 45K$$
$$K = 4$$

The specific equation is

$$y = 4xz^2$$

When $x = 2$ and $z = 4$, the last equation becomes

$$y = 4(2)(4)^2$$
$$y = 128$$ ▲

▼   **Example 6**   In electricity, the resistance of a cable is directly proportional to its length and inversely proportional to the square of the diameter. If a 100-foot cable 0.5 inch in diameter has a resistance of 0.2 ohm, what will be the resistance of a cable made from the same material if it is 200 feet long with a diameter of 0.25 inch?

Solution   Let R = resistance, l = length, and d = diameter. The equation is

$$R = \frac{Kl}{d^2}$$

When $R = 0.2$, $l = 100$, and $d = 0.5$, the equation becomes

$$0.2 = \frac{K(100)}{(0.5)^2}$$

or $K = 0.0005$

Using this value of K in our original equation, the result is

$$R = \frac{0.0005l}{d^2}$$

When $l = 200$ and $d = 0.25$, the equation becomes

$$R = \frac{0.0005(200)}{(0.25)^2}$$

$$R = 1.6$$

The resistance is 1.6 ohms.   ▲

Problem Set 7.5

For the following problems, y varies directly with x:

1. If y is 10 when x is 2, find y when x is 6.
2. If y is 20 when x is 5, find y when x is 3.
3. If y is -32 when x is 4, find x when y is -40.
4. If y is -50 when x is 5, find x when y is -70.
5. If y is 33 when x is 8, find y when x is 11.
6. If y is 26 when x is 3, find y when x is 4.

For the following problems, r is inversely proportional to s:

7. If r is -3 when s is 4, find r when s is 2.
8. If r is -10 when s is 6, find r when s is -5.
9. If r is 8 when s is 3, find s when r is 48.
10. If r is 12 when s is 5, find s when r is 30.
11. If r is 15 when s is 4, find r when s is 5.
12. If r is 21 when s is 10, find r when s is 8.

For the following problems, d varies directly with the square root of r:

13. If $d = 10$, when $r = 25$, find d when $r = 16$.
14. If $d = 12$, when $r = 36$, find d when $r = 49$.
15. If $d = 10\sqrt{2}$, when $r = 50$, find d when $r = 8$.
16. If $d = 6\sqrt{3}$, when $r = 12$, find d when $r = 75$.

For the following problems, y varies inversely with the square of x:

17. If $y = 45$ when $x = 3$, find y when x is 5.
18. If $y = 12$ when $x = 2$, find y when x is 6.
19. If $y = 10$ when $x = 6$, find x when y is 20.
20. If $y = 13$ when $x = 4$, find x when y is 30.

For the following problems, z varies jointly with x and the square of y:

21. If z is 54 when x and y are 3, find z when $x = 2$ and $y = 4$.
22. If z is 80 when x is 5 and y is 2, find z when $x = 2$ and $y = 5$.

23. If z is 64 when $x = 1$ and $y = 4$, find x when $z = 32$ and $y = 1$.

24. If z is 27 when $x = 6$ and $y = 3$, find x when $z = 50$ and $y = 4$.

25. The length a spring stretches is directly proportional to the force applied. If a force of 5 pounds stretches a spring 3 inches, how much force is necessary to stretch the same spring 10 inches?

26. The weight of a certain material varies directly with the surface area of that material. If 8 square feet weighs half a pound, how much will 10 square feet weigh?

27. The volume of a gas is inversely proportional to the pressure. If a pressure of 36 pounds per square inch corresponds to a volume of 25 cubic feet, what pressure is needed to produce a volume of 75 cubic feet?

28. The frequency of an electromagnetic wave varies inversely with the length. If a wave of length 200 meters has a frequency of 800 kilocycles per second, what frequency will be associated with a wave of length 500 meters?

29. The surface area of a hollow cylinder varies jointly with the height and radius of the cylinder. If a cylinder with radius 3 inches and height 5 inches has a surface area of 94 square inches, what is the surface area of a cylinder with radius 2 inches and height 8 inches?

30. The capacity of a cylinder varies jointly with the height and the square of the radius. If a cylinder with radius of 3 cm (centimeters) and a height of 6 cm has a capacity of 3 cm^3 (cubic centimeters), what will be the capacity of a cylinder with radius 4 cm and height 9 cm?

31. The resistance of a wire varies directly with the length and inversely with the square of the diameter. If 100 feet of wire with diameter 0.01 inch has a resistance of 10 ohms, what is the resistance of 60 feet of the same type of wire if its diameter is 0.02 inch?

32. The volume of a gas varies directly with its temperature and inversely with the pressure. If the volume of a certain gas is 30 cubic feet at a temperature of 300°K and a pressure of 20 pounds per square inch, what is the volume of the same gas at 340°K when the pressure is 30 pounds per square inch?

Chapter 7
Summary and Review

Linear Equations in Two Variables

A linear equation in two variables is any equation that can be put in the form $ax + by = c$. The graph of every linear equation is a straight line.

Intercepts

The x-intercept of an equation is the x-coordinate of the point where the graph crosses the x-axis. The y-intercept is the y-coordinate of the point where the graph crosses the y-axis. We find the y-intercept by substituting $x = 0$ into the equation and solving for y. The x-intercept is found by letting $y = 0$ and solving for x.

The Slope of a Line

The slope of the line containing points (x_1, y_1) and (x_2, y_2) is given by

$$\text{Slope} = m = \frac{\text{rise}}{\text{run}} = \frac{y_2 - y_1}{x_2 - x_1}$$

Horizontal lines have 0 slope, and vertical lines have no slope.
Parallel lines have equal slopes, and perpendicular lines have slopes which are negative reciprocals.

The Slope-Intercept Form of a Straight Line

The equation of a line with slope m and y-intercept b is given by

$$y = mx + b$$

The Point-Slope Form of a Straight Line

The equation of a line through (x_1, y_1) that has a slope of m can be written as

$$y - y_1 = m(x - x_1)$$

Linear Inequalities in Two Variables

An inequality of the form $ax + by < c$ is a linear inequality in two variables. The equation for the boundary of the solution set is given by $ax + by = c$. (This equation is found by simply replacing the inequality symbol with an equal sign.)

To graph a linear inequality, first graph the boundary. Next, choose any point off the boundary and substitute its coordinates into the original inequality. If the resulting statement is true, the graph lies on the same side of the boundary as the test point. A false statement indicates that the solution set lies on the other side of the boundary.

Variation

If y varies directly with x (y is directly proportional to x), then we say:

$$y = Kx$$

If y varies inversely with x (y is inversely proportional to x), then we say:

$$y = \frac{K}{x}$$

If z varies jointly with x and y (z is directly proportional to both x and y), then we say:

$$z = Kxy$$

In each case, K is called the constant of variation.

For each of the following straight lines, identify the x-intercept, y-intercept, and slope, and sketch the graph:

1. $2x + y = 6$
2. $3x - 2y = 5$
3. $y = -2x - 3$
4. $y = \frac{3}{2}x + 4$
5. $x = -2$
6. $y = 3$

Graph the following linear inequalities:

7. $x + y \geq 4$
8. $3x - 4y < 12$
9. $y \leq -x + 2$
10. $x > 5$

11. Give the equation of the line through $(-1, 3)$ that has slope $m = 2$.
12. Give the equation of the line through $(-3, 2)$ and $(4, -1)$.
13. Line l contains the point $(5, -3)$ and has a graph parallel to the graph of $2x - 5y = 10$. Find the equation for l.
14. Line l contains the point $(-1, -2)$ and has a graph perpendicular to the graph of $y = 3x - 1$. Find the equation for l.
15. Give the equation of the vertical line through $(4, -7)$.
16. Quantity y varies directly with the square of x. If y is 50 when x is 5, find y when x is 3.
17. Quantity z varies jointly with x and the square root of y. If z is 15 when x is 5 and y is 36, find z when x is 2 and y is 25.
18. The maximum load (L) a horizontal beam can safely hold varies jointly with the width (w) and the square of the depth (d) and inversely with the length (l). If a 10-foot beam with width 3 and depth 4 will safely hold up to 800 pounds load, how many pounds will a 12-foot beam with width 3 and depth 4 hold?

8

Systems of
Linear Equations

To the student:

In Chapter 7 we did some work with linear equations in two variables. In this chapter we will extend our work with linear equations to include systems of linear equations in two and three variables.

Systems of linear equations are used extensively in many different disciplines. Systems of linear equations can be used to solve multiple-loop circuit problems in electronics, kinship patterns in anthropology, genetics problems in biology, and profit-and-cost problems in economics. There are many other applications as well.

We will begin this chapter by looking at three different methods of solving linear systems in two variables. We will then extend two of these methods to include solutions to systems in three variables. A fourth method of solving linear systems involves what are known as determinants. The chapter ends with a look at some word problems whose solutions depend on linear systems.

To be successful in this chapter you should be familiar with the concepts in Chapter 7 as well as the process of solving a linear equation in one variable.

8.1
Systems of Linear Equations in Two Variables

In Chapter 7 we found the graph of an equation of the form $ax + by = c$ to be a straight line. Since the graph is a straight line, the equation is said to be a linear equation. Two linear equations considered together form a *linear system* of equations. For example,

$$3x - 2y = 6$$
$$2x + 4y = 20$$

252

is a linear system. The solution set to the system is the set of all ordered pairs that satisfy both equations. If we graph each equation on the same set of axes, we can see the solution set (see Figure 8-1).

The point (4, 3) lies on both lines and therefore must satisfy both equations. It is obvious from the graph that it is the only point that does so. The solution set for the system is $\{(4, 3)\}$.

More generally, if $a_1x + b_1y = c_1$ and $a_2x + b_2y = c_2$ are linear equations, then the solution set for the system

$$a_1x + b_1y = c_1$$
$$a_2x + b_2y = c_2$$

can be illustrated through one of the following graphs:

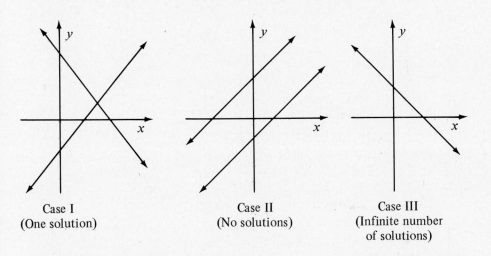

Case I   Case II   Case III
(One solution)   (No solutions)   (Infinite number
                                                          of solutions)

Case I. The two lines intersect at one and only one point. The coordinates of the point give the solution to the system. This is what usually happens.

Case II. The lines are parallel and therefore have no points in common. The solution set to the system is $\varnothing$. In this case, we say the equations are *inconsistent*.

Case III. The lines coincide. That is, their graphs represent the same line. The solution set consists of all ordered pairs that satisfy either equation. In this case, the equations are said to be *dependent*.

In the beginning of this section we found the solution set for the system

$$3x - 2y = 6$$
$$2x + 4y = 20$$

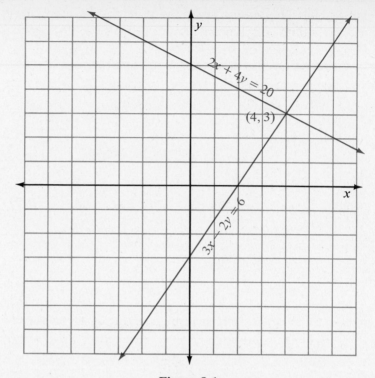

Figure 8-1

by graphing each equation and then reading the solution set from the graph. Solving a system of linear equations by graphing is the least accurate method. If the coordinates of the point of intersection are not integers, it can be very difficult to read the solution set from the graph. There is another method of solving a linear system that does not depend on the graph. It is called the *elimination method* and depends on the following theorem.

THEOREM 8.1 If the ordered pair (x, y) satisfies the equations in the system

$$a_1x + b_1y = c_1$$
$$a_2x + b_2y = c_2$$

then it will also satisfy the equation

$$A(a_1x + b_1y) + B(a_2x + b_2y) = Ac_1 + Bc_2$$

for any two real numbers A and B.

To prove this theorem we simply substitute c_1 for $a_1 x + b_1 y$ and c_2 for $a_2 x + b_2 y$ in the last equation to get the true statement

$$Ac_1 + Bc_2 = Ac_1 + Bc_2$$

Although Theorem 8.1 may look complicated, it simply indicates that the single equation that results from adding the left and right sides of the equations in our system has the same solution as the original system. We already know that the solution set for either equation in our system will not be changed by multiplying both sides of the equation by the same nonzero real number.

Here are some examples illustrating how we use Theorem 8.1 to find the solution set for a system of linear equations.

▼ **Example 1** Solve the system

$$4x + 3y = 10$$
$$2x + y = 4$$

Solution If we multiply the bottom equation by -3, the coefficients of y in the resulting equation and the top equation will be opposites:

$$
\begin{array}{lll}
4x + 3y = 10 & \xrightarrow{\text{No change}} & 4x + 3y = 10 \\
2x + y = 4 & \xrightarrow[\text{Multiply by } -3]{} & -6x - 3y = -12
\end{array}
$$

Adding the left and right sides of the resulting equations, according to Theorem 8.1, we have

$$
\begin{array}{rr}
4x + 3y = & 10 \\
-6x - 3y = & -12 \\
\hline
-2x = & -2
\end{array}
$$

The result is a linear equation in one variable. We have eliminated the variable y from the equations. (It is for this reason we call this method of solving a linear system the *elimination method*.) Solving $-2x = -2$ for x, we have

$$x = 1$$

This is the x-coordinate of the solution to our system. To find the y-coordinate, we substitute $x = 1$ into any of the equations containing both the variables x and y. Let's try the second equation in our original system:

$$2(1) + y = 4$$
$$2 + y = 4$$
$$y = 2$$

This is the y-coordinate of the solution to our system. The ordered pair $(1, 2)$ is the solution to the system.   ▲

▼   **Example 2**   Solve the system

$$3x - 5y = -2$$
$$2x - 3y = 1$$

Solution   We can eliminate either variable. Let's decide to eliminate the variable x. We can do so by multiplying the top equation by 2 and the bottom equation by -3, and then adding the left and right sides of the resulting equations:

$$
\begin{array}{ll}
3x - 5y = -2 & \xrightarrow{\text{Multiply by 2}} \\
2x - 3y = 1 & \xrightarrow[\text{Multiply by } -3]{}
\end{array}
\quad
\begin{array}{l}
6x - 10y = -4 \\
-6x + 9y = -3 \\
\hline
-y = -7 \\
y = 7
\end{array}
$$

The y-coordinate of the solution to the system is 7. Substituting this value of y into any of the equations with both x- and y-variables gives $x = 11$. The solution to the system is $(11, 7)$. It is the only ordered pair that satisfies both equations.   ▲

▼   **Example 3**   Solve the system

$$5x - 2y = 1$$
$$-10x + 4y = 3$$

Solution   We can eliminate y by multiplying the first equation by 2 and adding the result to the second equation:

$$
\begin{array}{ll}
5x - 2y = 1 & \xrightarrow{\text{Multiply by 2}} \\
-10x + 4y = 3 & \xrightarrow[\text{No change}]{}
\end{array}
\quad
\begin{array}{l}
10x - 4y = 2 \\
-10x + 4y = 3 \\
\hline
0 = 5
\end{array}
$$

The result is the false statement $0 = 5$, which indicates there is no solution to the system. The equations are said to be *inconsistent;*

their graphs are parallel lines. Whenever both variables have been eliminated and the resulting statement is false, the solution set for the system will be $\varnothing$.   ▲

▼ **Example 4** Solve the system

$$4x - 3y = 2$$
$$8x - 6y = 4$$

Solution Multiplying the top equation by -2 and adding, we can eliminate the variable x:

$$
\begin{array}{ll}
4x - 3y = 2 & \xrightarrow{\text{Multiply by } -2} \quad -8x + 6y = -4 \\
8x - 6y = 4 & \xrightarrow{\text{No change}} \quad \underline{8x - 6y = 4} \\
& \phantom{\xrightarrow{}} \qquad\qquad\quad 0 = 0
\end{array}
$$

Both variables have been eliminated and the resulting statement $0 = 0$ is true. In this case the lines coincide and the equations are said to be *dependent*. The solution set consists of all ordered pairs that satisfy either equation. We can write the solution set as $\{(x, y) | 4x - 3y = 2\}$ or $\{(x, y) | 8x - 6y = 4\}$.   ▲

The last two examples illustrate the two special cases in which the graphs of the equations in the system either coincide or are parallel. In both cases the left-hand sides of the equations were multiples of one another. In the case of the dependent equations, the right-hand sides were also multiples. We can generalize these observations as follows:
The equations in the system

$$a_1 x + b_1 y = c_1$$
$$a_2 x + b_2 y = c_2$$

will be inconsistent (their graphs are parallel lines) if

$$\frac{a_1}{a_2} = \frac{b_1}{b_2} \neq \frac{c_1}{c_2}$$

and will be dependent (their graphs will coincide) if

$$\frac{a_1}{a_2} = \frac{b_1}{b_2} = \frac{c_1}{c_2}$$

We end this section by considering another method of solving a linear system. The method is called the *substitution* method and is shown in the following example.

▼ **Example 5** Solve the system

$$2x - 3y = -6$$
$$y = 3x - 5$$

Solution The second equation tells us y *is* $3x - 5$. Substituting the expression $3x - 5$ for y in the first equation, we have

$$2x - 3(3x - 5) = -6$$

The result of the substitution is the elimination of the variable y. Solving the resulting linear equation in x as usual, we have

$$2x - 9x + 15 = -6$$
$$-7x + 15 = -6$$
$$-7x = -21$$
$$x = 3$$

Putting $x = 3$ into the second equation in the original system, we have

$$y = 3(3) - 5$$
$$= 9 - 5$$
$$= 4$$

The solution to the system is (3, 4).   ▲

▼ **Example 6** Solve by substitution:

$$2x + 3y = 5$$
$$x - 2y = 6$$

Solution In order to use the substitution method we must solve one of the two equations for x or y. We can solve for x in the second equation by adding $2y$ to both sides:

$$x - 2y = 6$$
$$x = 2y + 6 \qquad \text{Add } 2y \text{ to both sides}$$

Substituting the expression $2y + 6$ for x in the first equation of our system, we have

$$2(2y + 6) + 3y = 5$$
$$4y + 12 + 3y = 5$$
$$7y + 12 = 5$$
$$7y = -7$$
$$y = -1$$

Using $y = -1$ in either equation in the original system, we find $x = 4$. The solution is $(4, -1)$.   ▲

Solve each system by graphing both equations on the same set of axes and then reading the solution from the graph: *Problem Set 8.1*

1. $x + y = 3$
    $x - y = 3$
2. $x + y = -2$
    $x - y = 6$
3. $4x + 5y = 3$
    $2x - 5y = 9$
4. $6x - y = 4$
    $2x + y = 4$
5. $x + y = 4$
    $-2x - 2y = 3$
6. $2x - y = 5$
    $4x - 2y = 10$
7. $y = 2x + 3$
    $y = 3x + 4$
8. $y = 3x - 1$
    $y = 2x + 1$

Solve each of the following systems by the elimination method:

9. $x + y = 5$
    $3x - y = 3$
10. $x - y = 4$
     $-x + 2y = -3$
11. $3x + y = 4$
     $4x + y = 5$
12. $6x - 2y = -10$
     $6x + 3y = -1$
13. $3x - 2y = 6$
     $6x - 4y = 12$
14. $4x + 5y = -3$
     $-8x - 10y = 3$
15. $3x - 5y = 7$
     $-x + y = -1$
16. $4x + 2y = 32$
     $x + y = -2$
17. $x + 2y = 0$
     $2x - y = 0$
18. $x + 3y = 9$
     $2x - y = 4$
19. $2x + y = 5$
     $5x + 3y = 11$
20. $5x + 2y = 11$
     $7x - y = 10$
21. $4x + 3y = 14$
     $9x - 2y = 14$
22. $7x - 6y = 13$
     $6x - 5y = 11$
23. $2x - 5y = 3$
     $-4x + 10y = 3$
24. $3x - 2y = 1$
     $-6x + 4y = -2$
25. $2x + 3y = 8$
     $3x - 4y = -5$
26. $3x - 4y = -1$
     $6x - 5y = 10$
27. $\frac{1}{2}x + \frac{1}{3}y = 13$
     $\frac{1}{5}x + \frac{1}{8}y = 5$
28. $\frac{1}{2}x + \frac{1}{3}y = \frac{2}{3}$
     $\frac{1}{3}x + \frac{1}{5}y = \frac{7}{15}$
29. $\frac{1}{3}x + \frac{1}{5}y = 2$
     $\frac{1}{3}x - \frac{1}{2}y = -\frac{1}{3}$
30. $\frac{1}{2}x - \frac{1}{3}y = \frac{5}{6}$
     $-\frac{1}{5}x + \frac{1}{4}y = -\frac{9}{20}$

Solve each of the following systems by the substitution method:

31. $y = x + 3$
     $x + y = 3$
32. $y = x - 5$
     $2x - 6y = -2$
33. $x - y = 4$
     $2x - 3y = 6$
34. $x + y = 3$
     $2x + 3y = -4$

35. $y = 3x - 2$
    $y = 4x - 4$.
36. $y =\ \ \ 5x - 2$
    $y = -2x + 5$
37. $7x -\ \ y = 24$
    $x = 2y + 9$
38. $3x - y = -8$
    $y = 6x + 3$
39. $2x -\ \ y = 5$
    $4x - 2y = 10$
40. $\ \ \ \ 5x - 4y = 3$
    $-10x + 8y = -6$

Solve each of the following systems by letting $a = 1/x$ and $b = 1/y$. Solve by any method you choose.

41. $\dfrac{2}{x} - \dfrac{1}{y} = 1$

    $\dfrac{1}{x} + \dfrac{1}{y} = 2$

Hint: If we let $a = 1/x$ and $b = 1/y$, then this system becomes

$$2a - b = 1$$
$$a + b = 2$$

which can be solved very simply by either the elimination method or the substitution method.

42. $\dfrac{1}{x} + \dfrac{3}{y} = 9$

    $\dfrac{1}{x} - \dfrac{1}{y} = 1$

43. $\dfrac{2}{x} - \dfrac{3}{y} = 3$

    $\dfrac{2}{x} + \dfrac{2}{y} = 8$

44. $\dfrac{2}{x} - \dfrac{5}{y} = 10$

    $\dfrac{3}{x} + \dfrac{1}{y} = 15$

45. $\dfrac{1}{x} + \dfrac{3}{y} = 6$

    $\dfrac{2}{x} + \dfrac{6}{y} = 12$

46. $\dfrac{3}{x} - \dfrac{2}{y} = 5$

    $\dfrac{6}{x} - \dfrac{4}{y} = 3$

47. One telephone company charges 25 cents for the first minute and 15 cents for each additional minute for a certain long-distance phone call. If the number of additional minutes after the first minute is x and the cost, in cents, for the call is y, then the equation that gives the total cost, in cents, for the call is $y = 15x + 25$.

   a. Graph the equation $y = 15x + 25$.
   b. If a second phone company charges 22 cents for the first minute and 16 cents for each additional minute, write the equation that gives the total cost (y) of a call in terms of the number of additional minutes (x).

   c.   Graph the equation from part b on the same set of axes used in part a.

   d.   After how many additional minutes will the two companies charge an equal amount? (What is the x-coordinate of the point of intersection of the two lines?) To check the answer, solve the two equations by the elimination method or the substitution method.

A solution to an equation in three variables such as

$$2x + y - 3z = 6$$

is an ordered triple of numbers (x, y, z). For example, the ordered triples $(0, 0, -2)$, $(2, 2, 0)$, and $(0, 9, 1)$ are solutions to the equation $2x + y - 3z = 6$, since they produce a true statement when their coordinates are replaced for x, y, and z in the equation.

In this section we are concerned with solution sets to systems of linear equations in three variables.

DEFINITION   The solution set for a system of three linear equations in three variables is the set of ordered triples that satisfy all three equations.

We can solve a system in three variables by methods similar to those developed in the last section.

▼   **Example 1**   Solve the system

$$
\begin{aligned}
x + y + z &= 6 \qquad &(1)\\
2x - y + z &= 3 \qquad &(2)\\
x + 2y - 3z &= -4 \qquad &(3)
\end{aligned}
$$

Solution   We want to find the ordered triple (x, y, z) that satisfies all three equations. We have numbered the equations so it will be easier to keep track of where they are and what we are doing.

There are many ways to proceed. The main idea is to take two different pairs of equations and eliminate the same variable from each pair. We begin by adding equations (1) and (2) to eliminate the y-variable. The resulting equation is numbered (4):

$$
\begin{aligned}
x + y + z &= 6 \qquad &(1)\\
2x - y + z &= 3 \qquad &(2)\\
\hline
3x + 2z &= 9 \qquad &(4)
\end{aligned}
$$

Adding twice equation (2) to equation (3) will also eliminate the variable y. The resulting equation is numbered (5):

$$
\begin{array}{ll}
4x - 2y + 2z = 6 & \text{Twice (2)} \\
\underline{\;x + 2y - 3z = -4\;} & \text{(3)} \\
5x \qquad\;\; - z = 2 & \text{(5)}
\end{array}
$$

Equations (4) and (5) form a linear system in two variables. By multiplying equation (5) by 2 and adding the result to equation (4), we will succeed in eliminating the variable z from the new pair of equations:

$$
\begin{array}{ll}
3x + 2z = 9 & \text{(4)} \\
\underline{10x - 2z = 4\;} & \text{Twice (5)} \\
13x \qquad = 13 & \\
\;\;\; x = 1 &
\end{array}
$$

Substituting $x = 1$ into equation (4), we have

$$
\begin{aligned}
3(1) + 2z &= 9 \\
2z &= 6 \\
z &= 3
\end{aligned}
$$

Using $x = 1$ and $z = 3$ in equation (1) gives us

$$
\begin{aligned}
1 + y + 3 &= 6 \\
y + 4 &= 6 \\
y &= 2
\end{aligned}
$$

The solution set for the system is the ordered triple $\{(1, 2, 3)\}$. ▲

▼ **Example 2** Solve the system

$$
\begin{array}{ll}
2x + \;\;y - z = 3 & \text{(1)} \\
3x + 4y + z = 6 & \text{(2)} \\
2x - 3y + z = 1 & \text{(3)}
\end{array}
$$

Solution It is easiest to eliminate z from the equations. The equation produced by adding (1) and (2) is

$$
5x + 5y = 9 \qquad \text{(4)}
$$

The equation that results from adding (1) and (3) is

$$
4x - 2y = 4 \qquad \text{(5)}
$$

Equations (4) and (5) form a linear system in two variables. We can eliminate the variable y from this system as follows:

$$5x + 5y = 9 \xrightarrow{\text{Multiply by 2}} 10x + 10y = 18$$
$$4x - 2y = 4 \xrightarrow[\text{Multiply by 5}]{} \underline{20x - 10y = 20}$$
$$30x \qquad\quad = 38$$
$$x = \tfrac{38}{30}$$
$$x = \tfrac{19}{15}$$

Substituting $x = \tfrac{19}{15}$ into equation (5) or equation (4) and solving for y gives

$$y = \tfrac{8}{15}$$

Using $x = \tfrac{19}{15}$ and $y = \tfrac{8}{15}$ in equation (1), (2), or (3) and solving for z results in

$$z = \tfrac{1}{15}$$

The ordered triple that satisfies all three equations is $(\tfrac{19}{15}, \tfrac{8}{15}, \tfrac{1}{15})$.

▲

▼ **Example 3**   Solve the system

$$2x + 3y - z = 5 \qquad (1)$$
$$4x + 6y - 2z = 10 \qquad (2)$$
$$x - 4y + 3z = 5 \qquad (3)$$

Solution   Multiplying equation (1) by -2 and adding the result to equation (2) yields

$$-4x - 6y + 2z = -10 \qquad -2 \text{ times (1)}$$
$$\underline{4x + 6y - 2z = 10} \qquad (2)$$
$$0 = 0$$

All three variables have been eliminated, and we are left with a true statement. As was the case in Section 8.1, this implies that the two equations are dependent. There is no unique solution to the system.

▲

▼ **Example 4**   Solve the system

$$x - 5y + 4z = 8 \qquad (1)$$
$$3x + y - 2z = 7 \qquad (2)$$
$$-9x - 3y + 6z = 5 \qquad (3)$$

Solution   Multiplying equation (2) by 3 and adding the result to equation (3) produces

$$9x + 3y - 6z = 21 \qquad \text{3 times (2)}$$
$$\underline{-9x - 3y + 6z = 5} \qquad \text{(3)}$$
$$0 = 26$$

In this case all three variables have been eliminated, and we are left with a false statement. The two equations are inconsistent, there are no ordered triples that satisfy both equations. The solution set for the system is $\varnothing$. If equations (2) and (3) have no ordered triples in common, then certainly (1), (2), and (3) do not either. ▲

Note The graph of a linear equation in three variables is a plane in three-dimensional space. In Example 4 two of the equations are inconsistent, implying their graphs are parallel planes. In Example 3 we found two of the equations to be dependent, which implies that the two planes that represent the graphs of these equations coincide.

The graphs of dependent equations always coincide, while the graphs of inconsistent equations have no points in common.

▼ **Example 5**   Solve the system

$$x + 3y = 5 \qquad (1)$$
$$6y + \ z = 12 \qquad (2)$$
$$x - 2z = -10 \qquad (3)$$

Solution   It may be helpful to rewrite the system as

$$x + 3y \qquad\quad = 5 \qquad (1)$$
$$6y + \ z = 12 \qquad (2)$$
$$x \qquad\ - 2z = -10 \qquad (3)$$

Equation (2) does not contain the variable x. If we multiply equation (3) by -1 and add the result to equation (1), we will be left with another equation that does not contain the variable x:

$$x + 3y \qquad\quad = 5 \qquad (1)$$
$$\underline{-x \qquad\ + 2z = 10} \qquad \text{-1 times (3)}$$
$$3y + 2z = 15 \qquad (4)$$

Equations (2) and (4) form a linear system in two variables. Multiplying equation (2) by -2 and adding to equation (4) eliminates the variable z:

$$6y + z = 12 \xrightarrow{\text{Multiply by } -2} -12y - 2z = -24$$
$$3y + 2z = 15 \xrightarrow{\text{No change}} \underline{\quad 3y + 2z = \quad 15}$$
$$-9y \quad\quad = -9$$
$$y = 1$$

Using $y = 1$ in equation (4) and solving for z, we have

$$z = 6$$

Substituting $y = 1$ into equation (1) gives

$$x = 2$$

The ordered triple that satisfies all three equations is $(2, 1, 6)$. ▲

Solve the following systems: *Problem Set 8.2*

1. $x + y + z = 4$
$x - y + 2z = 1$
$x - y - z = -2$

2. $x - y - 2z = -1$
$x + y + z = 6$
$x + y - z = 4$

3. $x + y + z = 6$
$x - y + 2z = 7$
$2x - y - z = 0$

4. $x + y + z = 0$
$x + y - z = 6$
$x - y + z = -4$

5. $x + 2y + z = 3$
$2x - y + 2z = 6$
$3x + y - z = 5$

6. $2x + y - 3z = -14$
$x - 3y + 4z = 22$
$3x + 2y + z = 0$

7. $2x - y - 3z = 1$
$x + 2y + 4z = 3$
$4x - 2y - 6z = 2$

8. $3x + 2y + z = 3$
$x - 3y + z = 4$
$-6x - 4y - 2z = 1$

9. $2x - y + 3z = 4$
$x + 2y - z = 2$
$4x + 3y + z = 8$

10. $6x - 2y + z = 5$
$3x + y + 3z = 7$
$x + 4y - z = 4$

11. $x - 4y + 3z = 2$
$2x - 8y + 6z = -1$
$3x - y + z = 8$

12. $5x - 2y + z = 6$
$2x + 3y - 4z = 2$
$4x + 6y - 8z = 4$

13. $x + y = 9$
$y + z = 7$
$x - z = 2$

14. $x - y = -3$
$x + z = 2$
$y - z = 7$

15. $2x + y = 2$
$y + z = 3$
$4x - z = 0$

16. $2x + y = 6$
$3y - 2z = -8$
$x + z = 5$

17. $3x + 4y = 15$
$2x - 5z = -3$
$4y - 3z = 9$

18. $6x - 4y = 2$
$3y + 3z = 9$
$2x - 5z = -8$

19. $2x - y + 2z = -8$
$\ \ 3x - y - 4z = 3$
$\ \ \ x + 2y - 3z = 9$

20. $2x + 2y + 4z = 2$
$\ \ 2x - y - z = 0$
$\ \ 3x + y + 2z = 2$

21. In the following diagram of an electrical circuit, x, y, and z represent the amount of current (in amperes) flowing across the 5-ohm, 20-ohm, and 10-ohm resistors, respectively. (In circuit diagrams resistors are represented by $\wedge\!\wedge\!\wedge$ and potential differences by $-\!|\!\vdash\!-$.)

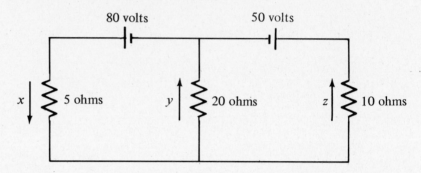

The system of equations used to find the three currents x, y, and z is

$$x - y - z = 0$$
$$5x + 20y = 80$$
$$20y - 10z = 50$$

Solve the system for all variables.

8.3
Introduction to
Determinants

In this section we will expand and evaluate determinants. The purpose of this section is simply to be able to find the value of a given determinant. As we will see in the next section, determinants are very useful in solving systems of linear equations. Before we apply determinants to systems of linear equations, however, we must practice calculating the value of some determinants. The problems in this section are rather mechanical in nature. We will have to wait until the next section to see the practical applications of the determinants developed in this section.

DEFINITION   The value of the 2×2 (2 by 2) determinant

$$\begin{vmatrix} a & c \\ b & d \end{vmatrix}$$

is given by

$$\begin{vmatrix} a & c \\ b & d \end{vmatrix} = ad - bc$$

From the definition above we see that a determinant is simply a square array of numbers with two vertical lines enclosing it. The value of a 2×2 determinant is found by cross-multiplying on the diagonals, a diagram of which looks like

$$\begin{vmatrix} a & c \\ b & d \end{vmatrix} = ad - bc$$

▼ **Example 1** Find the value of the following 2×2 determinants:

a. $\begin{vmatrix} 1 & 2 \\ 3 & 4 \end{vmatrix} = 1(4) - 3(2) = 4 - 6 = -2$

b. $\begin{vmatrix} 3 & 5 \\ -2 & 7 \end{vmatrix} = 3(7) - (-2)5 = 21 + 10 = 31$ ▲

▼ **Example 2** Solve for x if

$$\begin{vmatrix} -3 & x \\ 2 & x \end{vmatrix} = 20$$

Solution Applying the definition of a determinant to expand the left side, we have

$$-3(x) - 2(x) = 20$$
$$-5x = 20$$
$$x = -4$$ ▲

▼ **Example 3** Solve for x if

$$\begin{vmatrix} x^2 & 2 \\ x & 1 \end{vmatrix} = 8$$

Solution We expand the determinant on the left side to get

$$x^2(1) - x(2) = 8$$
$$x^2 - 2x = 8$$
$$x^2 - 2x - 8 = 0$$
$$(x - 4)(x + 2) = 0$$
$$x - 4 = 0 \quad \text{or} \quad x + 2 = 0$$
$$x = 4 \quad \text{or} \quad x = -2$$

The solution set is $\{4, -2\}$.   ▲

We now turn our attention to 3×3 determinants. A 3×3 determinant is also a square array of numbers, the value of which is given by the following definition.

DEFINITION The value of the 3×3 determinant

$$\begin{vmatrix} a_1 & b_1 & c_1 \\ a_2 & b_2 & c_2 \\ a_3 & b_3 & c_3 \end{vmatrix}$$

is given by

$$\begin{vmatrix} a_1 & b_1 & c_1 \\ a_2 & b_2 & c_2 \\ a_3 & b_3 & c_3 \end{vmatrix} = a_1 b_2 c_3 + a_3 b_1 c_2 + a_2 b_3 c_1 - a_3 b_2 c_1 - a_1 b_3 c_2 - a_2 b_1 c_3$$

At first glance, the expansion of a 3×3 determinant looks a little complicated. There are actually two different methods used to find the six products given above that simplify matters somewhat. The first method involves a process similar to the one used to find the value of a 2×2 determinant. It involves a cross-multiplication scheme.

Method 1

We begin by writing the determinant with the first two columns repeated on the right:

$$\begin{vmatrix} a_1 & b_1 & c_1 \\ a_2 & b_2 & c_2 \\ a_3 & b_3 & c_3 \end{vmatrix} \begin{matrix} a_1 & b_1 \\ a_2 & b_2 \\ a_3 & b_3 \end{matrix}$$

The positive products in the definition come from multiplying down the three full diagonals:

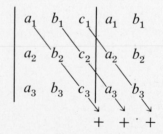

The negative products come from multiplying up the three full diagonals:

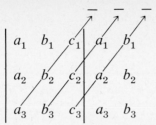

Check the products found by multiplying up and down the diagonals given here with the products given in the definition of a 3 × 3 determinant to see that they match.

▼   **Example 4** Find the value of

$$\begin{vmatrix} 1 & 3 & -2 \\ 2 & 0 & 1 \\ 4 & -1 & 1 \end{vmatrix}$$

Solution Repeating the first two columns and then finding the products up the diagonals and the products down the diagonals as given in Method 1, we have

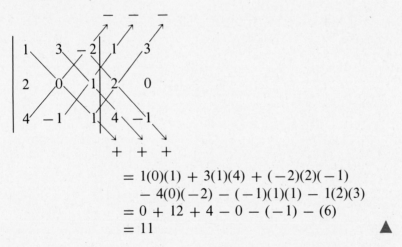

$$= 1(0)(1) + 3(1)(4) + (-2)(2)(-1)$$
$$- 4(0)(-2) - (-1)(1)(1) - 1(2)(3)$$
$$= 0 + 12 + 4 - 0 - (-1) - (6)$$
$$= 11 \qquad \blacktriangle$$

The second method of evaluating a 3 × 3 determinant is called *expansion by minors*.   *Method 2*

DEFINITION The *minor* for an element in a 3 × 3 determinant is the determinant consisting of the elements remaining when the row and column to which the element belongs are deleted. For example, in the determinant

$$\begin{vmatrix} a_1 & b_1 & c_1 \\ a_2 & b_2 & c_2 \\ a_3 & b_3 & c_3 \end{vmatrix}$$

$$\text{Minor for element } a_1 = \begin{vmatrix} b_2 & c_2 \\ b_3 & c_3 \end{vmatrix}$$

$$\text{Minor for element } b_2 = \begin{vmatrix} a_1 & c_1 \\ a_3 & c_3 \end{vmatrix}$$

$$\text{Minor for element } c_3 = \begin{vmatrix} a_1 & b_1 \\ a_2 & b_2 \end{vmatrix}$$

Before we can evaluate a 3 × 3 determinant by Method 2, we must first define what is known as the sign array for a 3 × 3 determinant.

DEFINITION   The *sign array* for a 3 × 3 determinant is a 3 × 3 array of signs in the following pattern:

$$\begin{vmatrix} + & - & + \\ - & + & - \\ + & - & + \end{vmatrix}$$

The sign array begins with a + sign in the upper left-hand corner. The signs then alternate between + and − across every row and down every column.

Note If you have read this far and are confused, hang on. After you have done a couple of examples you will find expansion by minors to be a fairly simple process. It just takes a lot of writing to explain it.

We can evaluate a 3 × 3 determinant by expanding across any row or down any column as follows:

Step 1. Choose a row or column to expand about.
Step 2. Write the product of each element in the row or column chosen in step 1 with its minor.
Step 3. Connect the three products in step 2 with the signs in the corresponding row or column in the sign array.

We will use the same determinant used in Example 4 to illustrate the procedure.

▼   **Example 5**   Expand across the first row:

$$\begin{vmatrix} 1 & 3 & -2 \\ 2 & 0 & 1 \\ 4 & -1 & 1 \end{vmatrix}$$

Solution   The products of the three elements in row 1 with their minors are

$$1\begin{vmatrix} 0 & 1 \\ -1 & 1 \end{vmatrix} \quad 3\begin{vmatrix} 2 & 1 \\ 4 & 1 \end{vmatrix} \quad (-2)\begin{vmatrix} 2 & 0 \\ 4 & -1 \end{vmatrix}$$

Connecting these three products with the signs from the first row of the sign array, we have

$$+1\begin{vmatrix} 0 & 1 \\ -1 & 1 \end{vmatrix} - 3\begin{vmatrix} 2 & 1 \\ 4 & 1 \end{vmatrix} + (-2)\begin{vmatrix} 2 & 0 \\ 4 & -1 \end{vmatrix}$$

We complete the problem by evaluating each of the three 2×2 determinants and then simplifying the resulting expression:

$$\begin{aligned} &+1[0 - (-1)] - 3(2 - 4) + (-2)(-2 - 0) \\ &= 1(1) - 3(-2) + (-2)(-2) \\ &= 1 + 6 + 4 \\ &= 11 \end{aligned}$$

The results of Examples 4 and 5 match. It makes no difference which method we use—the value of a 3×3 determinant is unique.   ▲

▼   **Example 6**   Expand down column 2:

$$\begin{vmatrix} 2 & 3 & -2 \\ 1 & 4 & 1 \\ 1 & 5 & -1 \end{vmatrix}$$

Solution   We connect the products of elements in column 2 and their minors with the signs from the second column in the sign array:

$$\begin{vmatrix} 2 & 3 & -2 \\ 1 & 4 & 1 \\ 1 & 5 & -1 \end{vmatrix} = -3\begin{vmatrix} 1 & 1 \\ 1 & -1 \end{vmatrix} + 4\begin{vmatrix} 2 & -2 \\ 1 & -1 \end{vmatrix} - 5\begin{vmatrix} 2 & -2 \\ 1 & 1 \end{vmatrix}$$

$$= -3(-1-1) + 4[-2-(-2)] - 5[2-(-2)]$$
$$= -3(-2) + 4(0) - 5(4)$$
$$= 6 + 0 - 20$$
$$= -14 \qquad \blacktriangle$$

Problem Set 8.3

Find the value of the following 2×2 determinants:

1. $\begin{vmatrix} 1 & 0 \\ 2 & 3 \end{vmatrix}$ 2. $\begin{vmatrix} 5 & 4 \\ 3 & 2 \end{vmatrix}$

3. $\begin{vmatrix} 2 & 1 \\ 3 & 4 \end{vmatrix}$ 4. $\begin{vmatrix} 4 & 1 \\ 5 & 2 \end{vmatrix}$

5. $\begin{vmatrix} 0 & 1 \\ 1 & 0 \end{vmatrix}$ 6. $\begin{vmatrix} 1 & 0 \\ 0 & 1 \end{vmatrix}$

7. $\begin{vmatrix} -3 & 2 \\ 6 & -4 \end{vmatrix}$ 8. $\begin{vmatrix} 8 & -3 \\ -2 & 5 \end{vmatrix}$

9. $\begin{vmatrix} -3 & -1 \\ 4 & -2 \end{vmatrix}$ 10. $\begin{vmatrix} 5 & 3 \\ 7 & -6 \end{vmatrix}$

Solve each of the following for x:

11. $\begin{vmatrix} 2x & 1 \\ x & 3 \end{vmatrix} = 10$ 12. $\begin{vmatrix} 3x & -2 \\ 2x & 3 \end{vmatrix} = 26$

13. $\begin{vmatrix} 1 & 2x \\ 2 & -3x \end{vmatrix} = 21$ 14. $\begin{vmatrix} -5 & 4x \\ 1 & -x \end{vmatrix} = 27$

15. $\begin{vmatrix} 2x & -4 \\ 2 & x \end{vmatrix} = -8x$ 16. $\begin{vmatrix} 3x & 2 \\ 2 & x \end{vmatrix} = -11x$

17. $\begin{vmatrix} x^2 & 3 \\ x & 1 \end{vmatrix} = 10$ 18. $\begin{vmatrix} x^2 & -2 \\ x & 1 \end{vmatrix} = 35$

Find the value of each of the following 3×3 determinants by using Method 1 of this section:

19. $\begin{vmatrix} 1 & 2 & 0 \\ 0 & 2 & 1 \\ 1 & 1 & 1 \end{vmatrix}$ 20. $\begin{vmatrix} -1 & 0 & 2 \\ 3 & 0 & 1 \\ 0 & 1 & 3 \end{vmatrix}$

21. $\begin{vmatrix} 1 & 2 & 3 \\ 3 & 2 & 1 \\ 1 & 1 & 1 \end{vmatrix}$ 22. $\begin{vmatrix} -1 & 2 & 0 \\ 3 & -2 & 1 \\ 0 & 5 & 4 \end{vmatrix}$

Find the value of each determinant by using Method 2 and expanding across the first row:

23. $\begin{vmatrix} 0 & 1 & 2 \\ 1 & 0 & 1 \\ -1 & 2 & 0 \end{vmatrix}$ **24.** $\begin{vmatrix} 3 & -2 & 1 \\ 0 & -1 & 0 \\ 2 & 0 & 1 \end{vmatrix}$

25. $\begin{vmatrix} 3 & 0 & 2 \\ 0 & -1 & -1 \\ 4 & 0 & 0 \end{vmatrix}$ **26.** $\begin{vmatrix} 1 & 1 & 1 \\ 1 & -1 & 1 \\ 1 & 1 & -1 \end{vmatrix}$

Find the value of each of the following determinants:

27. $\begin{vmatrix} 2 & -1 & 0 \\ 1 & 0 & -2 \\ 0 & 1 & 2 \end{vmatrix}$ **28.** $\begin{vmatrix} 5 & 0 & -4 \\ 0 & 1 & 3 \\ -1 & 2 & -1 \end{vmatrix}$

29. $\begin{vmatrix} 1 & 3 & 7 \\ -2 & 6 & 4 \\ 3 & 7 & -1 \end{vmatrix}$ **30.** $\begin{vmatrix} 2 & 1 & 5 \\ 6 & -3 & 4 \\ 8 & 9 & -2 \end{vmatrix}$

In this section we will use determinants to solve systems of linear equations in two and three variables. We begin by solving a general system of linear equations in two variables by the elimination method. The general system is

**8.4
Cramer's Rule**

$$a_1 x + b_1 y = c_1$$
$$a_2 x + b_2 y = c_2$$

We begin by eliminating the variable y from the system. To do so we multiply both sides of the first equation by b_2, and both sides of the second equation by $-b_1$:

$$a_1 b_2 x + b_1 b_2 y = c_1 b_2$$
$$-a_2 b_1 x - b_1 b_2 y = -c_2 b_1$$

We now add the two equations and solve the result for x:

$$(a_1 b_2 - a_2 b_1)x = c_1 b_2 - c_2 b_1$$
$$x = \frac{c_1 b_2 - c_2 b_1}{a_1 b_2 - a_2 b_1}$$

To solve for y we go back to the original system and eliminate the variable x. To do so we multiply the first equation by $-a_2$ and the second equation by a_1:

$$-a_1 a_2 x - a_2 b_1 y = -a_2 c_1$$
$$a_1 a_2 x + a_1 b_2 y = a_1 c_2$$

Adding these two equations and solving for y, we have

$$(a_1 b_2 - a_2 b_1)y = a_1 c_2 - a_2 c_1$$
$$y = \frac{a_1 c_2 - a_2 c_1}{a_1 b_2 - a_2 b_1}$$

The solutions we have obtained for x and y can be written in terms of 2×2 determinants as follows:

$$x = \frac{c_1 b_2 - c_2 b_1}{a_1 b_2 - a_2 b_1} = \frac{\begin{vmatrix} c_1 & b_1 \\ c_2 & b_2 \end{vmatrix}}{\begin{vmatrix} a_1 & b_1 \\ a_2 & b_2 \end{vmatrix}}$$

$$y = \frac{a_1 c_2 - a_2 c_1}{a_1 b_2 - a_2 b_1} = \frac{\begin{vmatrix} a_1 & c_1 \\ a_2 & c_2 \end{vmatrix}}{\begin{vmatrix} a_1 & b_1 \\ a_2 & b_2 \end{vmatrix}}$$

We summarize these results, known as Cramer's rule, into the following theorem:

THEOREM 8.2 (CRAMER'S RULE) The solution to the system

$$a_1 x + b_1 y = c_1$$
$$a_2 x + b_2 y = c_2$$

is given by

$$x = \frac{D_x}{D}, \qquad y = \frac{D_y}{D}$$

where

$$D = \begin{vmatrix} a_1 & b_1 \\ a_2 & b_2 \end{vmatrix} \qquad (D \neq 0)$$

$$D_x = \begin{vmatrix} c_1 & b_1 \\ c_2 & b_2 \end{vmatrix}$$

$$D_y = \begin{vmatrix} a_1 & c_1 \\ a_2 & c_2 \end{vmatrix}$$

The determinant D is made up of the coefficients of x and y in the original system. The terms D_x and D_y are found by replacing the coefficients of x or y by the constant terms in the original system. Notice also that Cramer's rule does not apply if $D = 0$. In this case the equations are either inconsistent or dependent.

▼ **Example 1** Use Cramer's rule to solve

$$2x - 3y = 4$$
$$4x + 5y = 3$$

Solution We begin by calculating the determinants D, D_x, and D_y:

$$D = \begin{vmatrix} 2 & -3 \\ 4 & 5 \end{vmatrix} = 2(5) - 4(-3) = 22$$

$$D_x = \begin{vmatrix} 4 & -3 \\ 3 & 5 \end{vmatrix} = 4(5) - 3(-3) = 29$$

$$D_y = \begin{vmatrix} 2 & 4 \\ 4 & 3 \end{vmatrix} = 2(3) - 4(4) = -10$$

$$x = \frac{D_x}{D} = \frac{29}{22} \quad \text{and} \quad y = \frac{D_y}{D} = \frac{-10}{22} = -\frac{5}{11}$$

The solution set for the system is $\{(\frac{29}{22}, -\frac{5}{11})\}$. ▲

Cramer's rule can be applied to systems of linear equations in three variables also.

THEOREM 8.3 (ALSO CRAMER'S RULE) The solution set to the system

$$a_1x + b_1y + c_1z = d_1$$
$$a_2x + b_2y + c_2z = d_2$$
$$a_3x + b_3y + c_3z = d_3$$

is given by

$$x = \frac{D_x}{D}, \quad y = \frac{D_y}{D}, \quad \text{and } z = \frac{D_z}{D}$$

where

$$D = \begin{vmatrix} a_1 & b_1 & c_1 \\ a_2 & b_2 & c_2 \\ a_3 & b_3 & c_3 \end{vmatrix} \qquad (D \neq 0)$$

$$D_x = \begin{vmatrix} d_1 & b_1 & c_1 \\ d_2 & b_2 & c_2 \\ d_3 & b_3 & c_3 \end{vmatrix}$$

$$D_y = \begin{vmatrix} a_1 & d_1 & c_1 \\ a_2 & d_2 & c_2 \\ a_3 & d_3 & c_3 \end{vmatrix}$$

$$D_z = \begin{vmatrix} a_1 & b_1 & d_1 \\ a_2 & b_2 & d_2 \\ a_3 & b_3 & d_3 \end{vmatrix}$$

Again the determinant D consists of the coefficients of x, y, and z in the original system. The determinants D_x, D_y, and D_z are found by replacing the coefficients of x, y, and z respectively with the constant terms from the original system. If $D = 0$, there is no unique solution to the system.

▼   **Example 2**   Use Cramer's rule to solve

$$\begin{aligned} x + y + z &= 6 \\ 2x - y + z &= 3 \\ x + 2y - 3z &= -4 \end{aligned}$$

Solution   This is the same system used in Example 1 in Section 8.2, so we can compare Cramer's rule with our previous methods of solving a system in three variables. We begin by setting up and evaluating D, D_x, D_y, and D_z. (Recall that there are a number of ways to evaluate a 3×3 determinant. Since we have four of these determinants, we can use both Methods 1 and 2 from the previous section.) We evaluate D using Method 1 from Section 8.3:

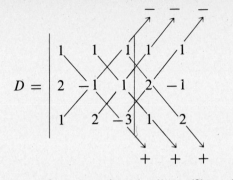

$$D = \begin{vmatrix} 1 & 1 & 1 \\ 2 & -1 & 1 \\ 1 & 2 & -3 \end{vmatrix} \begin{matrix} 1 & 1 \\ 2 & -1 \\ 1 & 2 \end{matrix}$$

$$= 3 + 1 + 4 - (-1) - (2) - (-6) = 13$$

We evaluate D_x using Method 2 from Section 8.3 and expanding across row 1:

$$D_x = \begin{vmatrix} 6 & 1 & 1 \\ 3 & -1 & 1 \\ -4 & 2 & -3 \end{vmatrix} = 6 \begin{vmatrix} -1 & 1 \\ 2 & -3 \end{vmatrix} - 1 \begin{vmatrix} 3 & 1 \\ -4 & -3 \end{vmatrix} + 1 \begin{vmatrix} 3 & -1 \\ -4 & 2 \end{vmatrix}$$

$$= 6(1) - 1(-5) + 1(2)$$
$$= 13$$

Find D_y by expanding across row 2:

$$D_y = \begin{vmatrix} 1 & 6 & 1 \\ 2 & 3 & 1 \\ 1 & -4 & -3 \end{vmatrix} = -2 \begin{vmatrix} 6 & 1 \\ -4 & -3 \end{vmatrix} + 3 \begin{vmatrix} 1 & 1 \\ 1 & -3 \end{vmatrix} - 1 \begin{vmatrix} 1 & 6 \\ 1 & -4 \end{vmatrix}$$

$$= -2(-14) + 3(-4) - 1(-10)$$
$$= 26$$

Find D_z by expanding down column 1:

$$D_z = \begin{vmatrix} 1 & 1 & 6 \\ 2 & -1 & 3 \\ 1 & 2 & -4 \end{vmatrix} = 1 \begin{vmatrix} -1 & 3 \\ 2 & -4 \end{vmatrix} - 2 \begin{vmatrix} 1 & 6 \\ 2 & -4 \end{vmatrix} + 1 \begin{vmatrix} 1 & 6 \\ -1 & 3 \end{vmatrix}$$

$$= 1(-2) - 2(-16) + 1(9)$$
$$= 39$$

$$x = \frac{D_x}{D} = \frac{13}{13} = 1, \quad y = \frac{D_y}{D} = \frac{26}{13} = 2, \quad z = \frac{D_z}{D} = \frac{39}{13} = 3$$

The solution set is $\{(1, 2, 3)\}$. ▲

▼   **Example 3**   Use Cramer's rule to solve

$$x + y = -1$$
$$2x - \ z = 3$$
$$y + 2z = -1$$

Solution   It is helpful to rewrite the system using zeros for the coefficients of those variables not shown:

$$x + y + 0z = -1$$
$$2x + 0y - z = 3$$
$$0x + y + 2z = -1$$

The four determinants used in Cramer's rule are

$$D = \begin{vmatrix} 1 & 1 & 0 \\ 2 & 0 & -1 \\ 0 & 1 & 2 \end{vmatrix} = -3$$

$$D_x = \begin{vmatrix} -1 & 1 & 0 \\ 3 & 0 & -1 \\ -1 & 1 & 2 \end{vmatrix} = -6$$

$$D_y = \begin{vmatrix} 1 & -1 & 0 \\ 2 & 3 & -1 \\ 0 & -1 & 2 \end{vmatrix} = 9$$

$$D_z = \begin{vmatrix} 1 & 1 & -1 \\ 2 & 0 & 3 \\ 0 & 1 & -1 \end{vmatrix} = -3$$

$$x = \frac{D_x}{D} = \frac{-6}{-3} = 2, \qquad y = \frac{D_y}{D} = \frac{9}{-3} = -3,$$

$$z = \frac{D_z}{D} = \frac{-3}{-3} = 1$$

The solution set is $\{(2, -3, 1)\}$.   ▲

Note   When solving a system of linear equations by Cramer's rule, it is best to find the determinant D first. If $D = 0$, Cramer's rule does not apply and there is no sense in calculating D_x, D_y, or D_z.

Solve each of the following systems using Cramer's rule: *Problem Set 8.4*

1. $2x - 3y = 3$
   $4x - 2y = 10$

2. $3x + y = -2$
   $-3x + 2y = -4$

3. $5x - 2y = 4$
   $-10x + 4y = 1$

4. $-4x + 3y = -11$
   $5x + 4y = 6$

5. $2x + 5y = 2$
   $4x + 10y = 4$

6. $3x + 5y = 2$
   $-6x - 10y = 6$

7. $4x - 7y = 3$
   $5x + 2y = -3$

8. $3x - 4y = 7$
   $6x - 2y = 5$

9. $9x - 8y = 4$
   $2x + 3y = 6$

10. $4x - 7y = 10$
    $-3x + 2y = -9$

11. $x + y - z = 2$
    $-x + y + z = 3$
    $x + y + z = 4$

12. $-x - y + z = 1$
    $x - y + z = 3$
    $x + y - z = 4$

13. $x + y + z = 4$
    $x - y - z = 2$
    $2x + 2y - z = 2$

14. $-x + y + 3z = 6$
    $x + y + 2z = 7$
    $2x + 3y + z = 4$

15. $3x - y + 2z = 4$
    $6x - 2y + 4z = 8$
    $x - 5y + 2z = 1$

16. $2x - 3y + z = 1$
    $3x - y - z = 4$
    $4x - 6y + 2z = 3$

17. $2x - y + 3z = 4$
    $x - 5y - 2z = 1$
    $-4x - 2y + z = 3$

18. $4x - y + 5z = 1$
    $2x + 3y + 4z = 5$
    $x + y + 3z = 2$

19. $-x - 2y = 1$
    $x + 2z = 11$
    $2y + z = 0$

20. $x + y = 2$
    $-x + z = 0$
    $2x + z = 3$

21. $x - y = 2$
    $3x + z = 11$
    $y - 2z = -3$

22. $4x + 5y = -1$
    $2y + 3z = -5$
    $x + 2z = -1$

23. If a company has fixed costs of $100 per week and each item it produces costs $10 to manufacture, then the total cost (y) per week to produce x items is

$$y = 10x + 100$$

If the company sells each item it manufactures for $12, then the total amount of money (y) the company brings in for selling x items is

$$y = 12x$$

Use Cramer's rule to solve the system

$$y = 10x + 100$$
$$y = 12x$$

for x to find the number of items the company must sell per week in order to break even.

8.5
Word Problems

Many times word problems involve more than one unknown quantity. If a problem is stated in terms of two unknowns and we represent each unknown quantity with a different variable, then we must write the relationship between the variables with two equations. The two equations written in terms of the two variables form a system of linear equations which we solve using the methods developed in this section. If we find a problem that relates three unknown quantities, then we need three different equations in order to form a linear system we can solve.

▼   **Example 1** One number is 2 more than 3 times another. Their sum is 26. Find the two numbers.

> **Solution** If we let x and y represent the two numbers, then the translation of the first sentence in the problem into an equation would be
>
> $$y = 3x + 2$$
>
> The second sentence gives us a second equation:
>
> $$x + y = 26$$
>
> The linear system that describes the situation is
>
> $$x + y = 26$$
> $$y = 3x + 2$$
>
> Substituting the expression for y from the second equation into the first and solving for x yields
>
> $$x + (3x + 2) = 26$$
> $$4x + 2 = 26$$
> $$4x = 24$$
> $$x = 6$$
>
> Using $x = 6$ in $y = 3x + 2$ gives the second number:
>
> $$y = 3(6) + 2$$
> $$y = 20$$
>
> The two numbers are 6 and 20. Their sum is 26, and the second is 2 more than 3 times the first.   ▲

▼   **Example 2**   Suppose 850 tickets were sold for the game for a total of $1,100. If adult tickets cost $1.50 and children's tickets cost $1.00, how many of each kind of ticket were sold?

Solution   If we let $x =$ the number of adult tickets and $y =$ the number of children's tickets, then

$$x + y = 850$$

since a total of 850 tickets were sold. Since each adult ticket costs $1.50 and each children's ticket costs $1.00 and the total amount of money paid for tickets was $1,100, a second equation is

$$1.50x + 1.00y = 1,100$$

which is equivalent to

$$15x + 10y = 11,000$$

The system that describes the situation is

$$
\begin{aligned}
x + \quad y &= 850 \\
15x + 10y &= 11,000
\end{aligned}
$$

Multiplying the first equation by -10 and adding the result to the second equation eliminates the variable y from the two systems:

$$
\begin{aligned}
-10x - 10y &= -8,500 \\
\underline{15x + 10y} &= \underline{11,000} \\
5x \quad\quad &= 2,500 \\
x &= 500
\end{aligned}
$$

The number of adult tickets sold was 500. To find the number of children's tickets, we substitute $x = 500$ into $x + y = 850$ to get

$$
\begin{aligned}
500 + y &= 850 \\
y &= 350
\end{aligned}
$$

The number of children's tickets sold was 350.   ▲

▼   **Example 3**   Suppose a person invests a total of $10,000 in two accounts. One account earns 5% annually and the other earns 6% annually. If the total interest earned from both accounts in a year is $560, how much is invested in each account?

Solution   The form of the solution to this problem is very similar to that of Example 2. We let x equal the amount invested at 6% and y

be the amount invested at 5%. Since the total investment is $10,000, one relationship between x and y can be written as

$$x + y = 10,000$$

The total interest earned from both accounts is $560. The amount of interest earned on x dollars at 6% is .06x, while the amount of interest earned on y dollars at 5% is .05y. This relationship is represented by the equation

$$.06x + .05y = 560$$

which is equivalent to

$$6x + 5y = 56,000$$

The original problem can be stated in terms of x and y with the system

$$x + y = 10,000$$
$$6x + 5y = 56,000$$

Eliminating x from the system and solving for y, we have

$$
\begin{array}{rcr}
-6x - 6y = & -60,000 \\
6x + 5y = & 56,000 \\
\hline
- y = & -4,000 \\
y = & 4,000
\end{array}
$$

The amount of money invested at 5% is $4,000. Since the total investment was $10,000, the amount invested at 6% must be $6,000.

▲

▼   **Example 4**   How much 20% alcohol solution and 50% alcohol solution must be mixed to get 12 gallons of 30% alcohol solution?

Solution   To solve this problem we must first understand that a 20% alcohol solution is 20% alcohol and 80% water.
   Let $x =$ the number of gallons of 20% alcohol solution needed, and $y =$ the number of gallons of 50% alcohol solution needed. Since we must end up with a total of 12 gallons of solution, one equation for the system is

$$x + y = 12$$

The amount of alcohol in the x gallons of 20% solution is .20x, while the amount of alcohol in the y gallons of 50% solution is .50y.

Since the total amount of alcohol in the 20% and 50% solutions must add up to the amount of alcohol in the 12 gallons of 30% solution, the second equation in our system can be written as

$$.20x + .50y = .30(12)$$

We eliminate the decimals by multiplying both sides by 10 to get

$$2x + 5y = 36$$

The system of equations that describes the situation is

$$x + y = 12$$
$$2x + 5y = 36$$

Multiplying the top equation by -2 to eliminate the x-variable, we have

$$
\begin{aligned}
-2x - 2y &= -24 \\
2x + 5y &= 36 \\
\hline
3y &= 12 \\
y &= 4
\end{aligned}
$$

Substituting $y = 4$ into $x + y = 12$, we solve for x:

$$x + 4 = 12$$
$$x = 8$$

It takes 8 gallons of 20% alcohol solution and 4 gallons of 50% alcohol solution to produce 12 gallons of 30% alcohol solution.

▲

▼   **Example 5**   It takes 2 hours for a boat to travel 28 miles downstream. The same boat can travel 18 miles upstream in 3 hours. What is the speed of the boat in still water and the speed of the current of the river?

Solution   Let $x = $ the speed of the boat in still water and $y = $ the speed of the current. Using a table as we did in Section 4.6, we have

	d	r	t
Upstream	18	$x - y$	3
Downstream	28	$x + y$	2

Since $d = r \cdot t$, the system we need to solve the problem is

$$18 = (x - y) \cdot 3$$
$$28 = (x + y) \cdot 2$$

which is equivalent to

$$6 = x - y$$
$$14 = x + y$$

Adding the two equations, we have

$$20 = 2x$$
$$x = 10$$

Substituting $x = 10$ into $14 = x + y$, we see that

$$y = 4$$

The speed of the boat in still water is 10 mph and the speed of the current is 4 mph.   ▲

Note   We solved each system in this section using either the elimination method or the substitution method. We could just as easily have used Cramer's rule to solve the systems. Cramer's rule applies to any system of linear equations with a unique solution.

Problem Set 8.5

1. One number is 3 more than twice another. The sum of the numbers is 18. Find the two numbers.
2. The sum of two numbers is 32. One of the numbers is 4 less than 5 times the other. Find the two numbers.
3. The difference of two numbers is 6. Twice the smaller is 4 more than the larger. Find the two numbers.
4. The larger of two numbers is 5 more than twice the smaller. If the smaller is subtracted from the larger, the result is 12. Find the two numbers.
5. The sum of three numbers is 8. Twice the smallest is 2 less than the largest, while the sum of the largest and smallest is 5. Use a linear system in three variables to find the three numbers.
6. The sum of three numbers is 14. The largest is 4 times the smallest, while the sum of the smallest and twice the largest is 18. Use a linear system in three variables to find the three numbers.
7. A total of 925 tickets were sold for the game for a total of $1150. If adult tickets sold for $2.00 and children's tickets sold for $1.00, how many of each kind of ticket were sold?
8. If tickets for the show cost $2.00 for adults and $1.50 for children, how

many of each kind of ticket were sold if a total of 300 tickets were sold for $525?

9. Bob has 20 coins totaling $1.40. If he has only dimes and nickels, how many of each coin does he have?

10. If Amy has 15 coins totaling $2.70, and the coins are quarters and dimes, how many of each coin does she have?

11. Mr. Jones has $20,000 to invest. He invests part at 6% and the rest at 7%. If he earns $1280 interest after one year, how much did he invest at each rate?

12. A man invests $17,000 in two accounts. One account earns 5% interest per year and the other 6.5%. If his yearly yield in interest is $970, how much does he invest at each rate?

13. Susan invests twice as much money at 7.5% as she does at 6%. If her total interest after a year is $840, how much does she have invested at each rate?

14. A woman earns $1350 interest from two accounts in a year. If she has three times as much invested at 7% as she does at 6%, how much does she have in each account?

15. How many gallons of 20% alcohol solution and 50% alcohol solution must be mixed to get 9 gallons of 30% alcohol solution?

16. How many ounces of 30% hydrochloric acid solution and 80% hydrochloric acid solution must be mixed to get 10 ounces of 50% hydrochloric acid solution?

17. A mixture of 16% disinfectant solution is to be made from 20% and 14% disinfectant solutions. How much of each solution should be used if 15 gallons of the 16% solution are needed?

18. How much 25% antifreeze and 50% antifreeze should be combined to give 40 gallons of 30% antifreeze?

19. It takes a boat 2 hours to travel 24 miles downstream and 3 hours to travel 18 miles upstream. What is the speed of the boat in still water and of the current of the river?

20. A boat on a river travels 20 miles downstream in only 2 hours. It takes the same boat 6 hours to travel 12 miles upstream. What are the speed of the boat and the speed of the current?

21. An airplane flying with the wind can cover a certain distance in 2 hours. The return trip against the wind takes $2\frac{1}{2}$ hours. How fast is the plane and what is the speed of the air, if the distance is 600 miles?

22. An airplane covers a distance of 1500 miles in 3 hours when it flies with the wind and $3\frac{1}{3}$ hours when it flies against the wind. What is the speed of the plane in still air?

A system of linear equations consists of two or more linear equations considered simultaneously. The solution set to a linear system in two variables is the set of ordered pairs that satisfies both equations. The

**Chapter 8
Summary and Review**
*Systems of Linear
Equations*

solution set to a linear system in three variables consists of all the ordered triples that satisfy each equation in the system.

Inconsistent and Dependent Equations

Two linear equations that have no solutions in common are said to be *inconsistent,* while two linear equations that have all their solutions in common are said to be *dependent.*

Determinants

The values of a 2 × 2 and a 3 × 3 determinant are as follows:

$$\begin{vmatrix} a & c \\ b & d \end{vmatrix} = ad - bc$$

$$\begin{vmatrix} a_1 & b_1 & c_1 \\ a_2 & b_2 & c_2 \\ a_3 & b_3 & c_3 \end{vmatrix} = a_1 b_2 c_3 + a_3 b_1 c_2 + a_2 b_3 c_1 \\ - a_3 b_2 c_1 - a_1 b_3 c_2 - a_2 b_1 c_3$$

There are two methods of finding the six products in the expansion of a 3 × 3 determinant. One method involves a cross-multiplication scheme. The other method involves expanding the determinant by minors.

Cramer's Rule for a Linear System in Two Variables

The solution to the system

$$a_1 x + b_1 y = c_1 \\ a_2 x + b_2 y = c_2$$

is given by

$$x = \frac{D_x}{D} \quad \text{and} \quad y = \frac{D_y}{D} \qquad (D \neq 0)$$

where

$$D = \begin{vmatrix} a_1 & b_1 \\ a_2 & b_2 \end{vmatrix}, \qquad D_x = \begin{vmatrix} c_1 & b_1 \\ c_2 & b_2 \end{vmatrix}, \qquad \text{and } D_y = \begin{vmatrix} a_1 & c_1 \\ a_2 & c_2 \end{vmatrix}$$

Cramer's Rule for a Linear System in Three Variables

The solution to the system

$$a_1 x + b_1 y + c_1 z = d_1 \\ a_2 x + b_2 y + c_2 z = d_2 \\ a_3 x + b_3 y + c_3 z = d_3$$

is given by

$$x = \frac{D_x}{D}, \qquad y = \frac{D_y}{D}, \qquad \text{and } z = \frac{D_z}{D} \qquad (D \neq 0)$$

where

$$D = \begin{vmatrix} a_1 & b_1 & c_1 \\ a_2 & b_2 & c_2 \\ a_3 & b_3 & c_3 \end{vmatrix} \qquad D_y = \begin{vmatrix} a_1 & d_1 & c_1 \\ a_2 & d_2 & c_2 \\ a_3 & d_3 & c_3 \end{vmatrix}$$

$$D_x = \begin{vmatrix} d_1 & b_1 & c_1 \\ d_2 & b_2 & c_2 \\ d_3 & b_3 & c_3 \end{vmatrix} \qquad D_z = \begin{vmatrix} a_1 & b_1 & d_1 \\ a_2 & b_2 & d_2 \\ a_3 & b_3 & d_3 \end{vmatrix}$$

Solve the following systems by the elimination method:

1. $2x - 5y = -8$
   $3x + y = 5$

2. $4x - 7y = -2$
   $-5x + 6y = -3$

Solve the following systems by the substitution method:

3. $2x - 5y = 14$
   $y = 3x + 8$

4. $6x - 3y = 0$
   $x + 2y = 5$

5. Solve the system

$$2x - y + z = 9$$
$$x + y - 3z = -2$$
$$3x + y - z = 6$$

Evaluate each determinant:

6. $\begin{vmatrix} 3 & -5 \\ -4 & 2 \end{vmatrix}$

7. $\begin{vmatrix} 1 & 0 & -3 \\ 2 & 1 & 0 \\ 0 & 5 & 4 \end{vmatrix}$

Use Cramer's rule to solve:

8. $5x - 4y = 2$
   $-2x + y = 3$

9. $2x + 4y = 3$
   $-4x - 8y = -6$

10. $2x - y + 3z = 2$
    $x - 4y - z = 6$
    $3x - 2y + z = 4$

11. John invests twice as much money at 6% as he does at 5%. If his investments earn a total of $680 in one year, how much does he have invested at each rate?

12. For a woman of average height and weight between the ages of 19 and 22, the Food and Nutrition Board of the National Academy of Sciences has determined the Recommended Daily Allowance (RDA) of ascorbic acid to be 45 mg (milligrams). They also determined the RDA for niacin to be 14 mg for the same woman.

Each ounce of cereal I contains 10 mg of ascorbic acid and 4 mg of niacin, while each ounce of cereal II contains 15 mg of ascorbic acid and 2 mg of niacin. How many ounces of each cereal must the average woman between the ages of 19 and 22 consume in order to have the RDAs for both ascorbic acid and niacin? (The following table is a summary of the information given.)

	Cereal I	Cereal II	Recommended Daily Allowance (RDA)
Ascorbic acid	10 mg	15 mg	45 mg
Niacin	4 mg	2 mg	14 mg

The Conic Sections 9

To the student:

This chapter is concerned with four special types of graphs and their associated equations. The four types of graphs are called parabolas, circles, ellipses, and hyperbolas. They are called *conic sections* because each can be found by slicing a cone with a plane as is shown in Figure 9-1.

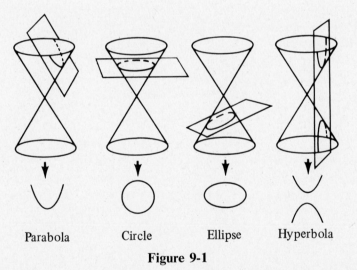

Parabola   Circle   Ellipse   Hyperbola

Figure 9-1

There are many applications associated with conic sections. The planets orbit the sun in elliptical orbits. Many of the comets that come in contact with the gravitational field surrounding the earth travel in parabolic or hyperbolic paths. Flashlight and searchlight mirrors have elliptical or parabolic shapes because of the way surfaces with those shapes reflect light. The arches of many bridges are in the shape of parabolas. Objects fired into the air travel in parabolic paths.

The equations associated with these graphs are all second-degree equations in two variables. The first four sections of this chapter deal with graphing the conic sections, given the appropriate equation. The chapter ends with a section on second-degree inequalities and a section on second-degree systems. A working knowledge of graphing will be useful throughout the chapter. Factoring is needed for the first section of the chapter.

9.1
Graphing Parabolas

The solution set to the equation

$$y = x^2 - 3$$

will consist of ordered pairs. One method of graphing the solution set is to find a number of ordered pairs that satisfy the equation and graph them. We can obtain some ordered pairs that are solutions to $y = x^2 - 3$ by use of a table as follows:

x	$y = x^2 - 3$	y	Solutions
-3	$y = (-3)^2 - 3 = 9 - 3 = 6$	6	$(-3, 6)$
-2	$y = (-2)^2 - 3 = 4 - 3 = 1$	1	$(-2, 1)$
-1	$y = (-1)^2 - 3 = 1 - 3 = -2$	-2	$(-1, -2)$
0	$y = 0^2 - 3 = 0 - 3 = -3$	-3	$(0, -3)$
1	$y = 1^2 - 3 = 1 - 3 = -2$	-2	$(1, -2)$
2	$y = 2^2 - 3 = 4 - 3 = 1$	1	$(2, 1)$
3	$y = 3^2 - 3 = 9 - 3 = 6$	6	$(3, 6)$

Graphing these solutions on a rectangular coordinate system indicates the shape of the graph of $y = x^2 - 3$. (See Figure 9-2.)

If we were to find more solutions by extending the table to include values of x between those we have already, we would see they follow the same pattern. Connecting the points on the graph with a smooth curve we have the graph of the second-degree equation $y = x^2 - 3$. (See Figure 9-3.)

This graph is an example of a parabola. All equations of the form $y = ax^2 + bx + c$, $a \neq 0$, have parabolas for graphs. Although it is always possible to graph the solution set for an equation by using a

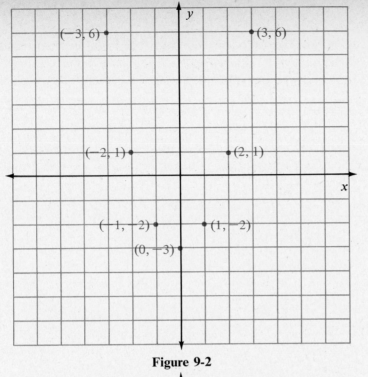

Figure 9-2

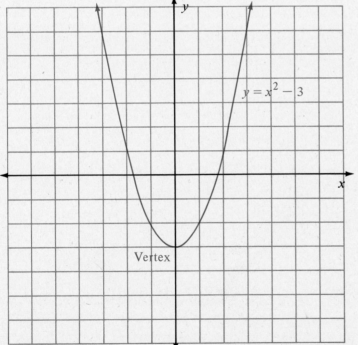

$y = x^2 - 3$

Vertex

Figure 9-3

table, we would like to find a method that is faster, less tedious, and more accurate.

The important points associated with the graph of a parabola are the highest (or lowest) point on the graph and the x-intercepts. The y-intercepts can also be useful.

Intercepts for Parabolas

The graph of the equation $y = ax^2 + bx + c$ will cross the y-axis at $y = c$, since substituting $x = 0$ into $y = ax^2 + bx + c$ yields $y = c$.

Since the graph will cross the x-axis when $y = 0$, the x-intercepts are those values of x that are solutions to the quadratic equation $0 = ax^2 + bx + c$. Applying the quadratic formula, if necessary, the x-intercepts can be written

$$x = \frac{-b \pm \sqrt{b^2 - 4ac}}{2a}$$

The x-intercepts can also be found by solving the equation $ax^2 + bx + c = 0$ by factoring, if the left side is factorable.

▼   **Example 1**   Find the x-intercepts for $y = x^2 - 5x - 6$.

Solution   To find the x-intercept, we let $y = 0$ and solve for x:

$$0 = x^2 - 5x - 6$$
$$0 = (x - 6)(x + 1)$$
$$x - 6 = 0 \quad \text{or} \quad x + 1 = 0$$
$$x = 6 \quad \text{or} \quad\quad x = -1$$

The x-intercepts are 6 and -1.   ▲

Not every parabola will cross the x-axis. If the equation $0 = ax^2 + bx + c$ does not have real solutions, there will be no x-intercepts for the graph of $y = ax^2 + bx + c$.

▼   **Example 2**   Find the intercepts for $y = 2x^2 + 3x + 5$.

Solution   Substituting $y = 0$ into the equation and solving for x, we have

$$0 = 2x^2 + 3x + 5$$

The right side does not factor so we must use the quadratic formula:

$$x = \frac{-3 \pm \sqrt{9 - 4(2)(5)}}{2(2)}$$

$$x = \frac{-3 \pm \sqrt{-31}}{4}$$

$$x = \frac{-3 \pm \sqrt{31}\, i}{4}$$

Since the solutions are complex numbers, the graph does not cross the x-axis. There are no x-intercepts.   ▲

The highest or lowest point on a parabola is called the *vertex*. The vertex for the graph of $y = ax^2 + bx + c$ will always occur when

The Vertex of a Parabola

$$x = \frac{-b}{2a}$$

That is, the x-coordinate of the vertex is $-b/2a$. To find the y-coordinate of the vertex, we substitute $x = -b/2a$ into $y = ax^2 + bx + c$ and solve for y.

If we complete the square on the first two terms on the right side of the equation $y = ax^2 + bx + c$, we can transform the equation as follows:

$$y = ax^2 + bx + c$$

$$y = a\left(x^2 + \frac{b}{a}x\right) + c$$

$$y = a\left[x^2 + \frac{b}{a}x + \left(\frac{b}{2a}\right)^2\right] + c - a\left(\frac{b}{2a}\right)^2$$

$$y = a\left(x + \frac{b}{2a}\right)^2 + \frac{4ac - b^2}{4a}$$

It may not look like it, but this last line indicates the vertex of the graph of $y = ax^2 + bx + c$ has an x-coordinate of $-b/2a$. Since a, b, and c are constants, the only quantity that is varying in the last expression is the x in $(x + b/2a)^2$. Since the quantity $(x + b/2a)^2$ is the square of $x + b/2a$, the smallest it will ever be is 0, and that will happen when $x = -b/2a$.

We can use the vertex point along with the x- and y-intercepts to sketch the graph of any equation of the form $y = ax^2 + bx + c$. Here is a summary of the information given above:

The graph of $y = ax^2 + bx + c$ will have

1. a y-intercept at $y = c$
2. x-intercepts (if they exist) at

$$x = \frac{-b \pm \sqrt{b^2 - 4ac}}{2a}$$

3. a vertex when $x = -b/2a$

▼ **Example 3** Sketch the graph of $y = x^2 - 4x - 5$.

Solution To find the x-intercepts, we let $y = 0$ and solve for x:

$$0 = x^2 - 4x - 5$$
$$0 = (x - 5)(x + 1)$$
$$x - 5 = 0 \quad \text{or} \quad x + 1 = 0$$
$$x = 5 \quad \text{or} \quad x = -1$$

The x-coordinate of the vertex will occur when

$$x = \frac{-b}{2a} = \frac{-(-4)}{2(1)}$$
$$= 2$$

To find the y-coordinate of the vertex we substitute 2 for x:

$$y = 2^2 - 4(2) - 5$$
$$= 4 - 8 - 5$$
$$= -9$$

The graph crosses the x-axis at 5 and -1 and has its vertex at $(2, -9)$. Plotting these points and connecting them with a smooth curve, we have the graph of $y = x^2 - 4x - 5$. (See Figure 9-4.) A parabola that opens up as this one does is said to be *concave-up*. The vertex is the lowest point on the graph. ▲

▼ **Example 4** Graph $y = -x^2 - 4x + 12$.

Solution Letting $y = 0$, the x-intercepts come from

$$0 = -x^2 - 4x + 12$$
$$0 = x^2 + 4x - 12$$
$$0 = (x + 6)(x - 2)$$
$$x + 6 = 0 \quad \text{or} \quad x - 2 = 0$$
$$x = -6 \quad \text{or} \quad x = 2$$

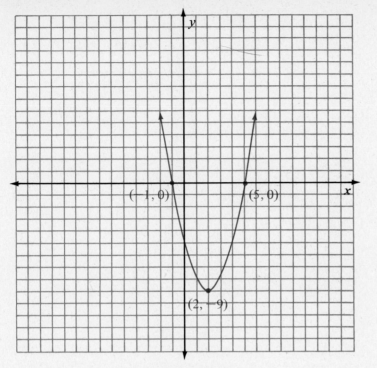

Figure 9-4

The x-coordinate of the vertex is given by

$$x = \frac{-b}{2a} = \frac{-(-4)}{2(-1)} = -2$$

We find the y-coordinate of the vertex by substituting $x = -2$ into the original equation:

$$y = -(-2)^2 - 4(-2) + 12$$
$$= -4 + 8 + 12$$
$$= 16$$

The graph crosses the x-axis at -6 and 2, and has its vertex at $(-2, 16)$. (See Figure 9-5.)

The graph is *concave-down*. The vertex in this case is the highest point on the graph. ▲

▼ **Example 5** Graph $y = 3x^2 - 6x + 1$.

Solution To find the x-intercepts, we let $y = 0$ and solve for x:

$$0 = 3x^2 - 6x + 1$$

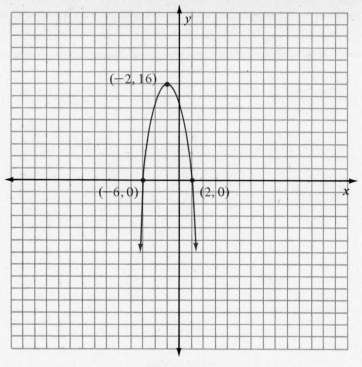

Figure 9-5

Since the right side of this equation will not factor, we must use the quadratic formula:

$$x = \frac{-(-6) \pm \sqrt{36 - 4(3)(1)}}{2(3)}$$

$$= \frac{6 \pm \sqrt{24}}{6}$$

$$= \frac{6 \pm 2\sqrt{6}}{6}$$

$$= \frac{3 \pm \sqrt{6}}{3}$$

The x-intercepts are

$$\frac{3 + \sqrt{6}}{3} \quad \text{and} \quad \frac{3 - \sqrt{6}}{3}$$

Since we are going to graph these points, it is a good idea to get a decimal approximation for them. Since $\sqrt{6} \approx 2.45$ ($\approx$ is read "approximately equals"), the x-intercepts are

$$\frac{3 + \sqrt{6}}{3} \approx \frac{3 + 2.45}{3} = \frac{5.45}{3} \approx 1.8$$

$$\frac{3 - \sqrt{6}}{3} \approx \frac{3 - 2.45}{3} = \frac{.55}{3} \approx .2$$

The x-coordinate of the vertex is

$$x = \frac{-b}{2a} = \frac{-(-6)}{2(3)} = 1$$

To find the y-coordinate of the vertex, we let $x = 1$:

$$\begin{aligned} y &= 3(1)^2 - 6(1) + 1 \\ &= 3 - 6 + 1 \\ &= -2 \end{aligned}$$

The graph crosses the x-axis at 1.8 and .2 (approximately) and has its vertex at $(1, -2)$. (See Figure 9-6.)

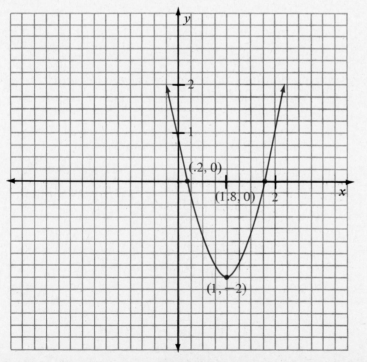

Figure 9-6

The graph is concave-up. The vertex is the lowest point on the graph.   ▲

Note It is apparent from these last three examples that the x-coordinate of the vertex will always fall halfway between the x-intercepts—*if* the graph has x-intercepts. If the x-intercepts are x_1 and x_2, then the value halfway between them will always be their average, or

$$\frac{x_1 + x_2}{2}$$

▼ **Example 6**   Graph $y = -2x^2 + 6x - 5$.

Solution   Letting $y = 0$, we have

$$0 = -2x^2 + 6x - 5$$

Since the right side does not factor, we must use the quadratic formula:

$$x = \frac{-6 \pm \sqrt{36 - 4(-2)(-5)}}{2(-2)}$$

$$= \frac{-6 \pm \sqrt{-4}}{-4}$$

$$= \frac{-6 \pm 2i}{-4}$$

The results are complex numbers, so the x-intercepts do not exist. The graph does not cross the x-axis. The x-coordinate of the vertex is

$$x = \frac{-b}{2a} = \frac{-6}{2(-2)} = \frac{6}{4} = \frac{3}{2}$$

To find the y-coordinate, we let $x = \frac{3}{2}$:

$$y = -2\left(\frac{3}{2}\right)^2 + 6\left(\frac{3}{2}\right) - 5$$

$$= \frac{-18}{4} + \frac{18}{2} - 5$$

$$= \frac{-18 + 36 - 20}{4}$$

$$= \frac{-2}{4}$$

$$= \frac{-1}{2}$$

The vertex is $(\frac{3}{2}, -\frac{1}{2})$. Since this is the only point we have so far, we must find two others. Let's let $x = 3$ and $x = 0$, since each point is the same distance from $x = \frac{3}{2}$ and on either side:

When $x = 3$,   When $x = 0$,
$$y = -2(3)^2 + 6(3) - 5 \qquad\qquad y = -2(0)^2 + 6(0) - 5$$
$$= -18 + 18 - 5 \qquad\qquad\qquad = 0 + 0 - 5$$
$$= -5 \qquad\qquad\qquad\qquad\quad = -5$$

The two additional points on the graph are $(3, -5)$ and $(0, -5)$. Figure 9-7 shows the graph.

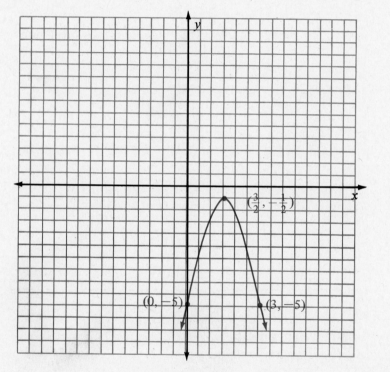

Figure 9-7

The graph is concave-down. The vertex is the highest point on the graph. ▲

We see from the examples of this section that the graph of $y = ax^2 + bx + c$ is concave-up whenever $a > 0$ and concave-down whenever $a < 0$.

For each of the following equations, give the x-intercepts and the coordinates of the vertex and sketch the graph:

1. $y = x^2 - 6x - 7$ 2. $y = x^2 + 6x - 7$
3. $y = x^2 + 2x - 3$ 4. $y = x^2 - 2x - 3$
5. $y = -x^2 - 4x + 5$ 6. $y = -x^2 + 4x - 5$
7. $y = -x^2 + 2x + 15$ 8. $y = -x^2 - 2x + 8$
9. $y = x^2 - 1$ 10. $y = x^2 - 4$
11. $y = -x^2 + 9$ 12. $y = -x^2 + 1$
13. $y = 3x^2 - 10x - 8$ 14. $y = 3x^2 + 10x + 8$
15. $y = x^2 - 2x - 11$ 16. $y = x^2 - 4x - 14$
17. $y = 4x^2 + 12x - 11$ 18. $y = 2x^2 + 8x - 14$
19. $y = -4x^2 - 4x + 31$ 20. $y = -9x^2 - 6x + 19$

Find the vertex and any two convenient points to sketch the graphs of the following:

21. $y = x^2 - 4x - 4$ 22. $y = x^2 - 2x + 3$
23. $y = -x^2 + 2x - 5$ 24. $y = -x^2 + 4x - 2$
25. $y = x^2 + 1$ 26. $y = x^2 + 4$
27. $y = -x^2 - 3$ 28. $y = -x^2 - 2$
29. $y = 3x^2 + 4x + 1$ 30. $y = 2x^2 + 4x + 3$
31. $y = 2x^2 + 3x + 2$ 32. $y = 3x^2 + 3x + 1$
33. $y = -4x^2 + x - 1$ 34. $y = -3x^2 + x - 1$

35. An arrow is shot straight up into the air with an initial velocity of 128 feet/second. If h is the height of the arrow at any time t, then the equation that gives h in terms of t is

$$h = 128t - 16t^2$$

Find the maximum height attained by the arrow.

9.2
The Circle

Before we find the general equation of a circle, we must first derive what is known as the *distance formula*.

Suppose (x_1, y_1) and (x_2, y_2) are any two points in the first quadrant. (Actually, we could choose the two points to be anywhere on the coordinate plane. It is just more convenient to have them in the first quadrant.) We can name the points P_1 and P_2, respectively, and draw the diagram shown in Figure 9-8.

Notice the coordinates of point Q. The x-coordinate is x_2 since Q is directly below point P_2. The y-coordinate of Q is y_1 since Q is directly across from point P_1. It is evident from the diagram that the length of P_2Q is $y_2 - y_1$ and the length of P_1Q is $x_2 - x_1$. Using the Pythagorean

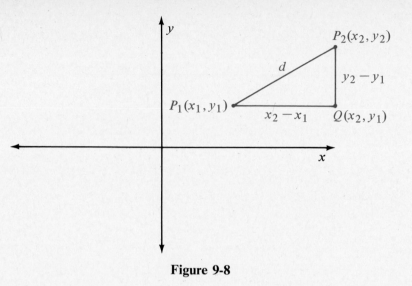

Figure 9-8

theorem, we have

$$(P_1P_2)^2 = (P_1Q)^2 + (P_2Q)^2$$

or

$$d^2 = (x_2 - x_1)^2 + (y_2 - y_1)^2$$

Taking the square root of both sides, we have

$$d = \sqrt{(x_2 - x_1)^2 + (y_2 - y_1)^2}$$

We know this is the positive square root, since d is the distance from P_1 to P_2 and must therefore be positive. This formula is called the *distance formula*.

▼ **Example 1** Find the distance between $(3, 5)$ and $(2, -1)$.

Solution If we let $(3, 5)$ be (x_1, y_1) and $(2, -1)$ be (x_2, y_2) and apply the distance formula, we have

$$
\begin{aligned}
d &= \sqrt{(2 - 3)^2 + (-1 - 5)^2} \\
&= \sqrt{(-1)^2 + (-6)^2} \\
&= \sqrt{1 + 36} \\
&= \sqrt{37}
\end{aligned}
$$

▲

Note The choice of $(3, 5)$ as (x_1, y_1) and $(2, -1)$ as (x_2, y_2) is arbitrary. We could just as easily have reversed them.

▼   **Example 2**   Find x if the distance from $(x, 5)$ to $(3, 4)$ is $\sqrt{2}$.

Solution   Using the distance formula, we have

$$\sqrt{2} = \sqrt{(x - 3)^2 + (5 - 4)^2}$$
$$2 = (x - 3)^2 + 1^2$$
$$2 = x^2 - 6x + 9 + 1$$
$$0 = x^2 - 6x + 8$$
$$0 = (x - 4)(x - 2)$$
$$x = 4 \quad \text{or} \quad x = 2$$

The two solutions are 4 and 2, which indicates there are two points, $(4, 5)$ and $(2, 5)$, which are $\sqrt{2}$ units from $(3, 4)$.   ▲

We can use the distance formula to derive the equation of a circle.

THEOREM 9.1   The equation of the circle with center at (a, b) and radius r is given by

$$(x - a)^2 + (y - b)^2 = r^2$$

Proof   By definition, all points on the circle are a distance r from the center (a, b). If we let (x, y) represent any point on the circle, then (x, y) is r units from (a, b). Applying the distance formula, we have

$$r = \sqrt{(x - a)^2 + (y - b)^2}$$

Squaring both sides of this equation gives the equation of the circle:

$$(x - a)^2 + (y - b)^2 = r^2$$

We can use Theorem 9.1 to find the equation of a circle given its center and radius, or to find its center and radius given the equation.

▼   **Example 3**   Find the equation of the circle with center at $(-3, 2)$ having a radius of 5.

Solution   We have $(a, b) = (-3, 2)$ and $r = 5$. Applying Theorem 9.1 yields

$$[x - (-3)]^2 + (y - 2)^2 = 5^2$$
$$(x + 3)^2 + (y - 2)^2 = 25$$ ▲

▼   **Example 4**   Give the equation of the circle with radius 3 whose center is at the origin.

Solution   The coordinates of the center are $(0, 0)$, and the radius is 3. The equation must be

$$(x - 0)^2 + (y - 0)^2 = 3^2$$
$$x^2 + y^2 = 9 \qquad \blacktriangle$$

We can see from Example 4 that the equation of any circle with its center at the origin and radius r will be

$$x^2 + y^2 = r^2$$

▼ **Example 5**   Find the center and radius, and sketch the graph, of the circle whose equation is

$$(x - 1)^2 + (y + 3)^2 = 4$$

Solution   Writing the equation in the form

$$(x - a)^2 + (y - b)^2 = r^2$$

we have

$$(x - 1)^2 + [y - (-3)]^2 = 2^2$$

The center is at $(1, -3)$ and the radius is 2. (See Figure 9-9.)

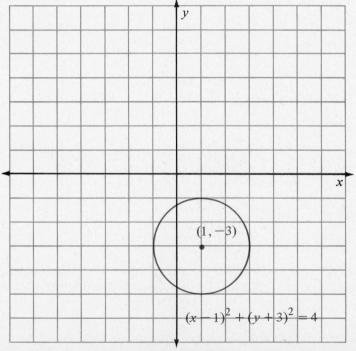

$(1, -3)$

$(x - 1)^2 + (y + 3)^2 = 4$

Figure 9-9   ▲

▼ **Example 6** Sketch the graph of $x^2 + y^2 = 9$.

Solution Since the equation can be written in the form

$$(x - 0)^2 + (y - 0)^2 = 3^2$$

it must have its center at $(0, 0)$ and a radius of 3. (See Figure 9-10.)

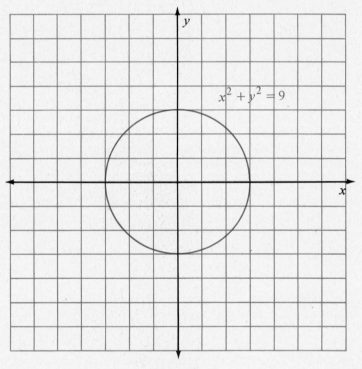

Figure 9-10

▼ **Example 7** Sketch the graph of $x^2 + y^2 + 6x - 4y - 12 = 0$.

Solution To sketch the graph we must find the center and radius. The center and radius can be identified if the equation has the form

$$(x - a)^2 + (y - b)^2 = r^2$$

The original equation can be written in this form by completing the squares on x and y:

$$x^2 + y^2 + 6x - 4y - 12 = 0.$$
$$x^2 + 6x + y^2 - 4y = 12$$
$$x^2 + 6x + 9 + y^2 - 4y + 4 = 12 + 9 + 4$$
$$(x + 3)^2 + (y - 2)^2 = 25$$
$$(x + 3)^2 + (y - 2)^2 = 5^2$$

From the last line it is apparent that the center is at $(-3, 2)$ and the radius is 5. (See Figure 9-11.)

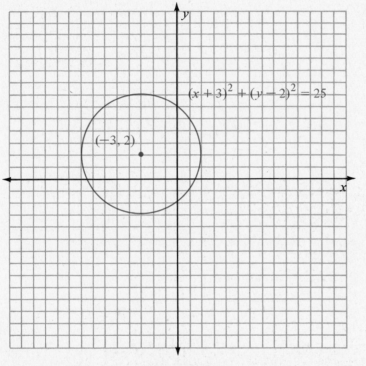

Figure 9-11

▼ **Example 8** Sketch the graph of $y = \sqrt{25 - x^2}$.

Solution In order to eliminate the radical from the equation, we square both sides. Recall that squaring both sides of an equation sometimes produces extraneous solutions. To avoid graphing any extraneous solutions we simply observe that y must be greater than or equal to 0, since it is equal to a positive square root. Keeping this in mind, we square both sides of the equation:

$$y = \sqrt{25 - x^2} \qquad y \geq 0$$
$$y^2 = 25 - x^2 \qquad y \geq 0$$
$$x^2 + y^2 = 25 \qquad y \geq 0$$

The graph of the equation is a circle with center at the origin and a radius of 5. Since $y \geq 0$, we graph only the top half of the circle. (See Figure 9-12.)

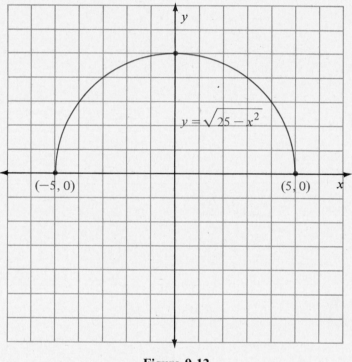

Figure 9-12

▼ **Example 9** Graph the solution set for $x = -\sqrt{9 - y^2}$.

Solution We observe that x is less than or equal to 0, and square both sides:

$$x = -\sqrt{9 - y^2} \qquad x \leq 0$$
$$x^2 = (-\sqrt{9 - y^2})^2 \qquad x \leq 0$$
$$x^2 = 9 - y^2 \qquad x \leq 0$$
$$x^2 + y^2 = 9 \qquad x \leq 0$$

The graph of the equation is a circle with center at the origin and a radius of 3. Since $x \leq 0$, we graph only that part of the circle that lies to the left of the y-axis. (See Figure 9-13.)

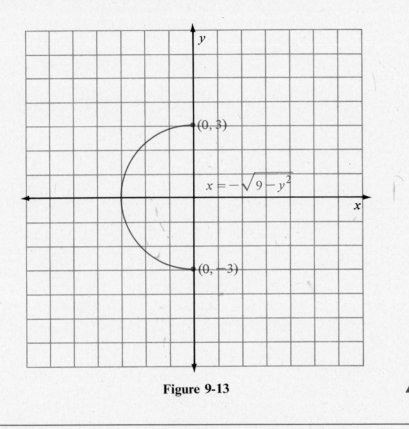

Figure 9-13

Find the distance between the following points: *Problem Set 9.2*

1. $(3, 7)$ and $(6, 3)$
2. $(4, 7)$ and $(8, 1)$
3. $(0, 9)$ and $(5, 0)$
4. $(-3, 0)$ and $(0, 4)$
5. $(3, -5)$ and $(-2, 1)$
6. $(-8, 9)$ and $(-3, -2)$
7. $(-1, -2)$ and $(-10, 5)$
8. $(-3, -8)$ and $(-1, 6)$

9. Find x so the distance between $(x, 2)$ and $(1, 5)$ is 13.
10. Find x so the distance between $(-2, 3)$ and $(x, 1)$ is 3.
11. Find y so the distance between $(7, y)$ and $(8, 3)$ is 1.
12. Find y so the distance between $(3, -5)$ and $(3, y)$ is 9.

Write the equation of the circle with the given center and radius:

13. Center $(2, 3)$; $r = 4$
14. Center $(3, -1)$; $r = 5$
15. Center $(3, -2)$; $r = 3$
16. Center $(-2, 4)$; $r = 1$

17. Center $(-5, -1)$; $r = \sqrt{5}$ 18. Center $(-7, -6)$; $r = \sqrt{3}$
19. Center $(0, -5)$; $r = 1$ 20. Center $(0, -1)$; $r = 7$
21. Center $(0, 0)$; $r = 2$ 22. Center $(0, 0)$; $r = 5$
23. Center $(-1, 0)$; $r = 2\sqrt{3}$ 24. Center $(3, 0)$; $r = 2\sqrt{2}$

Give the center and radius, and sketch the graph, of each of the following circles:

25. $x^2 + y^2 = 4$ 26. $x^2 + y^2 = 16$
27. $x^2 + y^2 = 5$ 28. $x^2 + y^2 = 3$
29. $(x - 1)^2 + (y - 3)^2 = 25$ 30. $(x - 4)^2 + (y - 1)^2 = 36$
31. $(x + 2)^2 + (y - 4)^2 = 8$ 32. $(x - 3)^2 + (y + 1)^2 = 12$
33. $(x + 1)^2 + (y + 1)^2 = 1$ 34. $(x + 3)^2 + (y + 2)^2 = 9$
35. $x^2 + y^2 - 6y = 7$ 36. $x^2 + y^2 - 4y = 5$
37. $x^2 + y^2 + 2x = 1$ 38. $x^2 + y^2 + 10x = 2$
39. $x^2 + y^2 - 4x - 6y = -12$ 40. $x^2 + y^2 - 4x + 2y = 4$
41. $x^2 + y^2 + 2x + y = 3$ 42. $x^2 + y^2 - 6x - y = 1$
43. $x^2 + y^2 - x - 3y = 2$ 44. $x^2 + y^2 + x - 5y = 1$

Graph each of the following:

45. $y = \sqrt{9 - x^2}$ 46. $y = -\sqrt{4 - x^2}$
47. $y = -\sqrt{9 - x^2}$ 48. $y = \sqrt{4 - x^2}$
49. $x = \sqrt{1 - y^2}$ 50. $x = \sqrt{25 - y^2}$
51. $x = -\sqrt{4 - y^2}$ 52. $x = -\sqrt{9 - y^2}$
53. $x = \sqrt{16 - y^2}$ 54. $x = \sqrt{49 - y^2}$
55. $y = -\sqrt{25 - x^2}$ 56. $y = -\sqrt{1 - x^2}$

57. Find the equation of the circle with center at the origin that contains the point $(3, 4)$.
58. Find the equation of the circle with center at the origin and x-intercepts $(3, 0)$ and $(-3, 0)$.
59. Find the equation of the circle with y-intercepts $(0, 4)$ and $(0, -4)$, and center at the origin.
60. A circle with center at $(-1, 3)$ passes through the point $(4, 3)$. Find the equation.
61. A circle with center at $(2, 5)$ passes through the point $(-1, 4)$. Find the equation.
62. A circle with center at $(-3, -4)$ contains the point $(2, -1)$. Find the equation.

9.3
Ellipses and Hyperbolas

This section is concerned with the graphs of ellipses and hyperbolas. To simplify matters somewhat we will only consider those graphs that are centered about the origin.

Suppose we want to graph the equation

$$\frac{x^2}{25} + \frac{y^2}{9} = 1$$

We can find the y-intercepts by letting $x = 0$, and the x-intercepts by letting $y = 0$:

When $x = 0$,   When $y = 0$,

$$\frac{0^2}{25} + \frac{y^2}{9} = 1 \qquad\qquad \frac{x^2}{25} + \frac{0^2}{9} = 1$$

$$y^2 = 9 \qquad\qquad\qquad x^2 = 25$$

$$y = \pm 3 \qquad\qquad\qquad x = \pm 5$$

The graph crosses the y-axis at $(0, 3)$ and $(0, -3)$ and the x-axis at $(5, 0)$ and $(-5, 0)$. Graphing these points and then connecting them with a smooth curve gives the graph shown in Figure 9-14.

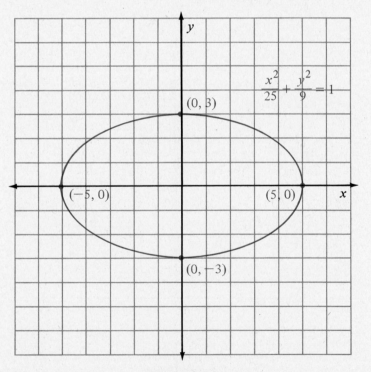

Figure 9-14

A graph of this type is called an *ellipse*. If we were to find some other ordered pairs that satisfy our original equation, we would find that their graphs lie on the ellipse. Also, the coordinates of any point on the ellipse will satisfy the equation. We can generalize these results as follows.

The graph of any equation of the form

$$\frac{x^2}{a^2} + \frac{y^2}{b^2} = 1$$

will be an ellipse. The ellipse will cross the x-axis at $(a, 0)$ and $(-a, 0)$. It will cross the y-axis at $(0, b)$ and $(0, -b)$. When a and b are equal, the ellipse will be a circle.

The most convenient method for graphing an ellipse is by locating the intercepts.

▼   **Example 1**   Graph the ellipse described by $\dfrac{x^2}{16} + \dfrac{y^2}{36} = 1$.

Solution   The x-intercepts are 4 and -4. The y-intercepts are 6 and -6. (See Figure 9-15.)

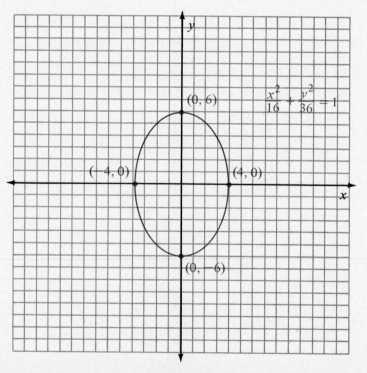

Figure 9-15   ▲

▼ **Example 2**   Sketch the graph of $4x^2 + 9y^2 = 36$.

Solution   To write the equation in the form

$$\frac{x^2}{a^2} + \frac{y^2}{b^2} = 1$$

we must divide both sides by 36:

$$\frac{4x^2}{36} + \frac{9y^2}{36} = \frac{36}{36}$$

$$\frac{x^2}{9} + \frac{y^2}{4} = 1$$

The graph crosses the x-axis at $(3, 0)$, $(-3, 0)$ and the y-axis at $(0, 2)$, $(0, -2)$. (See Figure 9-16.)

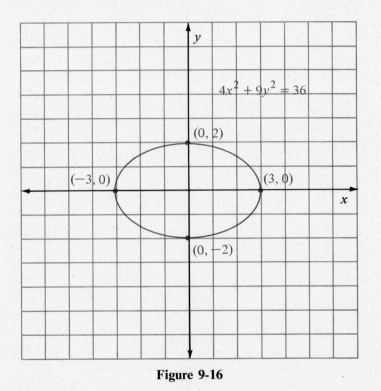

Figure 9-16   ▲

▼ **Example 3**   Sketch the graph of $x = \sqrt{16 - 4y^2}$.

Solution   Keeping in mind the fact that $x \geq 0$, we square both sides:

$$x = \sqrt{16 - 4y^2} \qquad x \geq 0$$
$$x^2 = 16 - 4y^2 \qquad x \geq 0$$
$$x^2 + 4y^2 = 16 \qquad x \geq 0$$
$$\frac{x^2}{16} + \frac{y^2}{4} = 1 \qquad x \geq 0$$

The graph lies to the right of the y-axis because $x \geq 0$. The intercepts are $(4, 0)$, $(0, 2)$, and $(0, -2)$. (We cannot include the intercept $(-4, 0)$ since x must be positive or zero.) The graph is shown in Figure 9-17.

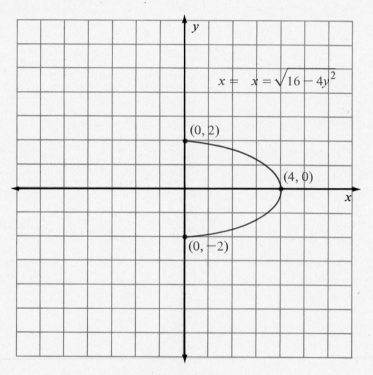

Figure 9-17

Consider the equation

$$\frac{x^2}{9} - \frac{y^2}{4} = 1$$

If we were to find a number of ordered pairs that are solutions to the equation and connect their graphs with a smooth curve, we would have Figure 9-18.

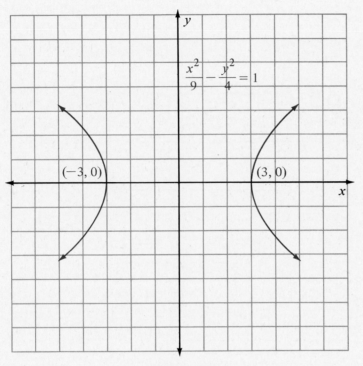

Figure 9-18

This graph is an example of a *hyperbola*. Notice that the graph has x-intercepts at $(3, 0)$ and $(-3, 0)$. The graph has no y-intercepts and hence does not cross the y-axis, since substituting $x = 0$ into the equation yields

$$\frac{0^2}{9} - \frac{y^2}{4} = 1$$

$$-y^2 = 4$$
$$y^2 = -4$$

for which there is no real solution. We can, however, use the number below y^2 to help sketch the graph. If we draw a rectangle that has its sides parallel to the x- and y-axes and that passes through the x-intercepts and the points on the y-axis corresponding to the square roots of the number below y, $+2$ and -2, it looks like Figure 9-19.

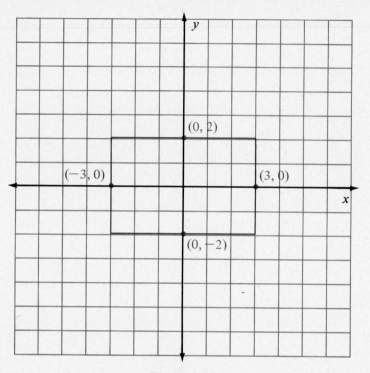

Figure 9-19

The lines that connect opposite corners of the rectangle are called *asymptotes*. The graph of the hyperbola

$$\frac{x^2}{9} - \frac{y^2}{4} = 1$$

will approach these lines. Figure 9-20 is the graph again, this time showing the asymptotes.

Although we won't attempt to prove so, it is true that there are two asymptotes associated with the graph of every hyperbola.

▼   **Example 4**   Graph the equation $\dfrac{y^2}{9} - \dfrac{x^2}{16} = 1.$

Solution   In this case the y-intercepts are 3 and -3, and the x-intercepts do not exist. We can use the square root of the number

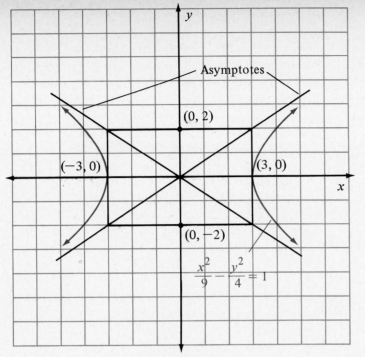

Figure 9-20

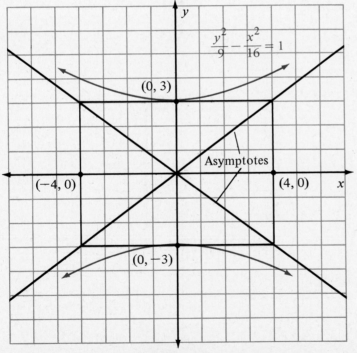

Figure 9-21

below x^2, however, to find the asymptotes associated with the graph. The sides of the rectangle used to draw the asymptotes must pass through 3 and -3 on the y-axis, and 4 and -4 on the x-axis. (See Figure 9-21.)

▲

▼ **Example 5** Sketch the graph of $5x^2 - y^2 = 25$.

Solution We begin by dividing both sides by 25:

$$\frac{5x^2}{25} - \frac{y^2}{25} = \frac{25}{25}$$

$$\frac{x^2}{5} - \frac{y^2}{25} = 1$$

The graph crosses the x-axis at $(\sqrt{5}, 0)$ and $(-\sqrt{5}, 0)$. The rectangle used to draw the asymptotes passes through these points and the points $(0, 5)$ and $(0, -5)$. (See Figure 9-22.)

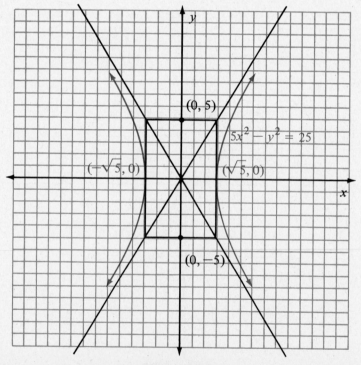

Figure 9-22

▲

▼   **Example 6**   Graph the equation $y = \sqrt{9 + x^2}$.

Solution   Keeping in mind the restriction $y \geq 0$, we square both sides:

$$
\begin{aligned}
y &= \sqrt{9 + x^2}   && y \geq 0 \\
y^2 &= (\sqrt{9 + x^2})^2   && y \geq 0 \\
y^2 &= 9 + x^2   && y \geq 0 \\
y^2 - x^2 &= 9   && y \geq 0 \\
\frac{y^2}{9} - \frac{x^2}{9} &= 1 && y \geq 0
\end{aligned}
$$

Because of the restriction $y \geq 0$, the only y-intercept is 3. (See Figure 9-23.)

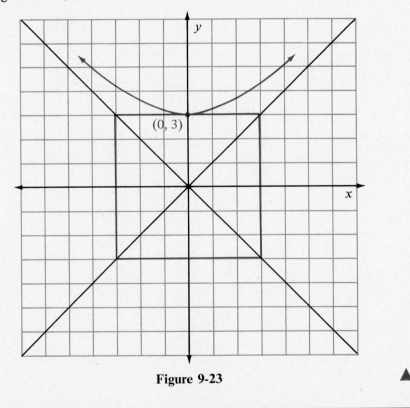

Figure 9-23   ▲

Graph each of the following. Be sure to label both the x- and y-intercepts.   *Problem Set 9.3*

1. $\dfrac{x^2}{9} + \dfrac{y^2}{16} = 1$ 2. $\dfrac{x^2}{25} + \dfrac{y^2}{4} = 1$

3. $\dfrac{x^2}{16} + \dfrac{y^2}{9} = 1$ 4. $\dfrac{x^2}{4} + \dfrac{y^2}{25} = 1$

5. $\dfrac{x^2}{3} + \dfrac{y^2}{4} = 1$ 6. $\dfrac{x^2}{4} + \dfrac{y^2}{3} = 1$

7. $4x^2 + 25y^2 = 100$ 8. $4x^2 + 9y^2 = 36$
9. $x^2 + 8y^2 = 16$ 10. $12x^2 + y^2 = 36$

Graph each of the following. Show all intercepts and the asymptotes in each case.

11. $\dfrac{x^2}{9} - \dfrac{y^2}{16} = 1$ 12. $\dfrac{x^2}{25} - \dfrac{y^2}{4} = 1$

13. $\dfrac{x^2}{16} - \dfrac{y^2}{9} = 1$ 14. $\dfrac{x^2}{4} - \dfrac{y^2}{25} = 1$

15. $\dfrac{y^2}{9} - \dfrac{x^2}{16} = 1$ 16. $\dfrac{y^2}{25} - \dfrac{x^2}{4} = 1$

17. $\dfrac{y^2}{36} - \dfrac{x^2}{4} = 1$ 18. $\dfrac{y^2}{4} - \dfrac{x^2}{36} = 1$

19. $x^2 - 4y^2 = 4$ 20. $y^2 - 4x^2 = 4$
21. $16y^2 - 9x^2 = 144$ 22. $4y^2 - 25x^2 = 100$

Graph each of the following equations:
23. $y = \sqrt{9 + x^2}$ 24. $y = \sqrt{4 + x^2}$
25. $y = -\sqrt{9 + x^2}$ 26. $y = -\sqrt{4 + x^2}$
27. $x = \sqrt{9 + y^2}$ 28. $x = \sqrt{4 + y^2}$
29. $2x = -\sqrt{36 + 9y^2}$ 30. $2y = \sqrt{36 + 9x^2}$
31. $3x = \sqrt{36 + 4y^2}$ 32. $3y = -\sqrt{36 + 4x^2}$
33. $y = -\sqrt{16 + 4x^2}$ 34. $y = \sqrt{16 + 4x^2}$

9.4
Second-Degree
Inequalities

In Section 7.4 we graphed linear inequalities by first graphing the boundary and then choosing a test point not on the boundary to indicate the region used for the solution set. The problems in this section are very similar. We will use the same general methods for graphing the inequalities in this section as we did in Section 7.4.

▼   **Example 1**   Graph $x^2 + y^2 < 16$.

Solution   The boundary is $x^2 + y^2 = 16$, which is a circle with center at the origin and a radius of 4. Since the inequality sign is $<$, the boundary is not included in the solution set and must therefore be represented with a broken line. The graph of the boundary is shown in Figure 9-24.

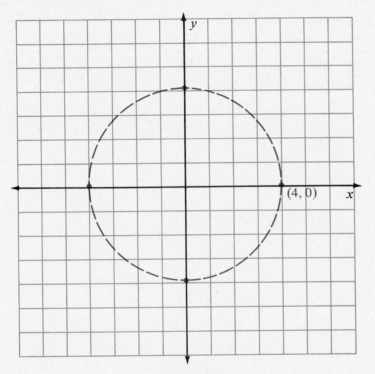

Figure 9-24

The solution set for $x^2 + y^2 < 16$ is either the region inside the circle or the region outside the circle. To see which region represents the solution set, we choose a convenient point not on the boundary and test it in the original inequality. The origin $(0, 0)$ is a convenient point. Since the origin satisfies the inequality $x^2 + y^2 < 16$, all points in the same region will also satisfy the inequality. The graph of the solution set is shown in Figure 9-25.   ▲

▼ **Example 2** Graph the inequality $y \leq x^2 - 2$.

Solution The parabola $y = x^2 - 2$ is the boundary and is included in the solution set. Using $(0, 0)$ as the test point, we see that

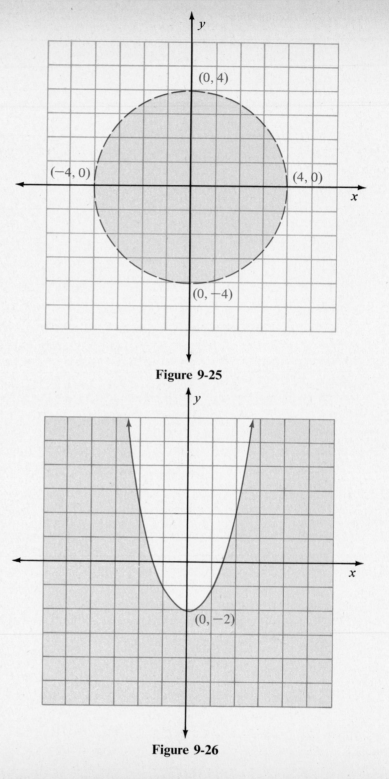

Figure 9-25

Figure 9-26

$0 \leq 0^2 - 2$ is a false statement, which means that the region containing $(0, 0)$ is not in the solution set. (See Figure 9-26.) ▲

▼ **Example 3** Graph $4y^2 - 9x^2 < 36$.

Solution The boundary is the hyperbola $4y^2 - 9x^2 = 36$ and is not included in the solution set. Testing $(0, 0)$ in the original inequality yields a true statement, which means that the region containing the origin is the solution set. (See Figure 9-27.)

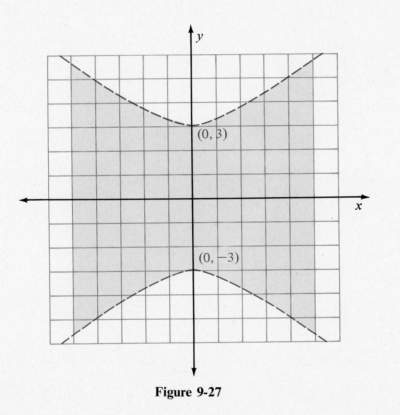

Figure 9-27   ▲

We now turn our attention to systems of inequalities. To solve a system of inequalities by graphing, we simply graph each inequality on the same set of axes. The solution set for the system is the region common to both graphs—the intersection of the individual solution sets.

▼ **Example 4** Graph the solution set for the system

$$x^2 + y^2 \leq 9$$

$$\frac{x^2}{4} + \frac{y^2}{25} \geq 1$$

Solution The boundary for the top equation is a circle with center at the origin and a radius of 3. The solution set lies inside the boundary. The boundary for the second equation is an ellipse. In this case the solution set lies outside the boundary. (See Figure 9-28.)

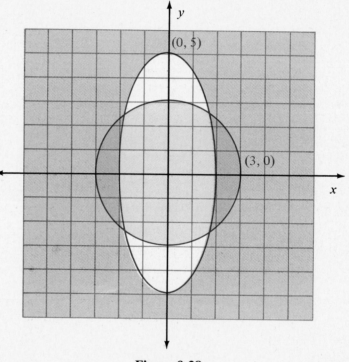

Figure 9-28

The solution set for the system is the intersection of the two individual solution sets.   ▲

Problem Set 9.4   Graph each of the following inequalities:

1. $x^2 + y^2 > 49$ 2. $x^2 + y^2 \geq 49$
3. $x^2 + y^2 \leq 49$ 4. $x^2 + y^2 < 49$

5. $(x - 2)^2 + (y + 3)^2 < 16$ **6.** $(x + 3)^2 + (y - 2)^2 \geq 25$

7. $y \leq x^2 - 4$ **8.** $y \geq x^2 + 3$

9. $y < x^2 - 6x + 7$ **10.** $y \geq x^2 + 2x - 8$

11. $\dfrac{x^2}{9} + \dfrac{y^2}{25} < 1$ **12.** $\dfrac{x^2}{9} - \dfrac{y^2}{25} > 1$

13. $\dfrac{x^2}{25} - \dfrac{y^2}{9} \geq 1$ **14.** $\dfrac{x^2}{25} - \dfrac{y^2}{9} \leq 1$

15. $4x^2 + 25y^2 \leq 100$ **16.** $25x^2 - 4y^2 > 100$

Graph the solution sets to the following systems:

17. $x^2 + y^2 < 9$ **18.** $x^2 + y^2 \leq 16$
    $\quad\quad y \geq x^2 - 1$ $\quad\quad y < x^2 + 2$

19. $\dfrac{x^2}{9} + \dfrac{y^2}{25} \leq 1$ **20.** $\dfrac{x^2}{4} + \dfrac{y^2}{16} \geq 1$

$\quad\quad \dfrac{x^2}{4} - \dfrac{y^2}{9} > 1$ $\quad\quad \dfrac{x^2}{9} - \dfrac{y^2}{25} < 1$

21. $4x^2 + 9y^2 \leq 36$ **22.** $9x^2 + 4y^2 \geq 36$
    $\quad\quad y > x^2 + 2$ $\quad\quad y < x^2 + 1$

23. $x^2 + y^2 \geq 25$ **24.** $x^2 + y^2 < 4$

$\quad\quad \dfrac{x^2}{36} - \dfrac{y^2}{25} \geq 1$ $\quad\quad x^2 - 4y^2 > 4$

Each system of equations in this section contains at least one second-degree equation. The most convenient method of solving a system that contains one or two second-degree equations is by substitution, although the elimination method can be used at times.

**9.5
Nonlinear Systems**

▼   **Example 1**   Solve the system

$$x^2 + y^2 = 4$$
$$x - 2y = 4$$

Solution   In this case the substitution method is the most convenient. Solving the second equation for x in terms of y, we have

$$x - 2y = 4$$
$$x = 2y + 4$$

We now substitute $2y + 4$ for x in the first equation in our original system and proceed to solve for y:

$$(2y + 4)^2 + y^2 = 4$$
$$4y^2 + 16y + 16 + y^2 = 4$$
$$5y^2 + 16y + 12 = 0$$
$$(5y + 6)(y + 2) = 0$$
$$5y + 6 = 0 \quad \text{or} \quad y + 2 = 0$$
$$y = -\tfrac{6}{5} \quad \text{or} \quad y = -2$$

These are the y-coordinates of the two solutions to the system. Substituting $y = -\tfrac{6}{5}$ into $x - 2y = 4$ and solving for x gives us $x = \tfrac{8}{5}$. Using $y = -2$ in the same equation yields $x = 0$. The two solutions to our system are $(\tfrac{8}{5}, -\tfrac{6}{5})$ and $(0, -2)$. Although graphing the system is not necessary, it does help us visualize the situation. (See Figure 9-29.)

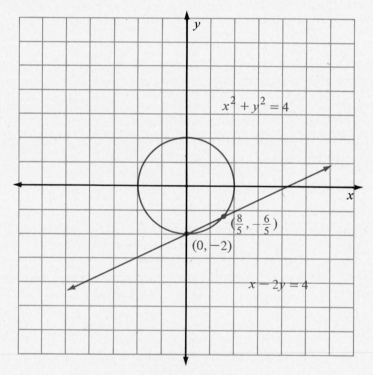

Figure 9-29

▼ **Example 2** Solve the system

$$16x^2 - 4y^2 = 64$$
$$x^2 + y^2 = 9$$

Solution   Since each equation is of the second degree in both x and y, it is easier to solve this system by eliminating one of the variables. To eliminate y we multiply the bottom equation by 4 and add the results to the top equation:

$$
\begin{aligned}
16x^2 - 4y^2 &= 64 \\
4x^2 + 4y^2 &= 36 \\
\hline
20x^2 \qquad\; &= 100
\end{aligned}
$$

$$x^2 = 5$$
$$x = \pm\sqrt{5}$$

The x-coordinates of the points of intersection are $\sqrt{5}$ and $-\sqrt{5}$. We substitute each back into the second equation in the original system and solve for y:

When $x = \sqrt{5}$,   When $x = -\sqrt{5}$,
$$(\sqrt{5})^2 + y^2 = 9 \qquad\qquad (-\sqrt{5})^2 + y^2 = 9$$
$$5 + y^2 = 9 \qquad\qquad\qquad 5 + y^2 = 9$$
$$y^2 = 4 \qquad\qquad\qquad\qquad y^2 = 4$$
$$y = \pm 2 \qquad\qquad\qquad\qquad y = \pm 2$$

The four points of intersection are $(\sqrt{5}, 2)$, $(\sqrt{5}, -2)$, $(-\sqrt{5}, 2)$, and $(-\sqrt{5}, -2)$. Graphically the situation is as shown in Figure 9-30.   ▲

▼ **Example 3**   Solve the system

$$
\begin{aligned}
x^2 - 2y &= 2 \\
y &= x^2 - 3
\end{aligned}
$$

Solution   We can solve this system using the substitution method. Replacing y in the first equation with $x^2 - 3$ from the second equation, we have

$$x^2 - 2(x^2 - 3) = 2$$
$$-x^2 + 6 = 2$$
$$x^2 = 4$$
$$x = \pm 2$$

Using either $+2$ or -2 in the equation $y = x^2 - 3$ gives us $y = 1$. The system has two solutions: $(2, 1)$ and $(-2, 1)$. The graph of the system supports our results. (See Figure 9-31.)   ▲

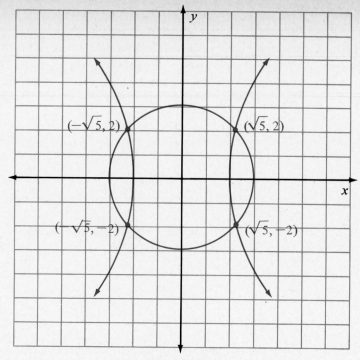

Figure 9-30

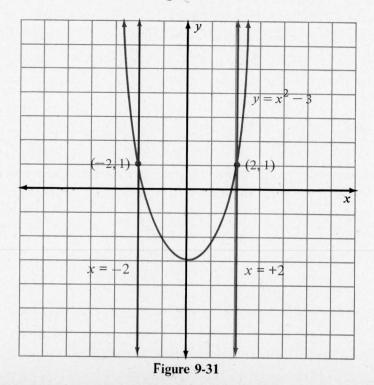

Figure 9-31

▼ **Example 4** The sum of the squares of two numbers is 34. The difference of their squares is 16. Find the two numbers.

Solution Let x and y be the two numbers. The sum of their squares is $x^2 + y^2$ and the difference of their squares is $x^2 - y^2$. (We can assume here that x^2 is the larger number.) The system of equations that describes the situation is

$$x^2 + y^2 = 34$$
$$x^2 - y^2 = 16$$

We can eliminate y by simply adding the two equations. The result of doing so is

$$2x^2 = 50$$
$$x^2 = 25$$
$$x = \pm 5$$

Substituting $x = 5$ into either equation in the system gives $y = \pm 3$. Using $x = -5$ gives the same results, $y = \pm 3$. The four pairs of numbers that are solutions to the original problem are

$$\{5, 3\} \qquad \{-5, 3\} \qquad \{5, -3\} \qquad \{-5, -3\} \qquad ▲$$

Solve each of the following systems of equations:

Problem Set 9.5

1. $x^2 + y^2 = 9$
    $2x + y = 3$
2. $x^2 + y^2 = 9$
    $x + 2y = 3$
3. $x^2 + y^2 = 16$
    $x + 2y = 8$
4. $x^2 + y^2 = 16$
    $x - 2y = 8$
5. $x^2 + y^2 = 25$
    $x^2 - y^2 = 25$
6. $x^2 - y^2 = 4$
    $2x^2 + y^2 = 5$
7. $x^2 + y^2 = 9$
    $y = x^2 - 3$
8. $x^2 + y^2 = 9$
    $y = x^2 + 3$
9. $x^2 + y^2 = 1$
    $y = x^2 + 1$
10. $x^2 + y^2 = 1$
    $y = x^2 - 1$
11. $4x^2 - 3y^2 = 12$
    $2x + 3y = 6$
12. $3x^2 + 4y^2 = 12$
    $3x - 2y = 6$
13. $3x + 2y = 10$
    $y = x^2 - 5$
14. $4x + 2y = 10$
    $y = x^2 - 10$
15. $4x^2 - 9y^2 = 36$
    $4x^2 + 9y^2 = 36$
16. $4x^2 + 25y^2 = 100$
    $4x^2 - 25y^2 = 100$
17. $2x^2 - 4y^2 = 8$
    $x^2 + 2y^2 = 10$
18. $4x^2 + 2y^2 = 8$
    $2x^2 - y^2 = 10$

19. $x - y = 4$
      $x^2 + y^2 = 16$

20. $x + y = 2$
      $x^2 - y^2 = 4$

21.   The sum of the squares of two numbers is 89. The difference of the numbers is 3. Find the numbers.

22.   The difference of the squares of two numbers is 35. The sum of their squares is 37. Find the numbers.

23.   One number is 3 less than the square of another. Their sum is 9. Find the numbers.

24.   The square of one number is 2 less than twice the square of another. The sum of the squares of the two numbers is 25. Find the numbers.

25.   Suppose the total weekly cost c to manufacture x items is given by the equation $c = x^2 - 90x + 1500$. If each item sells for $25, then the total revenue obtained from selling x items per week is $R = 25x$. The company manufacturing the item will break even when the cost of making the x items is equal to the revenue obtained from selling them. How many should be produced in order to break even?

Chapter 9
Summary and Review

Conic Sections

Each of the four conic sections can be obtained by slicing a cone with a plane at different angles as shown in Figure 9-32.

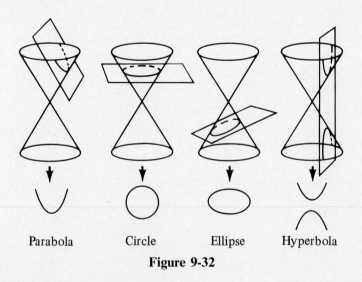

Parabola   Circle   Ellipse   Hyperbola

Figure 9-32

The Parabola

The graph of any equation of the form

$$y = ax^2 + bx + c \qquad a \neq 0$$

is a parabola. The graph is concave-up if $a > 0$, and concave-down if $a < 0$. The highest or lowest point on the graph is called the *vertex* and will always occur when $x = -b/2a$.

The graph of any equation of the form   *The Circle*

$$(x - a)^2 + (y - b)^2 = r^2$$

will be a circle having its center at (a, b) and a radius of r.

Any equation that can be put in the form   *The Ellipse*

$$\frac{x^2}{a^2} + \frac{y^2}{b^2} = 1$$

will have an ellipse for its graph. The x-intercepts will be at a and $-a$, the y-intercepts at b and $-b$.

The graph of an equation that can be put in either of the forms   *The Hyperbola*

$$\frac{x^2}{a^2} - \frac{y^2}{b^2} = 1 \quad \text{or} \quad \frac{y^2}{b^2} - \frac{x^2}{a^2} = 1$$

will be a hyperbola. The x-intercepts, if they exist, will be a and $-a$. The y-intercepts, if they occur, will be at b and $-b$. There are two straight lines called asymptotes associated with the graph of every hyperbola. Although the asymptotes are not part of the hyperbola, they are useful in sketching the graph.

---

1. Sketch the graph of each of the following. Give the coordinates of the   **Chapter 9**
   vertex in each case.   **Test**

   a. $y = x^2 - x - 12$ **b.** $y = -x^2 + 2x + 8$

2. Give the equation of the circle with center at $(-2, 4)$ and radius 3.
3. Give the equation of the circle with center at the origin that contains the point $(-3, -4)$.
4. Find the center, radius, and sketch the graph of the circle whose equation is $x^2 + y^2 - 10x + 6y = 5$.
5. Graph each of the following:

    a. $4x^2 - y^2 = 16$

    b. $3x^2 + 9y^2 = 27$

    c. $2y = -\sqrt{36 - 9x^2}$

    d. $x = \sqrt{16 - y^2}$

    e. $\dfrac{x^2}{4} + \dfrac{y^2}{25} \leq 1$

6. Solve the following systems:

    a. $x^2 + y^2 = 25$ **b.** $x^2 + y^2 = 16$

          $2x + y = 5$ $y = x^2 - 4$

Relations and Functions 10

To the student:

In this chapter we will study two main concepts, relations and functions. Relations and functions have many applications in the real world. The idea of a relation is already familiar to us on an intuitive level. When we say "the price of gasoline is increasing because there is more demand for it this year," we are expressing a relationship between the price of gasoline and the demand for it. We are implying the price of gasoline is a function of the demand for it. Mathematics becomes a part of this problem when we express, with an equation, the exact relationship between the two quantities.

Actually, we have been working with functions and relations for some time now, we just haven't said so. Any time we have used an equation containing two variables we have been using a relation. This chapter is really just a formalization and classification of concepts and equations that we have already encountered. The chapter begins with some basic definitions associated with functions and relations. We will develop a new notation associated with functions called function notation; we will consider combinations of functions and exponential functions. The chapter ends with a section on the inverse of a function.

We begin this section with the definition of a relation. It is apparent from the definition that we have worked with relations many times previously in this book.

**10.1
Relations and
Functions**

DEFINITION   A *relation* is any set of ordered pairs. The set of all first coordinates is called the *domain* of the relation, and the set of all second coordinates is said to be the *range* of the relation.

There are two ways to specify the ordered pairs in a relation. One method is to simply list them. The other method is to give the rule (equation) for obtaining them.

▼   **Example 1**   The set $\{(1, 2), (-3, \frac{1}{2}), (\pi, -4), (0, 1)\}$ is a relation. The domain for this relation is $\{1, -3, \pi, 0\}$ and the range is $\{2, \frac{1}{2}, -4, 1\}$.   ▲

▼   **Example 2**   The set of ordered pairs given by $\{(x, y) \mid x + y = 5\}$ is an example of a relation. In this case we have written the relation in terms of the equation used to obtain the ordered pairs in the relation. This relation is the set of all ordered pairs whose coordinates have a sum of 5. Some members of this relation are $(1, 4)$, $(0, 5)$, $(5, 0)$, $(-1, 6)$, and $(\frac{1}{2}, \frac{9}{2})$. It is impossible to list all the members of this relation since there are an infinite number of ordered pairs that are solutions to the equation $x + y = 5$. The domain, although not given directly, is the set of real numbers; the range is also. The graph of this relation is a straight line, shown in Figure 10-1.   ▲

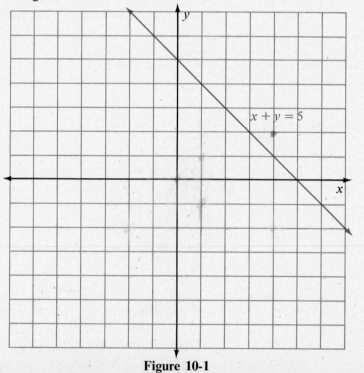

Figure 10-1

DEFINITION   A *function* is a relation in which no two different ordered pairs have the same first coordinates. The *domain* and *range* of a function are the sets of first and second coordinates, respectively.

   A function is simply a relation that does not repeat any first coordinates. The restriction on first coordinates may seem arbitrary at first. As it turns out, restricting ordered pairs to those that do not repeat first coordinates allows us to generalize and develop properties we would otherwise be unable to work with.

▼   **Example 3**   The relation $\{(2, 3), (5, 2), (6, 3)\}$ is also a function, since no two ordered pairs have the same first coordinates. The relation $\{(1, 7), (3, 7), (1, 5)\}$ is not a function since two of its ordered pairs, $(1, 7)$ and $(1, 5)$, have the same first coordinates.   ▲

If the ordered pairs of a relation are given in terms of an equation rather than a list, we can use the graph of the relation to determine if the relation is a function or not. Any two ordered pairs with the same first coordinates will lie along a vertical line parallel to the y-axis. Therefore, if a vertical line crosses the graph of a relation in more than one place, the relation cannot be a function. If no vertical line can be found that crosses the graph in more than one place, the relation must be a function. Testing to see if a relation is a function by observing if a vertical line crosses the graph of the relation in more than one place is called the *vertical line test*.

Vertical Line Test

▼   **Example 4**   Use the vertical line test to see which of the following are functions: (a) $y = x^2 - 2$; (b) $x^2 + y^2 = 9$.

   Solution   The graph of each relation is given in Figure 10-2. The equation $y = x^2 - 2$ is a function, since there are no vertical lines that cross its graph in more than one place. The equation $x^2 + y^2 = 9$ does not represent a function, since we can find a vertical line that crosses its graph in more than one place.   ▲

When a function (or relation) is given in terms of an equation, the domain is the set of all possible replacements for the variable x. If the domain of a function (or relation) is not specified, it is assumed to be all real numbers that do not give undefined terms in the equation. That is, we cannot use values of x in the domain that will produce 0 in a denominator or the square root of a negative number.

The Domain and Range of a Function

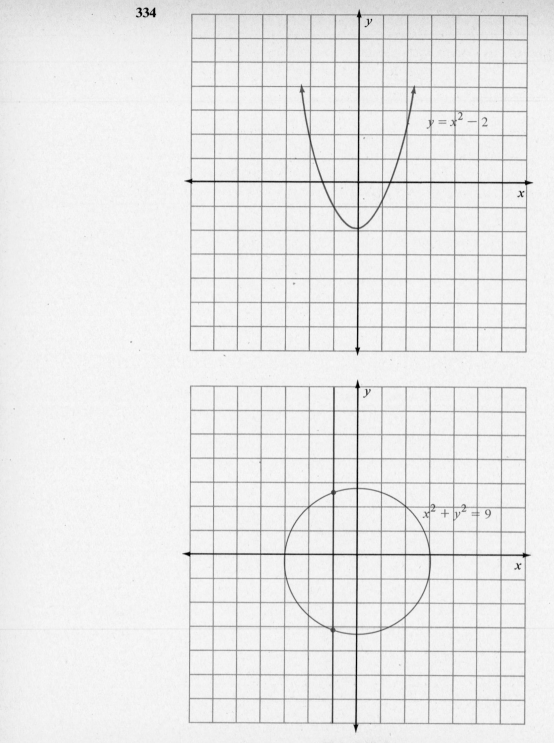

$y = x^2 - 2$

$x^2 + y^2 = 9$

Figure 10-2

▼   **Example 5**   Specify the domain for $y = \dfrac{1}{x - 3}$.

Solution   The domain can be any real number that does not produce an undefined term. If $x = 3$, the denominator on the right side will be 0. Hence, the domain is all real numbers except 3.   ▲

▼   **Example 6**   Give the domain for $y = \sqrt{x - 4}$.

Solution   Since the domain must consist of real numbers, the quantity under the radical will have to be greater than or equal to 0:

$$x - 4 \geq 0$$
$$x \geq 4$$

The domain in this case is $\{x \mid x \geq 4\}$.   ▲

   The graph of a function (or relation) is sometimes helpful in determining the domain and range.

▼   **Example 7**   Give the domain and range for $9x^2 + 4y^2 = 36$.

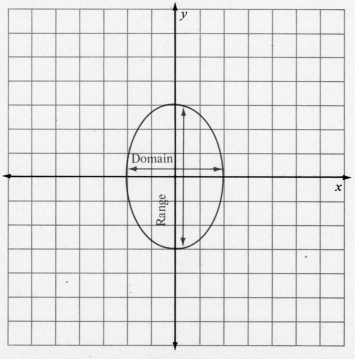

Figure 10-3

Solution The graph is the ellipse shown in Figure 10-3.

From the graph we have

$$\text{Domain} = \{x \mid -2 \leq x \leq 2\}$$
$$\text{Range} = \{y \mid -3 \leq y \leq 3\}$$

Note also that by applying the vertical line test we see the relation is not a function. ▲

▼ **Example 8** Give the domain and range for $y = x^2 - 3$.

Solution The graph is the parabola shown in Figure 10-4.

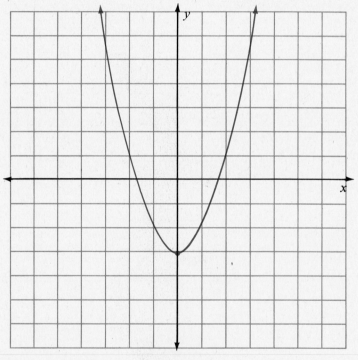

Figure 10-4

The domain is all real numbers, and the range is the set $\{y \mid y \geq -3\}$. Since no vertical line will cross the graph in more than one place, the graph represents a function. ▲

▼ **Example 9** Give the domain and range for $y = \sqrt{9 - x^2}$.

Solution   Since y is equal to a positive square root, y must be positive or zero. Keeping this in mind, we square both sides of the equation:

$$y^2 = 9 - x^2 \qquad y \geq 0$$
$$x^2 + y^2 = 9 \qquad y \geq 0$$

The graph of the last line is shown in Figure 10-5.

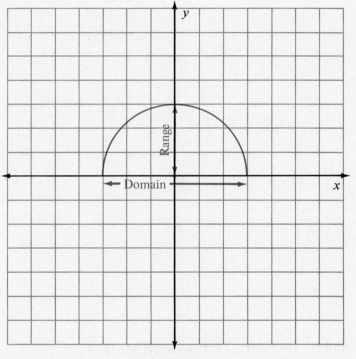

Figure 10-5

The graph is the graph of a function. The domain is $\{x \mid -3 \leq x \leq 3\}$ and the range is $\{y \mid 0 \leq y \leq 3\}$. ▲

For each of the following relations, give the domain and range and indicate which are also functions:

Problem Set 10.1

1. $\{(1, 3), (2, 5), (4, 1)\}$ **2.** $\{(3, 1), (5, 7), (2, 3)\}$
3. $\{(-1, 3), (1, 3), (2, -5)\}$ **4.** $\{(3, -4), (-1, 5), (3, 2)\}$
5. $\{(7, -1), (3, -1), (7, 4)\}$ **6.** $\{(5, -2), (3, -2), (5, -1)\}$
7. $\{(4, 3), (3, 4), (3, 5)\}$ **8.** $\{(4, 1), (1, 4), (-1, -4)\}$

9. $\{(5, -3), (-3, 2), (2, -3)\}$ **10.** $\{(2, 4), (3, 4), (4, 4)\}$

Which of the following graphs represent functions?

11.

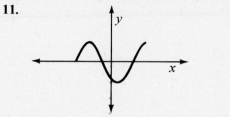

12.

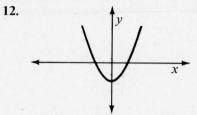

13.

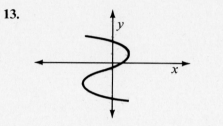

14.

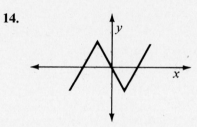

15.

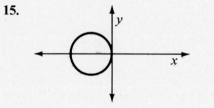

16.

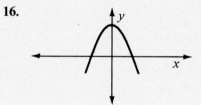

17.

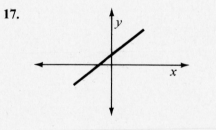

18.

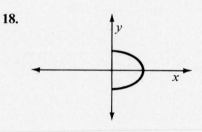

19.

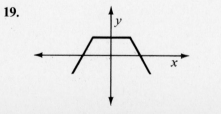

20.

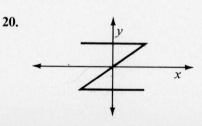

Give the domain for each of the following functions:

21. $y = \sqrt{x + 3}$ 22. $y = \sqrt{x + 4}$
23. $y = \sqrt{2x - 1}$ 24. $y = \sqrt{3x + 2}$
25. $y = \sqrt{1 - 4x}$ 26. $y = \sqrt{2 + 3x}$

27. $y = \dfrac{x + 2}{x - 5}$ 28. $y = \dfrac{x - 3}{x + 4}$

29. $y = \dfrac{3}{2x^2 + 5x - 3}$ 30. $y = \dfrac{-1}{3x^2 - 5x + 2}$

31. $y = \dfrac{-3}{x^2 - x - 6}$ 32. $y = \dfrac{4}{x^2 - 2x - 8}$

33. $y = \dfrac{4}{x^2 - 4}$ 34. $y = \dfrac{2}{x^2 - 9}$

Graph each of the following relations. Use the graph to find the domain and range, and indicate which relations are also functions.

35. $x^2 + 4y^2 = 16$ 36. $4x^2 + y^2 = 16$
37. $(x - 2)^2 + (y + 1)^2 = 9$ 38. $(x + 3)^2 + (y - 4)^2 = 25$
39. $x^2 + y^2 - 4x = 12$ 40. $x^2 + y^2 + 6x = 16$
41. $y = x^2 - x - 12$ 42. $y = x^2 + 2x - 8$
43. $y = 2x - 3$ 44. $y = 3x - 2$

45. $\dfrac{x^2}{4} - \dfrac{y^2}{9} = 1$ 46. $\dfrac{x^2}{9} - \dfrac{y^2}{4} = 1$

47. $\dfrac{x^2}{9} + \dfrac{y^2}{4} = 1$ 48. $\dfrac{x^2}{4} + \dfrac{y^2}{9} = 1$

49. $y = \sqrt{16 - x^2}$ 50. $y = -\sqrt{25 - x^2}$
51. $y = -\sqrt{9 + x^2}$ 52. $y = \sqrt{16 + x^2}$
53. $2y = \sqrt{36 - 9x^2}$ 54. $2y = -\sqrt{36 - 9x^2}$

55. A ball is thrown straight up into the air from ground level. The relationship between the height (h) of the ball at any time (t) is illustrated by the following graph:

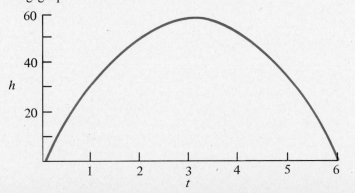

The horizontal axis represents time (t) and the vertical axis represents height (h).

a.   Is this graph the graph of a function?
b.   Identify the domain and range.
c.   At what time does the ball reach its maximum height?
d.   What is the maximum height of the ball?
e.   At what time does the ball hit the ground?

10.2
Function Notation

Consider the function

$$y = 3x - 2$$

Up to this point we have expressed the functions we have worked with as y in terms of x. There is an alternative to expressing y in terms of x called *function notation*. The notation $f(x)$ is read "f of x" and can be used instead of the letter y when writing functions. That is, the equations $y = 3x - 2$ and $f(x) = 3x - 2$ are equivalent. The symbols $f(x)$ and y are interchangeable when we are working with functions. If we wanted to find the value of y when x is 4 in the equation $y = 3x - 2$, we would have to say, "If $x = 4$, then $y = 3(4) - 2 = 10$." With function notation we simply write $f(4) = 3(4) - 2 = 10$. The following table illustrates the equivalence of the notations y and $f(x)$:

y in terms of x	Function notation
$y = x^2 - 3$	$f(x) = x^2 - 3$
If $x = 2$, then $y = 2^2 - 3 = 1$	$f(2) = (2)^2 - 3 = 1$
If $x = -4$, then $y = (-4)^2 - 3 = 13$	$f(-4) = (-4)^2 - 3 = 13$
If $x = 0$, then $y = 0^2 - 3 = -3$	$f(0) = 0^2 - 3 = -3$

It is apparent from the table that one advantage of using function notation is it takes less space to write some statements.

Note   The notation $f(x)$ does *not* mean "f times x." It is a special kind of notation that does not imply multiplication.

▼ **Example 1**   If $f(x) = 3x^2 + 2x - 1$, find $f(0), f(3), f(5)$, and $f(-2)$.

Solution   Since $f(x) = 3x^2 + 2x - 1$, we have

$$f(0) = 3(0)^2 + 2(0) - 1 = 0 + 0 - 1 = -1$$
$$f(3) = 3(3)^2 + 2(3) - 1 = 27 + 6 - 1 = 32$$
$$f(5) = 3(5)^2 + 2(5) - 1 = 75 + 10 - 1 = 84$$
$$f(-2) = 3(-2)^2 + 2(-2) - 1 = 12 - 4 - 1 = 7$$

▲

In the above example the function f is defined by the equation $f(x) = 3x^2 + 2x - 1$. We could just as easily have said $y = 3x^2 + 2x - 1$. That is, $y = f(x)$. Saying $f(-2) = 7$ is exactly the same as saying y is 7 when x is -2. If $f(-2) = 7$, then the ordered pair $(-2, 7)$ belongs to the function f; and conversely, if the ordered pair $(-2, 7)$ belongs to f, then $f(-2) = 7$. We can generalize this discussion by saying

$$(a, b) \in f \quad \text{if and only if} \quad f(a) = b$$

where $\in$ is read "belongs to."

▼   **Example 2**   If the function f is given by

$$f = \{(-2, 0), (3, -1), (2, 4), (3, 5)\}$$

then $f(-2) = 0$, $f(3) = -1$, $f(2) = 4$, and $f(3) = 5$.   ▲

▼   **Example 3**   If $f(x) = 4x - 1$ and $g(x) = x^2 + 2$, then

$$f(5) = 4(5) - 1 = 19 \qquad \text{and} \quad g(5) = 5^2 + 2 = 27$$
$$f(-2) = 4(-2) - 1 = -9 \quad \text{and} \quad g(-2) = (-2)^2 + 2 = 6$$
$$f(0) = 4(0) - 1 = -1 \qquad \text{and} \quad g(0) = 0^2 + 2 = 2$$
$$f(z) = 4z - 1 \qquad\qquad \text{and} \quad g(z) = z^2 + 2$$
$$f(a) = 4a - 1 \qquad\qquad \text{and} \quad g(a) = a^2 + 2$$

The ordered pairs $(5, 19)$, $(-2, -9)$, $(0, -1)(z, 4z - 1)$, and $(a, 4a - 1)$ belong to the function f, while the ordered pairs $(5, 27)(-2, 6)$, $(0, 2)$, $(z, z^2 + 2)$, and $(a, a^2 + 2)$ belong to the function g.   ▲

▼   **Example 4**   If $f(x) = 2x^2$ and $g(x) = 3x - 1$, find (a) $f[g(2)]$; (b) $g[f(2)]$; (c) $f[g(x)]$; and (d) $g[f(x)]$.

Solutions
a. Since $g(2) = 3(2) - 1 = 5$,
$$f[g(2)] = f(5) = 2(5)^2 = 50$$
b. Since $f(2) = 2(2)^2 = 8$,
$$g[f(2)] = g(8) = 3(8) - 1 = 23$$
c. Substituting $3x - 1$ for $g(x)$ in the expression $f[g(x)]$, we have

$$\begin{aligned} f[g(x)] &= f(3x - 1) \\ &= 2(3x - 1)^2 \\ &= 2(9x^2 - 6x + 1) \\ &= 18x^2 - 12x + 2 \end{aligned}$$

d.   Substituting $2x^2$ for $f(x)$ in the expression $g[f(x)]$, we have

$$\begin{aligned} g[f(x)] &= g(2x^2) \\ &= 3(2x^2) - 1 \\ &= 6x^2 - 1 \end{aligned}$$

The last two expressions in Example 4, $f[g(x)]$ and $g[f(x)]$, are called the *composition* of f with g and the *composition* of g with f, respectively.   ▲

▼   **Example 5**   If $f(x) = 2x - 3$, find $\dfrac{f(x + h) - f(x)}{h}$.

Solution   The expression $f(x + h)$ is given by

$$\begin{aligned} f(x + h) &= 2(x + h) - 3 \\ &= 2x + 2h - 3 \end{aligned}$$

Using this result gives us

$$\begin{aligned} \frac{f(x + h) - f(x)}{h} &= \frac{(2x + 2h - 3) - (2x - 3)}{h} \\ &= \frac{2h}{h} \\ &= 2 \end{aligned}$$

The expression

$$\frac{f(x + h) - f(x)}{h}$$

is a very important formula used in calculus. We are using it here just for practice.   ▲

▼   **Example 6**   If $f(x) = x^2 - 3x$, find $\dfrac{f(x + h) - f(x)}{h}$.

Solution

$$\begin{aligned} \frac{f(x + h) - f(x)}{h} &= \frac{[(x + h)^2 - 3(x + h)] - (x^2 - 3x)}{h} \\ &= \frac{x^2 + 2xh + h^2 - 3x - 3h - x^2 + 3x}{h} \\ &= \frac{2xh - 3h + h^2}{h} \\ &= 2x - 3 + h \end{aligned}$$

   ▲

▼ **Example 7** If $f(x) = 2x - 4$, find $\dfrac{f(x) - f(a)}{x - a}$.

Solution

$$\frac{f(x) - f(a)}{x - a} = \frac{(2x - 4) - (2a - 4)}{x - a}$$

$$= \frac{2x - 2a}{x - a}$$

$$= \frac{2(x - a)}{x - a}$$

$$= 2 \qquad\qquad\qquad ▲$$

Let $f(x) = 2x - 5$ and $g(x) = x^2 + 3x + 4$. Evaluate the following: *Problem Set 10.2*

1. $f(2)$ 2. $g(3)$
3. $f(-3)$ 4. $g(-2)$
5. $f(0)$ 6. $f(5)$
7. $g(-1)$ 8. $f(-4)$
9. $g(-3)$ 10. $g(2)$
11. $g(4) + f(4)$ 12. $f(2) - g(3)$
13. $f(3) - g(2)$ 14. $g(-1) + f(-1)$

If $f = \{(1, 4), (-2, 0), (3, \frac{1}{2}), (\pi, 0)\}$ and $g = \{(1, 1), (-2, 2), (\frac{1}{2}, 0)\}$, find each of the following values of f and g:

15. $f(1)$ 16. $g(1)$
17. $g(\frac{1}{2})$ 18. $f(3)$
19. $g(-2)$ 20. $f(\pi)$
21. $f(-2) + g(-2)$ 22. $g(1) + f(1)$
23. $f(-2) - g(1) + f(3)$ 24. $g(1) + g(-2) + f(-2)$

Let $f(x) = 3x^2 - 4x + 1$ and $g(x) = 2x - 1$. Evaluate each of the following:

25. $f(0)$ 26. $g(0)$
27. $g(-4)$ 28. $f(1)$
29. $f(a)$ 30. $g(z)$
31. $f(a + 3)$ 32. $g(a - 2)$
33. $f[g(2)]$ 34. $g[f(z)]$
35. $g[f(-1)]$ 36. $f[g(-2)]$
37. $g[f(0)]$ 38. $f[g(0)]$
39. $f[g(x)]$ 40. $g[f(x)]$

For each of the following functions, evaluate the quantity $\dfrac{f(x + h) - f(x)}{h}$:

41. $f(x) = 2x + 3$ **42.** $f(x) = 3x - 2$
43. $f(x) = x^2$ **44.** $f(x) = x^2 - 3$
45. $y = -4x - 1$ **46.** $y = -x + 4$
47. $y = 3x^2 - 2$ **48.** $y = 4x^2 + 3$
49. $f(x) = 2x^2 + 3x + 4$ **50.** $f(x) = 4x^2 + 3x + 2$

For each of the following functions evaluate the quantity $\dfrac{f(x) - f(a)}{x - a}$:

51. $f(x) = 3x$ **52.** $f(x) = -2x$
53. $f(x) = 4x - 5$ **54.** $f(x) = 3x + 1$
55. $f(x) = x^2$ **56.** $f(x) = 2x^2$
57. $y = 5x - 3$ **58.** $y = -2x + 7$
59. $y = x^2 + 1$ **60.** $y = x^2 - 1$

61.   Suppose the total cost C of manufacturing x items is given by the equation $C(x) = 2x^2 + 5x + 100$. If the notation $C(x)$ is used in the same way we have been using $f(x)$, then $C(10) = 2(10)^2 + 5(10) + 100 = 350$ is the total cost, in dollars, of manufacturing 10 items. Find the total cost of producing 1 item, 5 items, and 20 items.

**10.3
Algebra with
Functions**

If we are given two functions, f and g, with a common domain, we can define four other functions as follows.

DEFINITION

$(f + g)(x) = f(x) + g(x)$ The function $f + g$ is the sum of the functions f and g.

$(f - g)(x) = f(x) - g(x)$ The function $f - g$ is the difference of the functions f and g.

$(fg)(x) = f(x)g(x)$ The function fg is the product of the functions f and g.

$\dfrac{f}{g}(x) = \dfrac{f(x)}{g(x)}$ The function f/g is the quotient of the functions f and g, where $g(x) \neq 0$.

▼ **Example 1**   If $f(x) = 4x^2 + 3x + 2$ and $g(x) = 2x^2 - 5x - 6$, write the formula for the functions $f + g$, $f - g$, fg, and f/g.

Solution The function $f + g$ is defined by

$$(f + g)(x) = f(x) + g(x)$$
$$= (4x^2 + 3x + 2) + (2x^2 - 5x - 6)$$
$$= 6x^2 - 2x - 4$$

The function $f - g$ is defined by

$$(f - g)(x) = f(x) - g(x)$$
$$= (4x^2 + 3x + 2) - (2x^2 - 5x - 6)$$
$$= 4x^2 + 3x + 2 - 2x^2 + 5x + 6$$
$$= 2x^2 + 8x + 8$$

The function fg is defined by

$$(fg)(x) = f(x)g(x)$$
$$= (4x^2 + 3x + 2)(2x^2 - 5x - 6)$$
$$= 8x^4 - 20x^3 - 24x^2$$
$$+ 6x^3 - 15x^2 - 18x$$
$$+ 4x^2 - 10x - 12$$
$$= 8x^4 - 14x^3 - 35x^2 - 28x - 12$$

The function f/g is defined by

$$(f/g)(x) = f(x)/g(x)$$
$$= \frac{4x^2 + 3x + 2}{2x^2 - 5x - 6} \qquad \blacktriangle$$

▼ **Example 2**   Let $f(x) = 4x - 3$, $g(x) = 4x^2 - 7x + 3$, and $h(x) = x - 1$. Find $f + g$, fh, fg, and g/f.

Solution The function $f + g$, the sum of functions f and g, is defined by

$$(f + g)(x) = f(x) + g(x)$$
$$= (4x - 3) + (4x^2 - 7x + 3)$$
$$= 4x^2 - 3x$$

The function fh, the product of functions f and h, is defined by

$$(fh)(x) = f(x)h(x)$$
$$= (4x - 3)(x - 1)$$
$$= 4x^2 - 7x + 3$$
$$= g(x)$$

The product of the functions f and g, fg, is given by

$$(fg)(x) = f(x)g(x)$$
$$= (4x - 3)(4x^2 - 7x + 3)$$
$$= 16x^3 - 28x^2 + 12x - 12x^2 + 21x - 9$$
$$= 16x^3 - 40x^2 + 33x - 9$$

The quotient of the functions g and f, g/f, is defined as

$$\frac{g}{f}(x) = \frac{g(x)}{f(x)}$$

$$= \frac{4x^2 - 7x + 3}{4x - 3}$$

Factoring the numerator, we can reduce to lowest terms:

$$\frac{g}{f}(x) = \frac{(4x - 3)(x - 1)}{4x - 3}$$

$$= x - 1$$
$$= h(x)$$ ▲

▼   **Example 3**   If f, g, and h are the same functions defined in Example 2, evaluate $(f + g)(2)$, $(fh)(-1)$, $(fg)(0)$, and $(g/f)(5)$.

Solution   We use the formulas for $f + g$, fh, fg, and g/f found in Example 2:

$$(f + g)(2) = 4(2)^2 - 3(2)$$
$$= 16 - 6$$
$$= 10$$
$$(fh)(-1) = 4(-1)^2 - 7(-1) + 3$$
$$= 4 + 7 + 3$$
$$= 14$$
$$(fg)(0) = 16(0)^3 - 40(0)^2 + 33(0) - 9$$
$$= 0 - 0 + 0 - 9$$
$$= -9$$

$$\frac{g}{f}(5) = 5 - 1$$

$$= 4$$ ▲

Problem Set 10.3   Let $f(x) = 4x - 3$ and $g(x) = 2x + 5$. Write a formula for each of the following functions:

   1. $f + g$ **2.** $f - g$

3. $g - f$ **4.** $g + f$
5. fg **6.** f/g
7. g/f **8.** ff

If the functions f, g, and h are defined by $f(x) = 3x - 5$, $g(x) = x - 2$, and $h(x) = 3x^2 - 11x + 10$, write a formula for each of the following functions:

9. $g + f$ **10.** $f + h$
11. $g + h$ **12.** $f - g$
13. $g - f$ **14.** $h - g$
15. fg **16.** gf
17. fh **18.** gh
19. h/f **20.** h/g
21. f/h **22.** g/h
23. $f + g + h$ **24.** $h - g + f$
25. $h + fg$ **26.** $h - fg$

Let $f(x) = 2x + 1$, $g(x) = 4x + 2$, and $h(x) = 4x^2 + 4x + 1$, and find the following:

27. $(f + g)(2)$ **28.** $(f - g)(-1)$
29. $(fg)(3)$ **30.** $(f/g)(-3)$
31. $(h/g)(1)$ **32.** $(hg)(1)$
33. $(fh)(0)$ **34.** $(h - g)(-4)$
35. $(f + g + h)(2)$ **36.** $(h - f + g)(0)$
37. $(h + fg)(3)$ **38.** $(h - fg)(5)$

39. In California it costs 33¢ for the first minute and 24¢ for each additional minute to place a long-distance call out of state between 5 P.M. and 11 P.M. If x is the number of additional minutes and $f(x)$ is the cost of the call, then $f(x) = 24x + 33$.

    a. How much does it cost to talk for 10 minutes?
    b. What does $f(5)$ represent in this problem?
    c. If a call costs $1.29, how long was it?

The same phone company charges 52¢ for the first minute and 36¢ for each additional minute to place an out-of-state call between 8 A.M. and 5 P.M.

    d. Let $g(x)$ be the total cost of an out-of-state call between 8 A.M. and 5 P.M. and write an equation for $g(x)$.
    e. Find $g(5)$.
    f. Find the difference in price between a 10-minute call made between 8 A.M. and 5 P.M. and the cost of the same call made between 5 P.M. and 11 P.M.

10.4
Classification of
Functions

Much of the work we have done in previous chapters has involved functions. All linear equations in two variables, except those with vertical lines for graphs, are functions. The parabolas we worked with in Chapter 9 are graphs of functions. We will begin this section by looking at some of the major classifications of functions. We will then spend the remainder of this section working with exponential functions.

Constant Functions

Any function that can be written in the form

$$f(x) = c$$

where c is a real number, is called a *constant function*. The graph of every constant function is a horizontal line.

▼   **Example 1**   The function $f(x) = 3$ is an example of a constant function. Since all ordered pairs belonging to f have a y-coordinate of 3, the graph is the horizontal line given by $y = 3$. Remember, y and $f(x)$ are equivalent—that is, $y = f(x)$. (See Figure 10-6.)

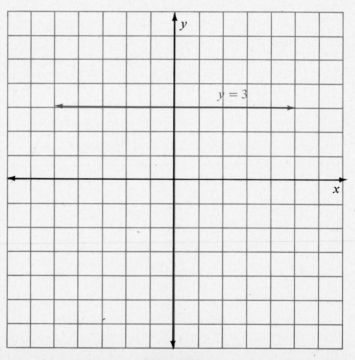

Figure 10-6

Any function that can be written in the form   *Linear Functions*

$$f(x) = ax + b$$

where a and b are real numbers, $a \neq 0$, is called a *linear function*. The graph of every linear function is a straight line. In the past we have written linear functions in the form

$$y = mx + b$$

▼   **Example 2**   The function $f(x) = 2x - 3$ is an example of a linear function. The graph of this function is a straight line with slope 2 and y-intercept -3. (See Figure 10-7.)

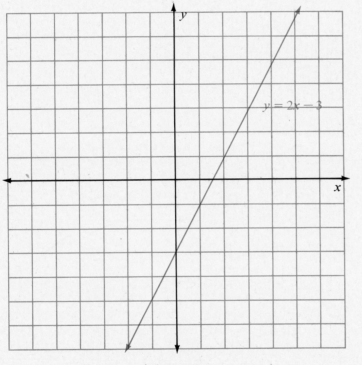

Figure 10-7   ▲

A *quadratic function* is any function that can be written in the form   *Quadratic Functions*

$$f(x) = ax^2 + bx + c$$

where a, b, and c are real numbers and $a \neq 0$. The graph of every quadratic function is a parabola. We considered parabolic graphs in

Section 9.1. At that time the quadratic functions were written as $y = ax^2 + bx + c$ using y instead of $f(x)$.

▼   **Example 3** The function $f(x) = x^2 + 3x - 4$ is an example of a quadratic function. Using the methods developed in Section 9.1, we sketch the graph shown in Figure 10-8.

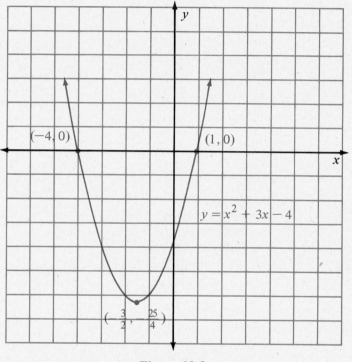

Figure 10-8   ▲

There are many other classifications of functions. One such classification we have not worked with previously is the exponential function. Our introduction to exponential functions, along with the material on inverse functions in the next section, will lay the groundwork necessary to get started in Chapter 11.

Exponential Functions   **DEFINITION**   An *exponential function* is any function that can be written in the form

$$f(x) = b^x$$

where b is a positive real number other than 1.

Each of the following is an exponential function:

$$f(x) = 2^x, \qquad y = 3^x, \qquad f(x) = (\tfrac{1}{4})^x$$

The first step in becoming familiar with exponential functions is to find some values for specific exponential functions.

▼ **Example 4** If the exponential functions f and g are defined by

$$f(x) = 2^x \quad \text{and} \quad g(x) = 3^x$$

then

$$f(0) = 2^0 = 1 \qquad\qquad g(0) = 3^0 = 1$$
$$f(1) = 2^1 = 2 \qquad\qquad g(1) = 3^1 = 3$$
$$f(2) = 2^2 = 4 \qquad\qquad g(2) = 3^2 = 9$$
$$f(3) = 2^3 = 8 \qquad\qquad g(3) = 3^3 = 27$$

$$f(-2) = 2^{-2} = \frac{1}{2^2} = \frac{1}{4} \qquad g(-2) = 3^{-2} = \frac{1}{3^2} = \frac{1}{9}$$

$$f(-3) = 2^{-3} = \frac{1}{2^3} = \frac{1}{8} \qquad g(-3) = 3^{-3} = \frac{1}{3^3} = \frac{1}{27} \qquad ▲$$

We will now turn our attention to the graphs of exponential functions. Since the notation y is easier to use when graphing, and $y = f(x)$, for convenience we will write the exponential functions as

$$y = b^x$$

▼ **Example 5** Sketch the graph of the exponential function

$$y = 2^x$$

Solution Using the results of Example 4 in addition to some other convenient values of x, we have the following table:

x	y
-3	$1/8$
-2	$1/4$
-1	$1/2$
0	1
1	2
2	4
3	8

Graphing the ordered pairs given in the table and connecting them with a smooth curve, we have the graph of $y = 2^x$ shown in Figure 10-9.

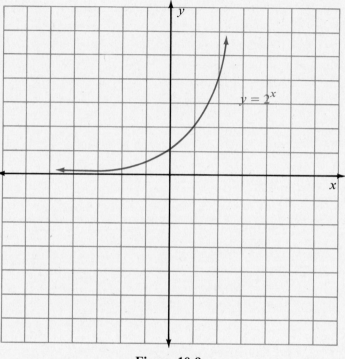

Figure 10-9

Notice that the graph does not cross the x-axis. It *approaches* the x-axis—in fact, we can get it as close to the x-axis as we want without its actually intersecting the x-axis. In order for the graph of $y = 2^x$ to intersect the x-axis, we would have to find a value of x that would make $2^x = 0$. Because no such value of x exists, the graph of $y = 2^x$ cannot intersect the x-axis.   ▲

▼   **Example 6**   Sketch the graph of $y = (\frac{1}{3})^x$.

Solution   We can make a table that will give some ordered pairs that satisfy the equation:

x	$y = \left(\dfrac{1}{3}\right)^x$	y
-3	$y = \left(\dfrac{1}{3}\right)^{-3} = 3^3 = 27$	27
-2	$y = \left(\dfrac{1}{3}\right)^{-2} = 3^2 = 9$	9
-1	$y = \left(\dfrac{1}{3}\right)^{-1} = 3^1 = 3$	3
0	$y = \left(\dfrac{1}{3}\right)^{0} = 1$	1
1	$y = \left(\dfrac{1}{3}\right)^{1} = \dfrac{1}{3}$	$\dfrac{1}{3}$
2	$y = \left(\dfrac{1}{3}\right)^{2} = \dfrac{1}{9}$	$\dfrac{1}{9}$
3	$y = \left(\dfrac{1}{3}\right)^{3} = \dfrac{1}{27}$	$\dfrac{1}{27}$

Using the ordered pairs from the table, we have the graph shown in Figure 10-10.

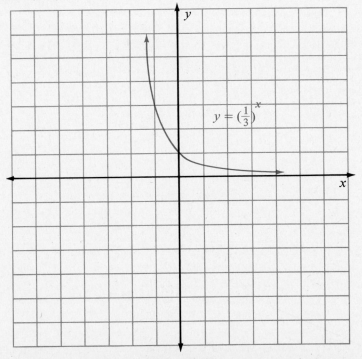

Figure 10-10

The graphs of all exponential functions have two things in common: (1) each crosses the y-axis at $(0, 1)$, since $b^0 = 1$; and (2) none can cross the x-axis, since $b^x = 0$ is impossible because of the restrictions on b.

Sketch the graph of each of the following functions. Identify each as a constant function, linear function, or quadratic function:

1. $f(x) = x^2 - 3$ **2.** $g(x) = 2x^2$
3. $g(x) = 4x - 1$ **4.** $f(x) = 3x + 2$
5. $f(x) = 5$ **6.** $f(x) = -3$
7. $g(x) = 0$ **8.** $g(x) = 1$
9. $f(x) = x$ **10.** $g(x) = \frac{1}{2}x + 1$
11. $f(x) = x^2 + 5x - 6$ **12.** $f(x) = -x^2 + 2x + 8$

Let $f(x) = 3^x$ and $g(x) = (\frac{1}{2})^x$ and evaluate each of the following:

13. $f(-2)$ **14.** $f(4)$
15. $g(2)$ **16.** $g(-2)$
17. $g(0)$ **18.** $f(0)$
19. $g(-1)$ **20.** $g(-4)$
21. $f(-3)$ **22.** $f(-1)$
23. $f(2) + g(-2)$ **24.** $f(2) - g(-2)$
25. $g(-3) + f(2)$ **26.** $g(2) + f(-1)$
27. $f(0) + g(0)$ **28.** $g(-3) + f(3)$
29. $f(-2) + g(3)$ **30.** $g(0) - f(0)$

Graph each of the following functions:

31. $y = 4^x$ **32.** $y = 2^{-x}$
33. $y = 3^{-x}$ **34.** $y = (\frac{1}{3})^{-x}$
35. $y = 2^{x+1}$ **36.** $y = 2^{x-3}$
37. $y = 3^{x-2}$ **38.** $y = 3^{x+1}$
39. $y = 2^{2x}$ **40.** $y = 3^{2x}$

The base b in an exponential function is restricted to positive numbers other than 1. To see why this is done, graph each of the following if possible:

41. $y = 1^x$ **42.** $y = 0^x$
43. $y = (-1)^x$ **44.** $y = (-2)^x$

45. Suppose it takes 1 day for a certain strain of bacteria to reproduce by dividing in half. If there are 100 bacteria present to begin with, then the total number present after x days will be

$$f(x) = 100 \cdot 2^x$$

Find the total number present after 1 day, 2 days, 3 days, and 4 days. How many days must elapse before there are over 100,000 bacteria present?

Suppose the function f is given by

$$f = \{(1, 4), (2, 5), (3, 6), (4, 7)\}$$

The inverse of f, written f^{-1}, is obtained by reversing the order of the coordinates in each ordered pair in f. The inverse of f is the relation given by

$$f^{-1} = \{(4, 1), (5, 2), (6, 3), (7, 4)\}$$

It is obvious that the domain of f is now the range of f^{-1}, and the range of f is now the domain of f^{-1}. Every function (or relation) has an inverse that is obtained from the original function by interchanging the components of each ordered pair. We can generalize this discussion by giving the following formal definition for the inverse of a function or relation.

DEFINITION   The inverse of the function (relation) f is written f^{-1} and is such that

$$(x, y) \in f \quad \text{if and only if} \quad (y, x) \in f^{-1}$$

The $^{-1}$ in the expression f^{-1} serves to denote the inverse of f. It is *not* an exponent—that is, $f^{-1}(x)$ is *not* the reciprocal of f. *In general,*

$$f^{-1}(x) \neq \frac{1}{f(x)}$$

Suppose a function f is defined with an equation instead of a list of ordered pairs. We can obtain the equation of the inverse f^{-1} by applying the above definition to interchange the role of x and y in the equation for f.

▼   **Example 1**   If the function f is defined by $f(x) = 2x - 3$, find the equation that represents the inverse of f.

Solution   Since the inverse of f is obtained by interchanging the components of all the ordered pairs belonging to f, and each ordered pair in f satisfies the equation $y = 2x - 3$, we simply exchange x and y in the equation $y = 2x - 3$ to get the formula for f^{-1}:

$$x = 2y - 3$$

We now solve this equation for y in terms of x:

$$x + 3 = 2y$$

$$\frac{x + 3}{2} = y$$

$$y = \frac{x + 3}{2}$$

The last line gives the equation that defines the function f^{-1}. We can write this equation with function notation as

$$f^{-1}(x) = \frac{x + 3}{2}$$

Let's compare the graphs of f and f^{-1} as given in Example 1. (See Figure 10-11.)

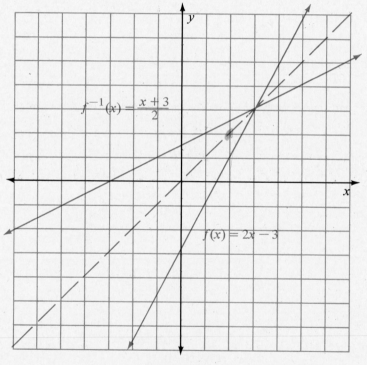

Figure 10-11

The graphs of f and f^{-1} have symmetry about the line $y = x$. This is a reasonable result, since the one function was obtained from the other by interchanging x and y in the equation. The ordered pairs (a, b) and (b, a) always have symmetry about the line $y = x$.   ▲

▼   **Example 2**   Graph the function $y = x^2 - 2$ and its inverse. Give
the equation for the inverse.

Solution   We can obtain the graph of the inverse of $y = x^2 - 2$ by
graphing $y = x^2 - 2$ by the usual methods, and then reflecting the
graph about the line $y = x$. The equation that corresponds to the
inverse of $y = x^2 - 2$ is obtained by interchanging x and y to get
$x = y^2 - 2$. (See Figure 10-12.)

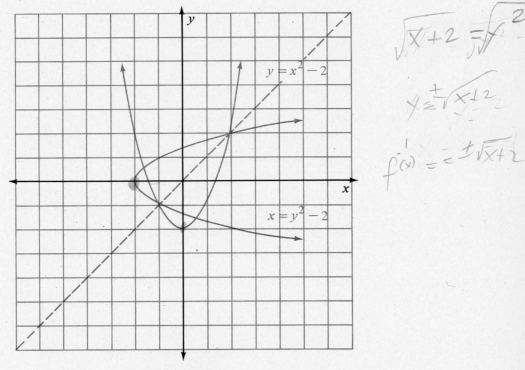

Figure 10-12

We can solve the equation $x = y^2 - 2$ for y in terms of x as follows:

$$x = y^2 - 2$$
$$x + 2 = y^2$$
$$y = \pm \sqrt{x + 2}$$

Using function notation, we can write the function and its inverse as

$$f(x) = x^2 - 2, \qquad f^{-1}(x) = \pm \sqrt{x + 2} \qquad \blacktriangle$$

Comparing the graphs from Examples 1 and 2, we observe that the inverse of a function is not always a function. In Example 1, both f and f^{-1} have graphs that are straight lines and therefore both represent functions. In Example 2, the inverse of function f is not a function, since a vertical line crosses the graph of f^{-1} in more than one place.

▼ **Example 3**   Graph the relation

$$\frac{x^2}{9} + \frac{y^2}{25} = 1$$

and its inverse. Write an equation for the inverse.

Solution   The graph of

$$\frac{x^2}{9} + \frac{y^2}{25} = 1$$

is an ellipse that crosses the x-axis at $(3, 0)$ and $(-3, 0)$ and the y-axis at $(0, 5)$ and $(0, -5)$. Exchanging x and y in the equation, we obtain the equation of the inverse shown in Figure 10-13:

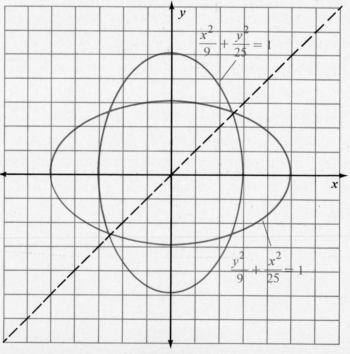

Figure 10-13

$$\frac{x^2}{9} + \frac{y^2}{25} = 1 \text{ and } \frac{y^2}{9} + \frac{x^2}{25} = 1 \text{ are inverse relations.}$$ ▲

▼   **Example 4**   Graph the function $y = 2^x$ and its inverse $x = 2^y$.

Solution   We graphed $y = 2^x$ in the preceding section. We simply reflect its graph about the line $y = x$ to obtain the graph of its inverse $x = 2^y$. (See Figure 10-14.)

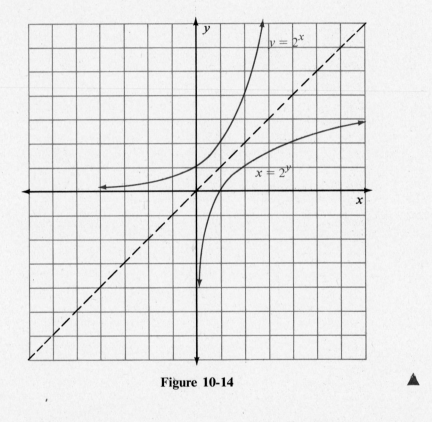

Figure 10-14   ▲

For each of the following functions, find the equation of the inverse. Write the inverse using the notation $f^{-1}(x)$.

Problem Set 10.5

1. $f(x) = 3x - 1$
2. $f(x) = 2x - 5$
3. $f(x) = 1 - 3x$
4. $f(x) = 3 - 4x$
5. $f(x) = x^2 + 4$
6. $f(x) = 3x^2$
7. $f(x) = \dfrac{x - 3}{4}$
8. $f(x) = \dfrac{x + 7}{2}$

9. $f(x) = \frac{1}{2}x - 3$ 10. $f(x) = \frac{1}{3}x + 1$
11. $f(x) = -x^2 + 3$ 12. $f(x) = 4 - x^2$

For each of the following relations, sketch the graph of the relation and its inverse, and write an equation for the inverse:

13. $y = 2x - 1$ 14. $y = 3x + 1$
15. $y = x^2 - 3$ 16. $y = x^2 + 1$
17. $y = x^2 - 2x - 8$ 18. $y = -x^2 + 2x + 8$
19. $y = 3^x$ 20. $y = (\frac{1}{2})^x$
21. $y = 4$ 22. $y = -2$
23. $4x^2 - 9y^2 = 36$ 24. $9x^2 + 4y^2 = 36$
25. $y = \frac{1}{2}x + 2$ 26. $y = \frac{1}{3}x - 1$
27. $x^2 + y^2 = 16$ 28. $x^2 - y^2 = 16$
29. $y = \sqrt{9 - x^2}$ 30. $y = -\sqrt{25 - x^2}$

31. $\dfrac{x^2}{4} + \dfrac{y^2}{25} = 1$ 32. $\dfrac{x^2}{25} + \dfrac{y^2}{9} = 1$

**Chapter 10
Summary and Review**

*Relations and
Functions*

A *relation* is any set of ordered pairs. The set of all first coordinates is called the *domain* of the relation. The set of all second coordinates is the *range* of the relation.

A *function* is a relation in which no two different ordered pairs have the same first coordinate.

If the domain for a relation or a function is not specified, it is assumed to be all real numbers for which the relation (or function) is defined. Since we are only concerned with real number functions, a function is not defined for those values of x that give 0 in the denominator or the square root of a negative number.

Vertical Line Test

If a vertical line crosses the graph of a relation in more than one place, then the relation is not a function. If no vertical line can cross the graph of a relation in more than one place, then the relation is a function.

Function Notation

The notation $f(x)$ is read "f of x." It is defined to be the value of the function f at x. The value of $f(x)$ is the value of y associated with a given value of x. The expressions $f(x)$ and y are equivalent. That is, $y = f(x)$.

Algebra for Functions

If f and g are any two functions with a common domain, then:

The sum of f and g, written $f + g$, is defined by

$$(f + g)(x) = f(x) + g(x)$$

The difference of f and g, written $f - g$, is defined by

$$(f - g)(x) = f(x) - g(x)$$

The product of f and g, written fg, is defined by

$$(fg)(x) = f(x)g(x)$$

The quotient of f and g, written f/g, is defined by

$$\left(\frac{f}{g}\right)(x) = \frac{f(x)}{g(x)} \qquad g(x) \neq 0$$

Constant function:	$f(x) = c$	$(c = \text{Constant})$	*Classification of*
Linear function:	$f(x) = ax + b$	$(a \neq 0)$	*Functions*
Quadratic function:	$f(x) = ax^2 + bx + c$	$(a \neq 0)$	
Exponential function:	$f(x) = b^x$	$(b > 0,\ b \neq 1)$	

If f is a function, then the inverse of f, written f^{-1}, is such that

$$(x, y) \in f \Leftrightarrow (y, x) \in f^{-1}$$

The Inverse of a Function

This is equivalent to saying that the inverse of a function is obtained by reversing the order of the coordinates of the ordered pairs belonging to the function.

The inverse of a function is not necessarily a function.

The graph of f^{-1} can be obtained from the graph of f by simply reflecting the graph of f across the line $y = x$. That is, the graphs of f and f^{-1} are symmetrical about the line $y = x$.

The Graph of the Inverse of a Function

---

1. The most common mistake made when working with functions is to interpret the notation $f(x)$ as meaning the product of f and x. The notation $f(x)$ does *not* mean f times x. It is the value of the function f at x and is equivalent to y.

2. Another common mistake occurs when the expression $f^{-1}(x)$ is interpreted as meaning the reciprocal of x. The notation $f^{-1}(x)$ is used to denote the *inverse* of $f(x)$:

COMMON MISTAKES

$$f^{-1}(x) \neq \frac{1}{f(x)}$$

$$f^{-1}(x) = \text{Inverse of } f$$

Specify the domain and range for the following relations and indicate which relations are also functions:

1. $\{(-3, 1)(2, 1)(3, 5)(0, 3)\}$ **2.** $\{(-2, 0)(-3, 0)(-2, 1)\}$

3. $y = x^2 - 9$ **4.** $4x^2 + 9y^2 = 36$

Indicate any restrictions on the domain of the following:

5. $y = \sqrt{x - 4}$ **6.** $y = \dfrac{4}{\sqrt{3 - x}}$

7. $y = \dfrac{6}{x + 1}$ **8.** $y = \dfrac{x - 1}{x^2 - 2x - 8}$

Let $f(x) = x - 2$, $g(x) = 3x + 4$, and $h(x) = 3x^2 - 2x - 8$, and find the following:

9. $f(3) + g(2)$ **10.** $h(0) + g(0)$

11. $f + g$ **12.** h/g

13. $(f + g + h)(-3)$ **14.** $(h + fg)(1)$

Graph each of the following exponential functions:

15. $y = 2^x$ **16.** $y = 3^{-x}$

Find an equation for the inverse of each of the following functions. Sketch the graph of f and f^{-1} on the same set of axes.

17. $f(x) = 2x - 3$ **18.** $f(x) = x^2 - 4$

Logarithms $\mathbf{11}$

To the student:

This chapter is mainly concerned with applications of a new notation for exponents. Logarithms are exponents. The properties of logarithms are actually the properties of exponents. As it turns out, writing exponents with the new notation, logarithms, allows us to solve some problems we would otherwise be unable to solve. Logarithms used to be used extensively to simplify tedious calculations. Since hand-held calculators are so common now, logarithms are seldom used in connection with computations. Nevertheless, there are many other applications of logarithms to both science and higher mathematics. For example, the pH of a liquid is defined in terms of logarithms. (That's the same pH that is given on the label of many hair conditioners.) The Richter scale for measuring earthquake intensity is a logarithmic scale, as is the decibel scale used for measuring the intensity of sound.

We will begin this chapter with the definition of logarithms and the three main properties of logarithms. The rest of the chapter involves applications of the definition and properties. Understanding the last two sections in Chapter 10 will be very useful in getting started in this chapter.

In Section 10.4 we considered some equations of the form

$$y = b^x \qquad (b > 0, b \neq 1)$$

called exponential functions. In the next section, Section 10.5, we found that the equation of the inverse of a function can be obtained by

**11.1
Logarithms Are
Exponents**

exchanging x and y in the equation of the original function. We can therefore exchange x and y in the equation of an exponential function to get the equation of its inverse. The equation of the inverse of an exponential function must have the form

$$x = b^y \qquad (b > 0, b \neq 1)$$

The problem with this expression is that y is not written explicitly in terms of x. That is, we would like to be able to write the equation $x = b^y$ as an equivalent equation with just y on the left side. One way to do so is with the following definition.

DEFINITION   The expression $y = \log_b x$ is read "y is the logarithm to the base b of x," and is equivalent to the expression

$$x = b^y \qquad (b > 0, b \neq 1)$$

We say y is the number we raise b to in order to get x.

It may seem rather strange to take an expression we already know something about, like $x = b^y$, and write it in another form, $y = \log_b x$, that we know nothing about. As you will see as we progress through the chapter, there are many advantages to using the notation $y = \log_b x$ in place of $x = b^y$, even though they say exactly the same thing about the relationship between x, y, and b. What we have to do now is realize that $y = \log_b x$ is *defined* to be equivalent to $x = b^y$. There isn't anything to prove about the definition—we can only accept it as it is.

NOTATION   When an expression is in the form $x = b^y$, it is said to be in *exponential form*. On the other hand, if an expression is in the form $y = \log_b x$, it is said to be in *logarithmic form*.

The following table illustrates the two forms:

Exponential form		Logarithmic form
$8 = 2^3$	$\Leftrightarrow$	$\log_2 8 = 3$
$25 = 5^2$	$\Leftrightarrow$	$\log_5 25 = 2$
$.1 = 10^{-1}$	$\Leftrightarrow$	$\log_{10} .1 = -1$
$\frac{1}{8} = 2^{-3}$	$\Leftrightarrow$	$\log_2 \frac{1}{8} = -3$
$r = z^s$	$\Leftrightarrow$	$\log_z r = s$

As the table indicates, logarithms are exponents. That is, $\log_2 8$ is 3 *because* 3 is the exponent to which we raise 2 in order to get 8. *Logarithms are exponents.*

The ability to write expressions in logarithmic form as equivalent expressions in exponential form allows us to solve some equations involving logarithms, as the following examples illustrate.

▼   **Example 1**   Solve for x: $\log_3 x = -2$.

Solution   In exponential form the equation looks like this:

$$x = 3^{-2}$$
$$\text{or}\quad x = \tfrac{1}{9}$$

The solution set is $\{\tfrac{1}{9}\}$.   ▲

▼   **Example 2**   Solve: $\log_x 4 = 3$.

Solution   Again, we use the definition of logarithms to write the expression in exponential form:

$$4 = x^3$$

Taking the cube root of both sides, we have

$$\sqrt[3]{4} = \sqrt[3]{x^3}$$
$$x = \sqrt[3]{4}$$

The solution set is $\{\sqrt[3]{4}\}$.   ▲

▼   **Example 3**   Solve: $\log_8 4 = x$.

Solution   We write the expression again in exponential form:

$$4 = 8^x$$

Since both 4 and 8 can be written as powers of 2, we write them in terms of powers of 2:

$$2^2 = (2^3)^x$$
$$2^2 = 2^{3x}$$

The only way the left and right sides of this last line can be equal is if the exponents are equal—that is, if

$$2 = 3x$$
$$\text{or}\quad x = \tfrac{2}{3}$$

The solution is $\tfrac{2}{3}$. We check as follows:

$$\log_8 4 = \tfrac{2}{3} \Longleftrightarrow 4 = 8^{2/3}$$
$$4 = (\sqrt[3]{8})^2$$
$$4 = 2^2$$
$$4 = 4$$

The solution checks when used in the original equation.   ▲

Graphing logarithmic functions can be done using the graphs of exponential functions and the fact that the graphs of inverse functions have symmetry about the line $y = x$. Here's an example to illustrate.

▼   **Example 4**   Graph the equation $y = \log_2 x$.

Solution   The equation $y = \log_2 x$ is, by definition, equivalent to the exponential equation

$$x = 2^y$$

which is the equation of the inverse of the function

$$y = 2^x$$

The graph of $y = 2^x$ was given in Figure 10-9. We simply reflect the graph of $y = 2^x$ about the line $y = x$ to get the graph of $x = 2^y$, which is also the graph of $y = \log_2 x$. (See Figure 11-1.)

It is apparent from the graph that $y = \log_2 x$ is a function, since no vertical line will cross its graph in more than one place. The same is true for all logarithmic equations of the form

$$y = \log_b x$$

where b is a positive number other than 1. All similar equations can be considered as functions.   ▲

Note   Since $y = \log_b x$ is equivalent to $x = b^y$ where $b > 0$ and $b \neq 1$, the only one of the quantities x, y, and b that can be negative is y. Both b and x in the expression $y = \log_b x$ must be positive.

We end this section with two special identities that arise from the definition of logarithms. Each is useful in proving the properties in the next section or for evaluating some simple logarithmic expressions.

Two Special Identities   If b is a positive real number other than 1, then each of the following is a consequence of the definition of a logarithm:

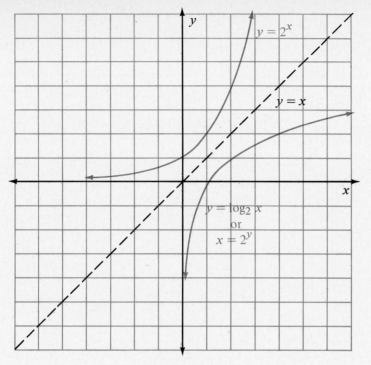

Figure 11-1

$$(1) \quad b^{\log_b x} = x \quad \text{and} \quad (2) \quad \log_b b^x = x$$

The justifications for these identities are similar. Let's consider only the first one. Consider the expression

$$y = \log_b x$$

By definition, it is equivalent to

$$x = b^y$$

Substituting $\log_b x$ for y in the last line gives us

$$x = b^{\log_b x}$$

Here are some examples that illustrate the use of identity 2 given above.

▼ **Example 5** Simplify $\log_2 8$.

Solution Substitute 2^3 for 8:

$$\log_2 8 = \log_2 2^3$$
$$= 3$$

▲

▼ **Example 6**   Simplify $\log_{10} 10,000$.

Solution   10,000 can be written as 10^4:

$$\log_{10} 10,000 = \log_{10} 10^4$$
$$= 4$$

▲

▼ **Example 7**   Simplify $\log_b b$ $(b > 0, b \neq 1)$.

Solution   Since $b^1 = b$, we have

$$\log_b b = \log_b b^1$$
$$= 1$$

▲

▼ **Example 8**   Simplify $\log_b 1$ $(b > 0, b \neq 1)$.

Solution   Since $1 = b^0$, we have

$$\log_b 1 = \log_b b^0$$
$$= 0$$

▲

▼ **Example 9**   Simplify $\log_4 (\log_5 5)$.

Solution   Since $\log_5 5 = 1$,

$$\log_4 (\log_5 5) = \log_4 1$$
$$= 0$$

▲

Problem Set 11.1   Write each of the following expressions in logarithmic form:

1. $2^4 = 16$ 2. $3^2 = 9$
3. $125 = 5^3$ 4. $16 = 4^2$
5. $.01 = 10^{-2}$ 6. $.001 = 10^{-3}$
7. $2^{-5} = \frac{1}{32}$ 8. $4^{-2} = \frac{1}{16}$
9. $(\frac{1}{2})^{-3} = 8$ 10. $(\frac{1}{3})^{-2} = 9$
11. $27 = 3^3$ 12. $81 = 3^4$

Write each of the following expressions in exponential form:

13. $\log_{10} 100 = 2$ 14. $\log_2 8 = 3$
15. $\log_2 64 = 6$ 16. $\log_2 32 = 5$
17. $\log_8 1 = 0$ 18. $\log_9 9 = 1$

19. $\log_{10} .001 = -3$
20. $\log_{10} .0001 = -4$
21. $\log_6 36 = 2$
22. $\log_7 49 = 2$
23. $\log_5 \frac{1}{25} = -2$
24. $\log_3 \frac{1}{81} = -4$

Solve each of the following equations for x:

25. $\log_3 x = 2$
26. $\log_4 x = 3$
27. $\log_5 x = -3$
28. $\log_2 x = -4$
29. $\log_2 16 = x$
30. $\log_3 27 = x$
31. $\log_8 2 = x$
32. $\log_{25} 5 = x$
33. $\log_x 4 = 2$
34. $\log_x 16 = 4$
35. $\log_x 5 = 3$
36. $\log_x 8 = 2$

Sketch the graph of each of the following logarithmic equations:

37. $y = \log_3 x$
38. $y = \log_{1/2} x$
39. $y = \log_{1/3} x$
40. $y = \log_4 x$
41. $y = \log_5 x$
42. $y = \log_{1/5} x$
43. $y = \log_{10} x$
44. $y = \log_{1/4} x$

Simplify each of the following:

45. $\log_2 16$
46. $\log_3 9$
47. $\log_{25} 125$
48. $\log_9 27$
49. $\log_{10} 1000$
50. $\log_{10} 10{,}000$
51. $\log_3 3$
52. $\log_4 4$
53. $\log_5 1$
54. $\log_{10} 1$
55. $\log_3 (\log_6 6)$
56. $\log_5 (\log_3 3)$
57. $\log_4 [\log_2 (\log_2 16)]$
58. $\log_4 [\log_3 (\log_2 8)]$
59. $\log_3 [\log_2 (\log_5 25)]$
60. $\log_4 [\log_3 (\log_4 64)]$

We begin this section by stating three properties of logarithms. Since the proofs for all three are very similar in form, we will prove only the first property.

For the following three properties, x, y, and b are all positive real numbers, $b \neq 1$, and r is any real number.

**11.2
Properties of
Logarithms**

PROPERTY 1 $\log_b (xy) = \log_b x + \log_b y$

In words: the logarithm of a *product* is the *sum* of the logarithms.

PROPERTY 2 $\log_b \left(\dfrac{x}{y} \right) = \log_b x - \log_b y$

In words: the logarithm of a *quotient* is the *difference* of the logarithms.

PROPERTY 3 $\log_b x^r = r \log_b x$

In words: the logarithm of a base raised to a *power* is the *product* of the power and the logarithm of the base.

Proof of Property 1   To prove property 1 we simply apply the first identity for logarithms given at the end of the preceding section:

$$b^{\log_b xy} = xy = (b^{\log_b x})(b^{\log_b y}) = b^{\log_b x + \log_b y}$$

Since the first and last expressions are equal and the bases are the same, the exponents $\log_b xy$ and $\log_b x + \log_b y$ must be equal. Therefore,

$$\log_b xy = \log_b x + \log_b y$$

The proofs of properties 2 and 3 proceed in much the same manner, so we will omit them here. The examples that follow show how the three properties can be used. These types of problems will be useful in applying logarithms to computations which we will do in the next section.

▼   **Example 1**   Expand using the properties of logarithms: $\log_5 \dfrac{3xy}{z}$.

Solution   Applying property 2, we can write the quotient of $3xy$ and z in terms of a difference:

$$\log_5 \frac{3xy}{z} = \log_5 3xy - \log_5 z$$

Applying property 1 to the product $3xy$, we write it in terms of addition:

$$\log_5 \frac{3xy}{z} = \log_5 3 + \log_5 x + \log_5 y - \log_5 z \qquad ▲$$

▼   **Example 2**   Expand using the properties of logarithms:

$$\log_2 \frac{x^4}{\sqrt{y} \cdot z^3}.$$

Solution   We write $\sqrt{y}$ as $y^{1/2}$ and apply the properties:

$$\log_2 \frac{x^4}{\sqrt{y} \cdot z^3} = \log_2 \frac{x^4}{y^{1/2} z^3}$$

$$= \log_2 x^4 - \log_2 (y^{1/2} \cdot z^3) \qquad \text{Property 2}$$

$$= \log_2 x^4 - (\log_2 y^{1/2} + \log_2 z^3) \qquad \text{Property 1}$$

$$= \log_2 x^4 - \log_2 y^{1/2} - \log_2 z^3 \qquad \begin{array}{l}\text{Remove} \\ \text{parentheses}\end{array}$$

$$= 4 \log_2 x - \frac{1}{2} \log_2 y - 3 \log_2 z \qquad \text{Property 3} \quad \blacktriangle$$

We can also use the three properties to write an expression in expanded form as just one logarithm.

▼ **Example 3** Write as a single logarithm: $2 \log_{10} a + 3 \log_{10} b - \frac{1}{3} \log_{10} c$.

Solution We begin by applying property 3:

$$2 \log_{10} a + 3 \log_{10} b - \frac{1}{3} \log_{10} c$$

$$= \log_{10} a^2 + \log_{10} b^3 - \log_{10} c^{1/3} \qquad \text{Property 3}$$
$$= \log_{10} (a^2 \cdot b^3) - \log_{10} c^{1/3} \qquad \text{Property 1}$$
$$= \log_{10} \frac{a^2 b^3}{c^{1/3}} \qquad \text{Property 2}$$

$$= \log_{10} \frac{a^2 b^3}{\sqrt[3]{c}} \qquad\qquad c^{1/3} = \sqrt[3]{c} \qquad \blacktriangle$$

The properties of logarithms along with the definition of logarithms are useful in solving equations that involve logarithms.

▼ **Example 4** Solve for x: $\log_2 (x + 2) + \log_2 x = 3$.

Solution Applying property 1 to the left side of the equation allows us to write it as a single logarithm:

$$\log_2 (x + 2) + \log_2 x = 3$$
$$\log_2 [(x + 2)(x)] = 3$$

The last line can be written in exponential form using the definition of logarithms:

$$(x + 2)(x) = 2^3$$

Solve as usual:

$$x^2 + 2x = 8$$
$$x^2 + 2x - 8 = 0$$
$$(x + 4)(x - 2) = 0$$
$$x + 4 = 0 \quad \text{or} \quad x - 2 = 0$$
$$x = -4 \quad \text{or} \quad x = 2$$

Earlier in this section we noted the fact that x in the expression $y = \log_b x$ cannot be a negative number. Since substitution of $x = -4$ into the original equation gives

$$\log_2 (-2) + \log_2 (-4) = 3$$

which contains logarithms of negative numbers, we cannot use -4 as a solution. The solution set is $\{2\}$.   ▲

Problem Set 11.2

Use the three properties of logarithms given in this section to expand each expression as much as possible:

1. $\log_3 4x$

2. $\log_2 5x$

3. $\log_6 \dfrac{5}{x}$

4. $\log_3 \dfrac{x}{5}$

5. $\log_2 y^5$

6. $\log_7 y^3$

7. $\log_9 \sqrt[3]{z}$

8. $\log_8 \sqrt{z}$

9. $\log_6 x^2 y^3$

10. $\log_{10} x^2 y^4$

11. $\log_5 \sqrt{x} \cdot y^4$

12. $\log_8 \sqrt[3]{xy^6}$

13. $\log_b \dfrac{xy}{z}$

14. $\log_b \dfrac{3x}{y}$

15. $\log_{10} \dfrac{4}{xy}$

16. $\log_{10} \dfrac{5}{4y}$

17. $\log_{10} \dfrac{x^2 y}{\sqrt{z}}$

18. $\log_{10} \dfrac{\sqrt{x} \cdot y}{z^3}$

19. $\log_{10} \dfrac{x^3 \sqrt{y}}{z^4}$

20. $\log_{10} \dfrac{x^4 \sqrt[3]{y}}{\sqrt{z}}$

21. $\log_b \sqrt[3]{\dfrac{x^2 y}{z^4}}$

22. $\log_b \sqrt[4]{\dfrac{x^4 y^3}{z^5}}$

Write each expression as a single logarithm:

23. $\log_b x + \log_b z$

24. $\log_b x - \log_b z$

25. $2 \log_3 x - 3 \log_3 y$

26. $4 \log_2 x + 5 \log_2 y$

27. $\frac{1}{2} \log_{10} x + \frac{1}{3} \log_{10} y$

28. $\frac{1}{3} \log_{10} x - \frac{1}{4} \log_{10} y$

29. $3 \log_2 x + \frac{1}{2} \log_2 y - \log_2 z$

30. $2 \log_3 x + 3 \log_3 y - \log_3 z$

31. $\frac{1}{2} \log_2 x - 3 \log_2 y - 4 \log_2 z$

32. $3 \log_{10} x - \log_{10} y - \log_{10} z$

33. $\frac{3}{2} \log_{10} x - \frac{3}{4} \log_{10} y - \frac{4}{5} \log_{10} z$

34. $3 \log_{10} x - \frac{4}{3} \log_{10} y - 5 \log_{10} z$

Solve each of the following equations:

35. $\log_2 x + \log_2 3 = 1$

36. $\log_2 x - \log_2 3 = 1$

37. $\log_3 x - \log_3 2 = 2$

38. $\log_3 x + \log_3 2 = 2$

39. $\log_3 x + \log_3 (x - 2) = 1$

40. $\log_6 x + \log_6 (x - 1) = 1$

41. $\log_3 (x + 3) - \log_3 (x - 1) = 1$

42. $\log_4 (x - 2) - \log_4 (x + 1) = 1$

43. $\log_2 x + \log_2 (x - 2) = 3$

44. $\log_4 x + \log_4 (x + 6) = 2$

In the past, logarithms have been very useful in simplifying calculations. With the widespread availability of hand-held calculators, the importance of logarithms in calculations has been decreased considerably. Working through arithmetic problems using logarithms is, however, good practice with logarithms. The problems in this section all involve the properties of logarithms developed in the preceding section and indicate how some fairly difficult computation problems can be simplified using logarithms.

**11.3
Common Logarithms
and Computations**

DEFINITION A *common logarithm* is a logarithm with a base of 10. Since common logarithms are used so frequently, it is customary, in order to save time, to omit notating the base. That is,

$$\log_{10} x = \log x$$

When the base is not shown, it is assumed to be 10.

The reason logarithms base 10 are so common is that our number system is a base-10 number system.

Common logarithms of powers of 10 are very simple to evaluate. We need only recognize that $\log 10 = \log_{10} 10 = 1$ and apply the third property of logarithms: $\log_b x^r = r \log_b x$.

$$\log 1000 = \log 10^3 = 3 \log 10 = 3(1) = 3$$
$$\log 100 = \log 10^2 = 2 \log 10 = 2(1) = 2$$

$$\begin{aligned}
\log 10 &= \log 10^1 &&= 1 \log 10 &&= 1(1) &&= 1 \\
\log 1 &= \log 10^0 &&= 0 \log 10 &&= 0(1) &&= 0 \\
\log .1 &= \log 10^{-1} &&= -1 \log 10 &&= -1(1) &&= -1 \\
\log .01 &= \log 10^{-2} &&= -2 \log 10 &&= -2(1) &&= -2 \\
\log .001 &= \log 10^{-3} &&= -3 \log 10 &&= -3(1) &&= -3
\end{aligned}$$

For common logarithms of numbers that are not powers of 10, we have to resort to a table. The table in Appendix D at the back of the book gives common logarithms of numbers between 1.00 and 9.99. To find the common logarithm of, say, 2.76, we read down the left-hand column until we get to 2.7, then across until we are below the 6 in the top row (or above the 6 in the bottom row):

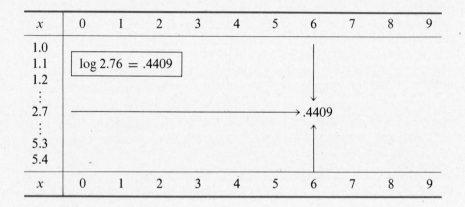

The table contains only logarithms of numbers between 1.00 and 9.99. Check the following logarithms in the table to be sure you know how to use the table:

$$\log 7.02 = .8463$$
$$\log 1.39 = .1430$$
$$\log 6.00 = .7782$$
$$\log 9.99 = .9996$$

To find the common logarithm of a number that is not between 1.00 and 9.99, we simply write the number in scientific notation, apply property 2 of logarithms, and use the table. The following examples illustrate the procedure.

Note A number is written in scientific notation when it is written as the product of a number between 1 and 10 and a power of 10. For example, $39,800 = 3.98 \times 10^4$ in scientific notation.

▼ **Example 1** Use the table in Appendix D to find $\log 2760$.

Solution
$$\begin{aligned} \log 2760 &= \log (2.76 \times 10^3) \\ &= \log 2.76 + \log 10^3 \\ &= .4409 + 3 \\ &= 3.4409 \end{aligned}$$

The 3 in the answer is called the *characteristic,* and its main function is to keep track of the decimal point. The decimal part of this logarithm is called the *mantissa.* It is found from the table. Here's more of the same: ▲

▼ **Example 2** Find $\log 843$.

Solution
$$\begin{aligned} \log 843 &= \log (8.43 \times 10^2) \\ &= \log 8.43 + \log 10^2 \\ &= .9258 + 2 \\ &= 2.9258 \end{aligned}$$ ▲

▼ **Example 3** Find $\log .0391$.

Solution
$$\begin{aligned} \log .0391 &= \log (3.91 \times 10^{-2}) \\ &= \log 3.91 + \log 10^{-2} \\ &= .5922 + (-2) \end{aligned}$$

Now there are two ways to proceed from here. We could add .5922 and -2 to get -1.4078. (If you were using a calculator to find $\log .0391$, this is the answer you would see.) The problem with -1.4078 is that the mantissa is negative. Our table contains only positive numbers. If we have to use $\log .0391$ again, and it is written as -1.4078, we will not be able to associate it with an entry in the table. It is most common, when using a table, to write the characteristic -2 as $8 + (-10)$ and proceed as follows:

$$\begin{aligned} \log .0391 &= .5922 + 8 + (-10) \\ &= 8.5922 - 10 \end{aligned}$$ ▲

Here is another example:

▼ **Example 4** Find $\log .00523$.

Solution
$$\begin{aligned} \log .00523 &= \log (5.23 \times 10^{-3}) \\ &= \log 5.23 + \log 10^{-3} \\ &= .7185 + (-3) \\ &= .7185 + 7 + (-10) \\ &= 7.7185 - 10 \end{aligned}$$ ▲

It is not necessary to show any of the steps we have shown in the examples above. As a matter of fact, it is better if we don't. With a little practice the steps can be eliminated. Once we have become familiar with using the table to find logarithms, we can go in the reverse direction and use the table to solve equations like log x = 3.8774.

▼   **Example 5**   Find x if log x = 3.8774.

Solution   We are looking for the number whose logarithm is 3.8774. The mantissa is .8774, which appears in the table across from 7.5 and under (or above) 4.

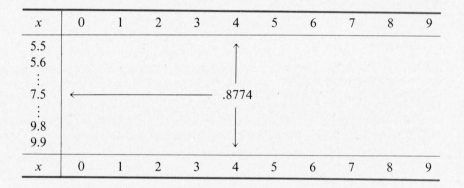

The characteristic is 3 and came from the exponent of 10. Putting these together, we have

$$\log x = 3.8774$$
$$\log x = .8774 + 3$$
$$x = 7.54 \times 10^3$$
$$x = 7540$$

The number 7540 is called the *antilogarithm* or just *antilog* of 3.8774. That is, 7540 is the number whose logarithm is 3.8774.   ▲

▼   **Example 6**   Find x if log x = 7.5821 − 10.

Solution   The mantissa is .5821, which comes from 3.82 in the table. The characteristic is $7 + (-10)$ or -3, which comes from a power of 10:

$$\log x = 7.5821 - 10$$
$$\log x = .5821 + 7 + (-10)$$
$$\log x = .5821 + (-3)$$
$$x = 3.82 \times 10^{-3}$$
$$x = .00382$$

The antilog of $7.5821 - 10$ is $.00382$. That is, the logarithm of $.00382$ is $7.5821 - 10$. ▲

The following examples illustrate how logarithms are used as an aid in computations.

▼ **Example 7** Use logarithms to find $(3780)(45,200)$.

Solution We will let n represent the answer to this problem:

$$n = (3780)(45,200)$$

Since these two numbers are equal, so are their logarithms. We therefore take the common logarithm of both sides:

$$\log n = \log (3780)(45,200)$$
$$= \log 3780 + \log 45,200$$
$$= 3.5775 + 4.6551$$
$$\log n = 8.2326$$

The number 8.2326 is the logarithm of the answer. The mantissa is .2326, which is not in the table. Since .2326 is closest to .2330, and .2330 is the logarithm of 1.71, we write

$$n = 1.71 \times 10^8$$ ▲

Note The answer 1.71×10^8 is not exactly equal to $(3780)(45,200)$. It is an approximation to it. If we want to be more accurate and have an answer with more significant digits, we would have to use a table with more significant digits.

▼ **Example 8** Use logarithms to find $\sqrt[3]{875}$.

Solution

$$\text{If} \quad n = \sqrt[3]{875},$$
$$\text{then} \quad n = (875)^{1/3}$$

$$\text{and} \quad \log n = \log 875^{1/3}$$
$$= \tfrac{1}{3} \log 875$$
$$= \tfrac{1}{3}(2.9420)$$
$$\log n = .9807$$
$$n = 9.56$$

A good approximation to the cube root of 875 is 9.56.   ▲

▼   **Example 9**   Use logarithms to find $35^{2.7}$.

Solution

$$\text{Let } n = 35^{2.7}.$$
$$\text{Then } \log n = \log 35^{2.7}$$
$$= 2.7 \log 35$$
$$= 2.7(1.5441)$$
$$\log n = 4.1691$$
$$n = 1.48 \times 10^4$$
$$= 14,800$$ ▲

▼   **Example 10**   Use logarithms to find $\dfrac{(34.5)^2 \sqrt{1080}}{(2.76)^3}$.

Solution

$$\text{If} \quad n = \frac{(34.5)^2 \sqrt{1080}}{(2.76)^3},$$

$$\text{then} \quad \log n = \log \left[\frac{(34.5)^2 \sqrt{1080}}{(2.76)^3} \right]$$

$$= 2 \log 34.5 + \frac{1}{2} \log 1080 - 3 \log 2.76$$

$$= 2(1.5378) + \frac{1}{2}(3.0334) - 3(.4409)$$

$$= 3.0756 + 1.5167 - 1.3227$$

$$\log n = 3.2696$$
$$n = 1.86 \times 10^3$$
$$= 1860$$ ▲

Problem Set 11.3   Use the table in Appendix D to find the following:

1. log 378   **2.** log 426

3. log 37.8   **4.** log 42,600

5. log 3780
7. log .0378
9. log 37,800
11. log 600
13. log 2010
15. log .00971
17. log .0314
19. log .399

6. log .4260
8. log .0426
10. log 4900
12. log 900
14. log 10,200
16. log .0312
18. log .00052
20. log .111

Find x in the following equations:

21. $\log x = 2.8802$
23. $\log x = 7.8802 - 10$
25. $\log x = 3.1553$
27. $\log x = 4.6503 - 10$
29. $\log x = 2.9628 - 10$

22. $\log x = 4.8802$
24. $\log x = 6.8802 - 10$
26. $\log x = 5.5911$
28. $\log x = 8.4330 - 10$
30. $\log x = 5.8000 - 10$

Use logarithms to evaluate the following:

31. $(378)(24.5)$

32. $(921)(2630)$

33. $\dfrac{496}{391}$

34. $\dfrac{512}{216}$

35. $\sqrt{401}$
37. $\sqrt[3]{1030}$

36. $\sqrt[3]{92.3}$
38. $\sqrt{.641}$

39. $\dfrac{(2390)(28.4)}{176}$

40. $\dfrac{(32.9)(5760)}{11.1}$

41. $(296)^2(459)$
43. $\sqrt{526}\,(1080)$
45. $45^{-3.1}$

42. $(3250)(24.2)^3$
44. $(32.7)\sqrt{.580}$
46. $72^{1.8}$

47. $\dfrac{(895)^3\sqrt{41.1}}{17.8}$

48. $\dfrac{(925)^2\sqrt{1.99}}{243}$

49. $\sqrt{\dfrac{24.1(32)^2}{25}}$

50. $\sqrt[3]{\dfrac{(895)^2(256)}{958}}$

Logarithms are very important in solving equations in which the variable appears as an exponent. The equation

$$5^x = 12$$

is an example of one such equation. Equations of this form are called *exponential equations*. Since the quantities 5^x and 12 are equal, so are their common logarithms. We begin our solution by taking the logarithm of both sides:

$$\log 5^x = \log 12$$

11.4
Exponential Equations and Change of Base

We now apply property 3 for logarithms, $\log x^r = r \log x$, to turn x from an exponent into a coefficient:

$$x \log 5 = \log 12$$

Dividing both sides by $\log 5$ gives us

$$x = \frac{\log 12}{\log 5}$$

If we want a decimal approximation to the solution, we can find $\log 12$ and $\log 5$ in the table in Appendix D and divide:

$$x = \frac{1.0792}{.6990}$$

$$= 1.5439$$

The complete problem looks like this:

$$5^x = 12$$
$$\log 5^x = \log 12$$
$$x \log 5 = \log 12$$
$$x = \frac{\log 12}{\log 5}$$
$$= \frac{1.0792}{.6990}$$
$$= 1.5439$$

Note   A very common mistake can occur in the third-from-the-last step. Many times the expression

$$\frac{\log 12}{\log 5}$$

is mistakenly simplified as $\log 12 - \log 5$. There is no property of logarithms that allows us to do this. The only property of logarithms that deals with division is property 2, which is *this:*

$$\log_b \frac{x}{y} = \log_b x - \log_b y \qquad \text{(Right)}$$

not this:

$$\frac{\log_b x}{\log_b y} = \log_b x - \log_b y \qquad \text{(Wrong)}$$

The second statement is simply not a property of logarithms, although it is sometimes mistaken for one.

Here is another example of solving an exponential equation using logarithms.

▼ **Example 1** Solve for x: $25^{2x+1} = 15$.

Solution Taking the logarithm of both sides and then writing the exponent $(2x + 1)$ as a coefficient, we proceed as follows:

$$25^{2x+1} = 15$$
$$\log 25^{2x+1} = \log 15$$
$$(2x + 1) \log 25 = \log 15$$
$$2x + 1 = \frac{\log 15}{\log 25}$$
$$2x = \frac{\log 15}{\log 25} - 1$$
$$x = \frac{1}{2}\left(\frac{\log 15}{\log 25} - 1\right)$$

Using the table of common logarithms (Appendix D), we can write a decimal approximation to the answer:

$$x = \frac{1}{2}\left(\frac{1.1761}{1.3979} - 1\right)$$
$$= \frac{1}{2}(.8413 - 1)$$
$$= \frac{1}{2}(-.1587)$$
$$= -.0793 \qquad \blacktriangle$$

There is a fourth property of logarithms we have not yet considered. This last property allows us to change from one base to another and is therefore called the *change-of-base property*.

PROPERTY 4 (Change of Base) If a and b are both positive numbers other than 1, and if $x > 0$, then

$$\log_a x = \frac{\log_b x}{\log_b a}$$

$$\underset{\text{Base } a}{\uparrow} \qquad \underset{\text{Base } b}{\uparrow}$$

The logarithm on the left side has a base of a, while both logarithms on the right side have a base of b. This allows us to change from base a to any other base b that is a positive number other than 1. Here is a proof of property 4 for logarithms:

Proof   We begin by writing the identity

$$a^{\log_a x} = x$$

Taking the logarithm base b of both sides and setting the exponent $\log_a x$ as a coefficient, we have

$$\log_b a^{\log_a x} = \log_b x$$
$$\log_a x \log_b a = \log_b x$$

Dividing both sides by $\log_b a$, we have the desired result:

$$\frac{\log_a x \log_b a}{\log_b a} = \frac{\log_b x}{\log_b a}$$

$$\log_a x = \frac{\log_b x}{\log_b a}$$

We can use this property to find logarithms we do not have a table for. The next two examples illustrate the use of this property.

▼   **Example 2**   Find $\log_8 24$.

Solution   Since we do not have a table for base-8 logarithms, we can change this expression to an equivalent expression that only contains base-10 logarithms:

$$\log_8 24 = \frac{\log 24}{\log 8}$$

Don't be confused. We did not just drop the base, we changed to base 10. We could have written the last line like this:

$$\log_8 24 = \frac{\log_{10} 24}{\log_{10} 8}$$

Looking up log 24 and log 8 in the table in Appendix D, we write

$$\log_8 24 = \frac{1.3802}{.9031}$$

$$= 1.5283 \qquad \blacktriangle$$

▼   **Example 3**   Find $\log_{20} 342$.

Solution   Changing to base 10 using property 4, we proceed as follows:

$$\log_{20} 342 = \frac{\log 342}{\log 20}$$

$$= \frac{2.5340}{1.3010}$$

$$= 1.9477$$

The decimal approximation to $\log_{20} 342$ is 1.9477.   ▲

Solve each exponential equation. Use the table in Appendix D to write the answer in decimal form.

Problem Set 11.4

1. $3^x = 5$
2. $4^x = 3$
3. $5^x = 3$
4. $3^x = 4$
5. $5^{-x} = 12$
6. $7^{-x} = 8$
7. $12^{-x} = 5$
8. $8^{-x} = 7$
9. $8^{x+1} = 4$
10. $9^{x+1} = 3$
11. $4^{x-1} = 4$
12. $3^{x-1} = 9$
13. $3^{2x+1} = 2$
14. $2^{2x+1} = 3$
15. $3^{1-2x} = 2$
16. $2^{1-2x} = 3$
17. $15^{3x-4} = 10$
18. $10^{3x-4} = 15$
19. $6^{5-2x} = 4$
20. $9^{7-3x} = 5$

Use the change-of-base property and the table in Appendix D to find a decimal approximation to each of the following logarithms:

21. $\log_8 16$
22. $\log_9 27$
23. $\log_{16} 8$
24. $\log_{27} 9$
25. $\log_7 15$
26. $\log_3 12$
27. $\log_{15} 7$
28. $\log_{12} 3$
29. $\log_{12} 11$
30. $\log_{14} 15$
31. $\log_{11} 12$
32. $\log_{15} 14$
33. $\log_8 240$
34. $\log_6 180$
35. $\log_4 321$
36. $\log_5 462$

There are many practical applications of exponential equations and logarithms. Many times an exponential equation will describe a situation arising in nature. Logarithms are then used to solve the exponential equation. Many of the problems in this section deal with expressions

11.5
Word Problems

and equations that have been derived or defined in some other discipline. We will not attempt to derive them here. We will simply accept them as they are and use them to solve some problems.

In chemistry the pH of a solution is defined in terms of logarithms as

$$pH = -\log (H_3O^+)$$

where (H_3O^+) is the concentration of the hydronium ion, H_3O^+, in solution. An acid solution has a pH lower than 7, and a basic solution has a pH above 7. (This is from chemistry. We don't need to worry here about *why* it's this way. We just want to work some problems involving logarithms.)

▼ **Example 1**   Find the pH of a solution in which (H_3O^+) = 5.1 × 10^{-3}.

Solution   Using the definition given above, we write

$$pH = -\log (H_3O^+)$$
$$= -\log (5.1 \times 10^{-3})$$
$$= -[.7076 + (-3)]$$
$$= -.7076 + 3$$
$$\approx 2.29$$

According to our previous discussion, this solution would be considered an acid solution, since its pH is less than 7.   ▲

▼ **Example 2**   What is (H_3O^+) for a solution with a pH of 9.3?

Solution   We substitute 9.3 for pH in the definition of pH and proceed as follows:

$$9.3 = -\log (H_3O^+)$$
$$\log (H_3O^+) = -9.3$$

It is impossible to look up −9.3 in the table in Appendix D because it is negative. We can get around this problem by adding and subtracting 10 to the −9.3:

$$\log (H_3O^+) = -9.3 + 10 - 10 \qquad \text{(Actually adding 0)}$$
$$= .7 - 10$$

The closest thing to .7 in the table is .6998, which is the logarithm of 5.01:

$$(H_3O^+) = 5.01 \times 10^{-10}$$

The concentration of H_3O^+ is 5.01×10^{-10}. (This quantity is usually given in moles/liter. We have left off the units to make things simpler.) ▲

Carbon-14 dating is used extensively in science to find the age of fossils. If at one time a nonliving substance contains an amount A_0 of carbon-14, then t years later it will contain an amount A of carbon-14, where

$$A = A_0 \cdot 2^{-t/5600}$$

▼ **Example 3** If a nonliving substance has 3 grams of carbon-14, how much carbon-14 will be present 500 years later?

Solution The original amount of carbon 14, A_0, is 3 grams. The number of years, t, is 500. Substituting these quantities into the equation given above, we have the expression

$$A = (3)2^{-500/5600}$$
$$= (3)2^{-5/56}$$

In order to evaluate this expression we must use logarithms. We begin by taking the logarithm of both sides:

$$\log A = \log [(3)2^{-5/56}]$$
$$= \log 3 - \tfrac{5}{56} \log 2$$
$$= .4771 - \tfrac{5}{56}(.3010)$$
$$= .4771 - .0269$$
$$\log A = .4502$$
$$A = 2.82$$

The amount remaining after 500 years is 2.82 grams. ▲

If an amount of money P (P for principal) is invested in an account that pays a rate r of interest compounded annually, then the total amount of money A (the original amount P plus all the interest) in the account after t years is given by the equation

$$A = P(1 + r)^t$$

▼ **Example 4** How much money will accumulate over 20 years if a person invests $5000 in an account that pays 6% interest per year?

Solution The original amount P is \$5000, the rate of interest r is .06 (6% = .06), and the length of time t is 20 years. We substitute these values into the equation and solve for A:

$$A = 5000(1 + .06)^{20}$$
$$A = 5000(1.06)^{20}$$
$$\log A = \log [5000(1.06)^{20}]$$
$$\log A = \log 5000 + 20 \log 1.06$$
$$= 3.6990 + 20(.0253)$$
$$= 3.6990 + .5060$$
$$\log A = 4.2050$$
$$A = 1.60 \times 10^4$$
$$= 16{,}000$$

The original \$5000 will more than triple to become \$16,000 in 20 years at 6% annual interest.   ▲

▼   **Example 5**   How long does it take for \$5000 to double if it is deposited in an account that yields 5% per year?

Solution The original amount P is \$5000, *the total amount after t years* is $A = \$10{,}000$, and the interest rate r is .05. Substituting into

$$A = P(1 + r)^t$$

we have

$$10{,}000 = 5000(1 + .05)^t$$
$$= 5000(1.05)^t$$

This is an exponential equation. We solve by taking the logarithm of both sides:

$$\log 10{,}000 = \log [(5000)(1.05)^t]$$
$$\log 10{,}000 = \log 5000 + t \log 1.05$$
$$4 = 3.6990 + t(.0212)$$

Subtract 3.6990 from both sides:

$$.3010 = .0212t$$

Dividing both sides by .0212, we have

$$t = 14.20$$

It takes a little over 14 years for \$5000 to double if it earns 5% per year.   ▲

For problems 1–8 find the pH of the solution using the formula $pH = -\log(H_3O^+)$.

Problem Set 11.5

1. $(H_3O^+) = 4 \times 10^{-3}$
2. $(H_3O^+) = 3 \times 10^{-4}$
3. $(H_3O^+) = 5 \times 10^{-6}$
4. $(H_3O^+) = 6 \times 10^{-5}$
5. $(H_3O^+) = 4.2 \times 10^{-5}$
6. $(H_3O^+) = 2.7 \times 10^{-7}$
7. $(H_3O^+) = 8.6 \times 10^{-2}$
8. $(H_3O^+) = 5.3 \times 10^{-4}$

For problems 9–12 find (H_3O^+) for solutions with the given pH.

9. $pH = 3.4$
10. $pH = 5.7$
11. $pH = 6.5$
12. $pH = 2.1$

13. A nonliving substance contains 3 micrograms of carbon-14. How much carbon-14 will be left at the end of

    a. 5000 years?
    b. 10,000 years?
    c. 56,000 years?
    d. 112,000 years?

14. A nonliving substance contains 5 micrograms of carbon-14. How much carbon-14 will be left at the end of

    a. 500 years?
    b. 5000 years?
    c. 56,000 years?
    d. 112,000 years?

15. At one time a certain nonliving substance contained 10 micrograms of carbon-14. How many years later did the same substance contain only 5 micrograms of carbon-14?

16. At one time a certain nonliving substance contained 20 micrograms of carbon-14. How many years later did the same substance contain only 5 micrograms of carbon-14?

17. How much money is in an account after 10 years if $5000 was deposited originally at 6% per year?

18. If you put $2000 in an account that earns 5% interest annually, how much will you have in the account at the end of 10 years?

19. If you deposit $2000 in an account that earns 7.5% annually, how much will you have in the account after 10 years?

20. Suppose $15,000 is deposited in an account that yields 10% per year. How much is in the account 5 years later?

21. If $4000 is in an account that earns 5% per year, how long does it take before the account has $8000 in it?

22. If $4000 is deposited in an account that yields 7% annually, how long does it take the account to reach $8000?

23. How long does it take to double $10,000 if it is in an account that earns 6% per year?

24. How long does it take to double $10,000 if it is in an account that earns 10% per year?

Chapter 11
Summary and Review

Definition of Logarithms

If b is a positive number not equal to 1, then the expression

$$y = \log_b x$$

is equivalent to $x = b^y$. That is, in the expression $y = \log_b x$, y is the number to which we raise b in order to get x. Expressions written in the form $y = \log_b x$ are said to be in *logarithmic form*. Expressions like $x = b^y$ are in *exponential form*.

Two Special Identities

For $b > 0, b \neq 1$, the following two expressions hold for all positive real numbers x:

1. $b^{\log_b x} = x$
2. $\log_b b^x = x$

Properties of Logarithms

If x, y, and b are positive real numbers, $b \neq 1$, and r is any real number, then

1. $\log_b (xy) = \log_b x + \log_b y$

2. $\log_b \left(\dfrac{x}{y}\right) = \log_b x - \log_b y$

3. $\log_b x^r = r \log_b x$

Common Logarithms

Common logarithms are logarithms with a base of 10. To save time in writing, we omit the base when working with common logarithms. That is,

$$\log x = \log_{10} x$$

Notation

In the expression

$$\log 4240 = 3.6274$$

the 3 is called the *characteristic* and the decimal .6274 is called the *mantissa*.

Change of Base

If x, a, and b are positive real numbers, $a \neq 1$ and $b \neq 1$, then

$$\log_a x = \frac{\log_b x}{\log_b a}$$

The most common mistakes that occur with logarithms come from trying to apply the three properties of logarithms to situations in which they don't apply. For example, a very common mistake looks like this:

$$\frac{\log 3}{\log 2} = \log 3 - \log 2$$

This is not a property of logarithms. In order to write the expression $\log 3 - \log 2$, we would have to start with

$$\log \frac{3}{2} \quad \text{not} \quad \frac{\log 3}{\log 2}$$

There is a difference.

Solve for x:

Chapter 11 Test

1. $\log_4 x = 3$

2. $\log_x 5 = 2$

Graph each of the following:

3. $y = \log_2 x$

4. $y = \log_{1/2} x$

Evaluate each of the following:

5. $\log_8 4$

6. $\log_7 21$

7. $\log 23{,}400$

8. $\log .0123$

Use logarithms to find an approximate answer for the following computations:

9. $(2.34)(6080)$

10. $3^{2.5}$

11. $\sqrt[4]{23}$

12. $\dfrac{(352)(41.5)^2}{(2.31)^3}$

Use the properties of logarithms as an aid in solving the following:

13. $5 = 3^x$

14. $4^{2x-1} = 8$

15. $\log_5 x - \log_5 3 = 1$

16. $\log_2 x + \log_2 (x - 7) = 3$

17. Find the total amount of money in an account if $4000 was deposited 5 years ago at 5% annual interest.

18. Give the pH of a solution in which $(H_3O^+) = 3 \times 10^{-4}$.

12 Sequences and Series

To the student:

This chapter begins with a look at sequences and series in general. In the first section we will use a new kind of notation, summation notation, that allows us to write long sums of numbers with few symbols. We will classify two main types of sequences, arithmetic and geometric sequences, and look at what is called the nth partial sum of a sequence. The section on the binomial expansion is very useful as it gives a formula for expanding $(a + b)^n$ without actually having to multiply it out. Each section in this chapter deals in some way with sequences of numbers. Sequences are used frequently in many different branches of mathematics and science. Many real-life situations can be described in terms of arithmetic and geometric sequences. For example, the yearly amount of money in a savings account that compounds interest at regular intervals will form a geometric progression, as will the value of any object that increases or decreases in value by a specified percentage each year. The binomial theorem has some useful applications which are not limited to mathematics. It is used in statistics, biology, and physics as well as other disciplines. Many of the problems we have previously worked in terms of functions can also be worked using sequences and series.

12.1 Sequences

This section serves as a general introduction to sequences and series. We begin with the definition of a sequence.

DEFINITION A *sequence* is a function whose domain is the set of positive integers $\{1, 2, 3, \ldots\}$.

We usually use the letter *a* with a subscript to denote specific terms in a sequence, instead of the usual function notation. That is, instead of writing $a(3)$ for the third term of a sequence, we use a_3 to denote the third term. For example,

$$a_1 = \text{First term of the sequence}$$
$$a_2 = \text{Second term of the sequence}$$
$$a_3 = \text{Third term of the sequence}$$
$$\vdots$$
$$a_n = n\text{th term of the sequence}$$
$$\vdots$$

The *n*th term is also called the *general term* of the sequence. The general term is used to define the other terms of the sequence. That is, if we are given the formula for the general term, a_n, we can find any other term in the sequence. The following examples illustrate.

▼   **Example 1**   Find the first four terms of the sequence whose general term is given by $a_n = 2n - 1$.

Solution   The subscript notation a_n works the same way function notation works. To find the first, second, third, and fourth terms of this sequence, we simply substitute 1, 2, 3, and 4 for *n* in the formula $2n - 1$:

$$\text{If} \quad \text{General term} = a_n = 2n - 1,$$
$$\text{then} \quad \text{First term} = a_1 = 2(1) - 1 = 1$$
$$\text{Second term} = a_2 = 2(2) - 1 = 3$$
$$\text{Third term} = a_3 = 2(3) - 1 = 5$$
$$\text{Fourth term} = a_4 = 2(4) - 1 = 7$$

The first four terms of this sequence are the odd numbers 1, 3, 5, and 7. The whole sequence can be written as

$$1, 3, 5, \ldots, 2n - 1, \ldots$$

Since each term in this sequence is larger than the preceding term, we say the sequence is an *increasing sequence*.   ▲

▼   **Example 2**   Write the first four terms of the sequence defined by

$$a_n = \frac{1}{n + 1}.$$

Solution   Replacing n with 1, 2, 3, and 4, we have, respectively, the first four terms:

$$\text{First term} = a_1 = \frac{1}{1+1} = \frac{1}{2}$$

$$\text{Second term} = a_2 = \frac{1}{2+1} = \frac{1}{3}$$

$$\text{Third term} = a_3 = \frac{1}{3+1} = \frac{1}{4}$$

$$\text{Fourth term} = a_4 = \frac{1}{4+1} = \frac{1}{5}$$

The sequence defined by

$$a_n = \frac{1}{n+1}$$

can be written as

$$\frac{1}{2}, \frac{1}{3}, \frac{1}{4}, \ldots, \frac{1}{n+1}, \ldots$$

Since each term in the sequence is smaller than the term preceding it, the sequence is said to be a *decreasing sequence*.   ▲

▼   **Example 3**   Find the fifth and sixth terms of the sequence whose general term is given by $a_n = \frac{(-1)^n}{n^2}$.

Solution   For the fifth term we replace n with 5. For the sixth term we replace n with 6:

$$\text{Fifth term} = a_5 = \frac{(-1)^5}{5^2} = \frac{-1}{25}$$

$$\text{Sixth term} = a_6 = \frac{(-1)^6}{6^2} = \frac{1}{36}$$ ▲

In the first three examples we found some terms of a sequence after being given the general term. In the next two examples we will do the reverse. That is, given some terms of a sequence, we will find the formula for the general term.

▼ **Example 4** Find a formula for the nth term of the sequence 2, 8, 18, 32, ...

Solution Solving a problem like this involves some guessing. Looking over the first four terms we see each is twice a perfect square:

$$2 = 2(1)$$
$$8 = 2(4)$$
$$18 = 2(9)$$
$$32 = 2(16)$$

If we write each square with an exponent of 2, the formula for the nth term becomes obvious:

$$a_1 = 2 \ = 2(1)^2$$
$$a_2 = 8 \ = 2(2)^2$$
$$a_3 = 18 = 2(3)^2$$
$$a_4 = 32 = 2(4)^2$$
$$\vdots$$
$$a_n = \qquad 2(n)^2 = 2n^2$$

The general term of the sequence 2, 8, 18, 32, . . . is $a_n = 2n^2$.

▲

▼ **Example 5** Find the general term for the sequence $2, \frac{3}{8}, \frac{4}{27}, \frac{5}{64}, \ldots$.

Solution The first term can be written as $\frac{2}{1}$. The denominators are all perfect cubes. The numerators are all 1 more than the base of the cubes in the denominator:

$$a_1 = \frac{2}{1} \ = \frac{1 + 1}{1^3}$$

$$a_2 = \frac{3}{8} \ = \frac{2 + 1}{2^3}$$

$$a_3 = \frac{4}{27} = \frac{3 + 1}{3^3}$$

$$a_4 = \frac{5}{64} = \frac{4 + 1}{4^3}$$

Observing this pattern, we recognize the general term to be

$$a_n = \frac{n + 1}{n^3}$$

▲

Note Finding the nth term of a sequence from the first few terms is not always automatic. That is, it sometimes takes a while to recognize the pattern. Don't be afraid to guess at the formula for the general term. Many times an incorrect guess leads to the correct formula.

Problem Set 12.1

Write the first five terms of the sequences with the following general terms:

1. $a_n = 3n + 1$

2. $a_n = 2n + 3$

3. $a_n = 4n - 1$

4. $a_n = n + 4$

5. $a_n = n$

6. $a_n = -n$

7. $a_n = n^2 + 3$

8. $a_n = n^3 + 1$

9. $a_n = \dfrac{n}{n + 3}$

10. $a_n = \dfrac{n}{n + 2}$

11. $a_n = \dfrac{n + 1}{n + 2}$

12. $a_n = \dfrac{n + 3}{n + 4}$

13. $a_n = \dfrac{1}{n^2}$

14. $a_n = \dfrac{1}{n^3}$

15. $a_n = 2^n$

16. $a_n = 3^n$

17. $a_n = 3^{-n}$

18. $a_n = 2^{-n}$

19. $a_n = 1 + \dfrac{1}{n}$

20. $a_n = 1 - \dfrac{1}{n}$

21. $a_n = n - \dfrac{1}{n}$

22. $a_n = n + \dfrac{1}{n}$

23. $a_n = (-2)^n$

24. $a_n = (-3)^n$

Determine the general term for each of the following sequences:

25.   2, 3, 4, 5, . . .

26.   3, 6, 9, 12, . . .

27.   4, 8, 12, 16, 20, . . .

28.   3, 4, 5, 6, . . .

29.   7, 10, 13, 16, . . .

30.   4, 9, 14, 19, . . .

31.   1, 4, 9, 16, . . .

32.   1, 8, 27, 64, . . .

33.   3, 12, 27, 48, . . .

34.   2, 16, 54, 128, . . .

35.   4, 8, 16, 32, . . .

36.   3, 9, 27, 81, . . .

37. $-2, 4, -8, 16, \ldots$

38. $-3, 9, -27, 81, \ldots$

39. $\frac{1}{4}, \frac{1}{8}, \frac{1}{16}, \frac{1}{32}, \ldots$

40. $\frac{1}{3}, \frac{1}{9}, \frac{1}{27}, \frac{1}{81}, \ldots$

41. $\frac{1}{4}, \frac{2}{9}, \frac{3}{16}, \frac{4}{25}, \ldots$

42. $\frac{1}{4}, \frac{2}{10}, \frac{3}{28}, \frac{4}{82}, \ldots$

43.   The value of a home in California is said to increase 1% per month. If a home were currently worth \$50,000, then in 1 month it would increase in value 1% of \$50,000 or .01(50,000) = \$500. It would then be worth \$50,500. Write a sequence that gives the value of this home every month for 6 months. What is the general term of the sequence? How much should the house be worth in 2 years? (Give answers to nearest dollar.)

We begin this section with the definition of a series.

DEFINITION   The sum of a number of terms in a sequence is called a *series*.

A sequence can be finite or infinite depending on whether or not the sequence ends at the *n*th term. For example,

$$1, 3, 5, 7, 9$$

is a finite sequence, while

$$1, 3, 5, \ldots$$

is an infinite sequence. Associated with each of the above sequences is a series found by adding the terms of the sequence:

$$1 + 3 + 5 + 7 + 9 \qquad \text{Finite series}$$
$$1 + 3 + 5 + \ldots \qquad \text{Infinite series}$$

We will consider only finite series in this book. We can introduce a new kind of notation here that is a compact way of indicating a finite series. The notation is called *summation notation,* or *sigma notation* since it is written using the Greek letter sigma. The expression

$$\sum_{i=1}^{4} (3i + 2)$$

is an example of summation notation. The expression is interpreted as indicating we are to add all the numbers $3i + 2$, where i is replaced by all the consecutive integers from 1 up to and including 4:

$$\sum_{i=1}^{4} (3i + 2) = [3(1) + 2] + [3(2) + 2] + [3(3) + 2] + [3(4) + 2]$$
$$= 5 + 8 + 11 + 14$$
$$= 38$$

The letter i as used here is called the *index of summation,* or just *index* for short.

Here are some examples illustrating the use of summation notation.

▼   **Example 1**   Expand and simplify $\displaystyle\sum_{i=1}^{5} (i^2 - 1)$.

Solution   We replace i in the expression $i^2 - 1$ with all consecutive integers from 1 up to 5, including 1 and 5:

$$\sum_{i=1}^{5} (i^2 - 1) = (1^2 - 1) + (2^2 - 1) + (3^2 - 1) + (4^2 - 1) + (5^2 - 1)$$

$$= 0 + 3 + 8 + 15 + 24$$
$$= 50 \qquad \blacktriangle$$

▼   **Example 2**   Expand and simplify $\sum_{i=3}^{6} (-2)^i$.

Solution   We replace i in the expression $(-2)^i$ with the consecutive integers beginning at 3 and ending at 6:

$$\sum_{i=3}^{6} (-2)^i = (-2)^3 + (-2)^4 + (-2)^5 + (-2)^6$$

$$= -8 + 16 + (-32) + 64$$
$$= 40 \qquad \blacktriangle$$

▼   **Example 3**   Expand $\sum_{i=2}^{5} (x^i - 3)$.

Solution   We must be careful here not to confuse the letter x with i. The index i is the quantity we replace by the consecutive integers from 1 to 5, not x.

$$\sum_{i=2}^{5} (x^i - 3) = (x^2 - 3) + (x^3 - 3) + (x^4 - 3) + (x^5 - 3) \qquad \blacktriangle$$

   In the first three examples we were given an expression with summation notation and asked to expand it. The next examples in this section illustrate how we can write an expression in expanded form as an expression involving summation notation.

▼   **Example 4**   Write with summation notation: $1 + 3 + 5 + 7 + 9$.

Solution   A formula that gives us the terms of this sum is

$$a_i = 2i - 1$$

where i ranges from 1 up to and including 5. Notice we are using the subscript i here in exactly the same way we used the subscript n in the last section—to indicate the general term. Writing the sum

$$1 + 3 + 5 + 7 + 9$$

with summation notation looks like this:

$$\sum_{i=1}^{5} (2i - 1) \qquad \blacktriangle$$

▼   **Example 5**   Write with summation notation: $3 + 12 + 27 + 48$.

Solution   We need a formula, in terms of i, that will give each term in the sum. Writing the sum as

$$3 \cdot 1^2 + 3 \cdot 2^2 + 3 \cdot 3^2 + 3 \cdot 4^2$$

we see the formula

$$a_i = 3 \cdot i^2$$

where i ranges from 1 up to and including 4. Using this formula and summation notation, we can represent the sum

$$3 + 12 + 27 + 48$$

as

$$\sum_{i=1}^{4} 3i^2 \qquad \blacktriangle$$

▼   **Example 6**   Write with summation notation:

$$\frac{x + 3}{x^3} + \frac{x + 4}{x^4} + \frac{x + 5}{x^5} + \frac{x + 6}{x^6}.$$

Solution   A formula that gives each of these terms is

$$a_i = \frac{x + i}{x^i}$$

where i assumes all integer values between 3 and 6, including 3 and 6. The sum can be written as

$$\sum_{i=3}^{6} \frac{x + i}{x^i} \qquad \blacktriangle$$

Problem Set 12.2   Expand and simplify each of the following:

1. $\displaystyle\sum_{i=1}^{4}(2i + 4)$ 2. $\displaystyle\sum_{i=1}^{5}(3i - 1)$

3. $\displaystyle\sum_{i=1}^{3}(2i - 1)$ 4. $\displaystyle\sum_{i=1}^{4}(2i - 1)$

5. $\displaystyle\sum_{i=2}^{5}(i^2 - 1)$ 6. $\displaystyle\sum_{i=3}^{6}(i^2 + 1)$

7. $\displaystyle\sum_{i=1}^{4}\frac{i}{1 + i}$ 8. $\displaystyle\sum_{i=1}^{4}\frac{i^2}{1 + i}$

9. $\displaystyle\sum_{i=1}^{5}\frac{i^2}{2i - 1}$ 10. $\displaystyle\sum_{i=3}^{5}(i^3 + 4)$

11. $\displaystyle\sum_{i=1}^{4}(-3)^i$ 12. $\displaystyle\sum_{i=1}^{4}\left(-\frac{1}{3}\right)^i$

13. $\displaystyle\sum_{i=3}^{6}(-2)^i$ 14. $\displaystyle\sum_{i=4}^{6}\left(-\frac{1}{2}\right)^i$

Expand the following:

15. $\displaystyle\sum_{i=1}^{5}(x + i)$ 16. $\displaystyle\sum_{i=3}^{6}(x - i)$

17. $\displaystyle\sum_{i=2}^{7}(x + 1)^i$ 18. $\displaystyle\sum_{i=1}^{4}(x + 3)^i$

19. $\displaystyle\sum_{i=1}^{5}\frac{x + i}{x - 1}$ 20. $\displaystyle\sum_{i=1}^{6}\frac{x - 3i}{x + 3i}$

21. $\displaystyle\sum_{i=3}^{8}(x + i)^i$ 22. $\displaystyle\sum_{i=4}^{7}(x - 2i)^i$

23. $\displaystyle\sum_{i=1}^{5}(x + i)^{i+1}$ 24. $\displaystyle\sum_{i=2}^{6}(x + i)^{i-1}$

Write each of the following sums with summation notation:

25. $2 + 4 + 8 + 16$ 26. $3 + 5 + 7 + 9 + 11$
27. $4 + 8 + 16 + 32 + 64$ 28. $1 + 3 + 5$

29. $5 + 10 + 17 + 26 + 37$

30. $3 + 8 + 15 + 24$

31. $\frac{3}{4} + \frac{4}{5} + \frac{5}{6} + \frac{6}{7} + \frac{7}{8}$

32. $\frac{1}{2} + \frac{2}{3} + \frac{3}{4} + \frac{4}{5}$

33. $\frac{1}{3} + \frac{2}{5} + \frac{3}{7} + \frac{4}{9}$

34. $\frac{3}{1} + \frac{5}{3} + \frac{7}{5} + \frac{9}{7}$

35. $(x - 3) + (x - 4) + (x - 5) + (x - 6)$

36. $x^2 + x^3 + x^4 + x^5 + x^6$

37. $\dfrac{x}{x + 3} + \dfrac{x}{x + 4} + \dfrac{x}{x + 5}$

38. $\dfrac{x - 3}{x^3} + \dfrac{x - 4}{x^4} + \dfrac{x - 5}{x^5} + \dfrac{x - 6}{x^6}$

39. $x^2(x + 2) + x^3(x + 3) + x^4(x + 4)$

40. $x(x + 2)^2 + x(x + 3)^3 + x(x + 4)^4$

41. A skydiver jumps from a plane and falls 16 feet the first second, 48 feet the second second, and 80 feet the third second. If he continues to fall in the same manner, how far will he fall the seventh second? What is the total distance he falls in 7 seconds?

In this and the following section we will classify two major types of sequences: arithmetic progressions and geometric progressions.

12.3 Arithmetic Progressions

DEFINITION An *arithmetic progression* is a sequence of numbers in which each term is obtained from the preceding term by adding the same amount each time.

The sequence

$$2, 7, 12, 17, \ldots$$

is an example of an arithmetic progression, since each term is obtained from the preceding term by adding 5 each time. The amount we add each time—in this case, 5—is called the *common difference,* since it can be obtained by subtracting any two consecutive terms. (The term with the larger subscript must be written first.) The common difference is denoted by d.

▼ **Example 1** Give the common difference d for the arithmetic progression 4, 10, 16, 22, ...

Solution Since each term can be obtained from the preceding term by adding 6, the common difference is 6. That is, $d = 6$. ▲

▼ **Example 2** Give the common difference for 100, 93, 86, 79, ...

Solution   The common difference in this case is $d = -7$, since adding -7 to any term always produces the next consecutive term.
▲

▼   **Example 3**   Give the common difference for $\frac{1}{2}$, 1, $\frac{3}{2}$, 2, ...

Solution   The common difference is $d = \frac{1}{2}$.   ▲

The general term, a_n, of an arithmetic progression can always be written in terms of the first term a_1 and the common difference d. Consider the sequence from Example 1:

$$2, 7, 12, 17, \ldots$$

We can write each term in terms of the first term 2 and the common difference 5:

$$
\begin{array}{cccc}
2, & 2 + (1 \cdot 5), & 2 + (2 \cdot 5), & 2 + (3 \cdot 5), \ldots \\
a_1, & a_2, & a_3, & a_4, \quad \ldots
\end{array}
$$

Observing the relationship between the subscript on the terms in the second line and the coefficients of the 5's in the first line, we write the general term for the sequence as

$$a_n = 2 + (n - 1)5$$

We generalize this result to include the general term of any arithmetic sequence.

The General Term

The *general term* of an arithmetic progression with the first term a_1 and common difference d is given by

$$a_n = a_1 + (n - 1)d$$

▼   **Example 4**   Find the general term for the sequence

$$7, 10, 13, 16, \ldots .$$

Solution   The first term is $a_1 = 7$, and the common difference is $d = 3$. Substituting these numbers into the formula given above, we have

$$a_n = 7 + (n - 1)3$$

which we can simplify, if we choose, to

$$
\begin{aligned}
a_n &= 7 + 3n - 3 \\
&= 3n + 4
\end{aligned}
$$
▲

▼ **Example 5** Find the general term of the arithmetic progression whose third term, a_3, is 7 and whose eighth term, a_8, is 17.

Solution According to the formula for the general term, the third term can be written as $a_3 = a_1 + 2d$, and the eighth term can be written as $a_8 = a_1 + 7d$. Since these terms are also equal to 7 and 17, respectively, we can write

$$a_3 = a_1 + 2d = 7$$
$$a_8 = a_1 + 7d = 17$$

To find a_1 and d, we simply solve the system on the right above:

$$a_1 + 2d = 7$$
$$a_1 + 7d = 17$$

We add the opposite of the top equation to the bottom equation. The result is

$$5d = 10$$
$$d = 2$$

To find a_1 we simply substitute 2 for d in either of the original equations and get

$$a_1 = 3$$

The general term for this progression is

$$a_n = 3 + (n - 1)2$$

which we can simplify to

$$a_n = 2n + 1 \qquad\qquad ▲$$

The sum of the first n terms of an arithmetic progression is denoted by S_n. The following theorem gives the formula for finding S_n, which is sometimes called the nth *partial sum*.

THEOREM 12.1 The sum of the first n terms of an arithmetic progression whose first term is a_1 and whose nth term is a_n is given by

$$S_n = \frac{n}{2}(a_1 + a_n)$$

Proof We can write S_n in expanded form as

$$S_n = a_1 + [a_1 + d] + [a_1 + 2d] + \cdots + [a_1 + (n - 1)d]$$

This sum can also be written in reverse order as

$$S_n = a_n + [a_n - d] + [a_n - 2d)] + \cdots + [a_n - (n-1)d]$$

If we add the preceding two expressions term by term, we have

$$2S_n = (a_1 + a_n) + (a_1 + a_n) + (a_1 + a_n) + \cdots + (a_1 + a_n)$$
$$2S_n = n(a_1 + a_n)$$
$$S_n = \frac{n}{2}(a_1 + a_n)$$

▼ **Example 6** Find the sum of the first 10 terms of the arithmetic progression 2, 10, 18, 26, ...

Solution The first term is 2 and the common difference is 8. The tenth term is

$$a_{10} = 2 + 9(8)$$
$$= 2 + 72$$
$$= 74$$

Substituting $n = 10$, $a_1 = 2$, and $a_{10} = 74$ into the formula

$$S_n = \frac{n}{2}(a_1 + a_n)$$

we have

$$S_{10} = \frac{10}{2}(2 + 74)$$
$$= 5(76)$$
$$= 380$$

The sum of the first 10 terms is 380.   ▲

Problem Set 12.3

Determine which of the following sequences are arithmetic progressions. For those that are arithmetic progressions, identify the common difference d.

1. 1, 2, 3, 4, . . .   2. 4, 6, 8, 10, . . .
3. 1, 2, 4, 7, . . .   4. 1, 2, 4, 8, . . .
5. 50, 45, 40, . . .   6. $1, \frac{1}{2}, \frac{1}{4}, \frac{1}{8}, \ldots$
7. 1, 4, 9, 16, . . .   8. 5, 7, 9, 11, . . .
9. $\frac{1}{3}, 1, \frac{5}{3}, \frac{7}{3}, \ldots$ 10. 5, 11, 17, . . .

Each of the following problems gives some information about a specific arithmetic progression. In each case use the given information to find:

    a. the general term a_n
    b. the tenth term a_{10}
    c. the twenty-fourth term a_{24}
    d. the sum of the first 10 terms S_{10}
    e. the sum of the first 24 terms S_{24}

11. $a_1 = 3$, $d = 4$ **12.** $a_1 = 5$, $d = 10$
13. $a_1 = 6$, $d = -2$ **14.** $a_1 = 7$, $d = -1$
15. 5, 9, 13, 17, . . .   **16.** 8, 11, 14, 17, . . .
17. 12, 7, 2, -3, . . .   **18.** 25, 20, 15, 10, . . .
19. $\frac{1}{2}$, 1, $\frac{3}{2}$, 2, . . .   **20.** $-\frac{1}{3}$, 0, $\frac{1}{3}$, $\frac{2}{3}$
21. $a_3 = 16$, $a_8 = 46$ **22.** $a_3 = 16$, $a_8 = 51$
23. $a_6 = 17$, $a_{12} = 29$ **24.** $a_5 = 23$, $a_{10} = 48$

25. Suppose a woman earns \$15,000 a year and gets a raise of \$850 per year. Write a sequence that gives her salary for each of the first 5 years she works. What is the general term of this sequence? How much will she be making annually if she stays on the job for 10 years?

This section is concerned with the second major classification of sequences, called geometric progressions. The problems in this section are very similar to the problems in the preceding section.

**12.4
Geometric
Progressions**

DEFINITION A sequence of numbers in which each term is obtained from the previous term by multiplying by the same amount each time is called a *geometric sequence*.

    The sequence

$$3, 6, 12, 24, \ldots$$

is an example of a geometric progression. Each term is obtained from the previous term by multiplying by 2. The amount by which we multiply each time—in this case, 2—is called the *common ratio*. The common ratio is denoted by r, and can be found by taking the ratio of any two consecutive terms. (The term with the larger subscript must be in the numerator.)

▼   **Example 1** Find the common ratio for the geometric progression

$$\tfrac{1}{2}, \tfrac{1}{4}, \tfrac{1}{8}, \tfrac{1}{16}, \ldots.$$

Solution Since each term can be obtained from the term before it by multiplying by $\frac{1}{2}$, the common ratio is $\frac{1}{2}$. That is, $r = \frac{1}{2}$.   ▲

▼ **Example 2** Find the common ratio for $\sqrt{3}$, 3, $3\sqrt{3}$, 9, ...

Solution If we take the ratio of the second and third terms, we have

$$\frac{3\sqrt{3}}{3} = \sqrt{3}$$

The common ratio is $r = \sqrt{3}$.   ▲

The General Term

The *general term, a_n,* of a geometric progression with a first term a_1 and common ratio r is given by

$$a_n = a_1 r^{n-1}$$

To see how we arrive at this formula, consider the following geometric progression whose common ratio is 3:

$$2, \ 6, \ 18, \ 54, \ . . .$$

We can write each term of the sequence in terms of the first term 2 and the common ratio 3:

$$\begin{array}{cccc} 2\cdot 3^0, & 2\cdot 3^1, & 2\cdot 3^2, & 2\cdot 3^3, \ . . . \\ a_1, & a_2, & a_3, & a_4, \ . . . \end{array}$$

Observing the relationship between the two lines written above, we find we can write the general term of this progression as

$$a_n = 2\cdot 3^{n-1}$$

Since the first term can be designated by a_1 and the common ratio by r, the formula

$$a_n = 2\cdot 3^{n-1}$$

coincides with the formula

$$a_n = a_1 r^{n-1}$$

▼ **Example 3** Find the general term for the geometric progression

$$5, \ 10, \ 20, \ . . . .$$

Solution The first term is $a_1 = 5$, and the common ratio is $r = 2$. Using these values in the formula

$$a_n = a_1 r^{n-1}$$

we have

$$a_n = 5 \cdot 2^{n-1}$$
▲

▼   **Example 4**   Find the tenth term of the sequence $3, \frac{3}{2}, \frac{3}{4}, \frac{3}{8}, \ldots.$

Solution   The sequence is a geometric progression with first term $a_1 = 3$ and common ratio $r = \frac{1}{2}$. The tenth term is

$$a_{10} = 3(\tfrac{1}{2})^9$$
▲

▼   **Example 5**   Find the general term for the geometric progression whose fourth term is 16 and whose seventh term is 128.

Solution   The fourth term can be written as $a_4 = a_1 r^3$, and the seventh term can be written as $a_7 = a_1 r^6$:

$$a_4 = a_1 r^3 = 16$$
$$a_7 = a_1 r^6 = 128$$

We can solve for r by using the ratio a_7/a_4:

$$\frac{a_7}{a_4} = \frac{a_1 r^6}{a_1 r^3} = \frac{128}{16}$$
$$r^3 = 8$$
$$r = 2$$

The common ratio is 2. To find the first term we substitute $r = 2$ into either of the original two equations. The result is

$$a_1 = 2$$

The general term for this progression is

$$a_n = 2 \cdot 2^{n-1}$$

which we can simplify by adding exponents, since the bases are equal:

$$a_n = 2^n$$
▲

As was the case in the preceding section, the sum of the first n terms of a geometric progression is denoted by S_n, which is called the nth *partial sum* of the progression.

THEOREM 12.2 The sum of the first n terms of a geometric progression with first term a_1 and common ratio r is given by the formula

$$S_n = \frac{a_1(r^n - 1)}{r - 1}$$

Proof We write the sum of the first n terms in expanded form:

$$S_n = a_1 + a_1 r + a_1 r^2 + \cdots + a_1 r^{n-1} \tag{1}$$

Then multiplying both sides by r, we have

$$r S_n = a_1 r + a_1 r^2 + a_1 r^3 + \cdots + a_1 r^n \tag{2}$$

If we subtract the left side of equation (1) from the left side of equation (2) and do the same for the right sides, we end up with

$$r S_n - S_n = a_1 r^n - a_1$$

We factor S_n from both terms on the left side and a_1 from both terms on the right side of this equation:

$$S_n(r - 1) = a_1(r^n - 1)$$

Dividing both sides by $r - 1$ gives the desired result:

$$S_n = \frac{a_1(r^n - 1)}{r - 1}$$

▼ **Example 6** Find the sum of the first 10 terms of the geometric progression 5, 15, 45, 135, ...

Solution The first term is $a_1 = 5$, and the common ratio is $r = 3$. Substituting these values into the formula for S_{10}, we have the sum of the first 10 terms of the sequence:

$$S_{10} = \frac{5(3^{10} - 1)}{3 - 1}$$

$$= \frac{5(3^{10} - 1)}{2}$$

The answer can be left in this form, since it would be too time-consuming (and boring) to calculate 3^{10} by hand.

Identify those sequences that are geometric progressions. For those that are geometric give the common ratio r.

1. 1, 5, 25, 125, . . .
2. 6, 12, 24, 48
3. $\frac{1}{2}, \frac{1}{6}, \frac{1}{18}, \frac{1}{54}$
4. 5, 10, 15, 20, . . .
5. 4, 9, 16, 25, . . .
6. $-1, \frac{1}{3}, -\frac{1}{9}, \frac{1}{27}$
7. $-2, 4, -8, 16, . . .$
8. 1, 8, 27, 64, . . .
9. 4, 6, 8, 10, . . .
10. $1, -3, 9, -27, . . .$

Each of the following problems gives some information about a specific geometric progression. In each case use the given information to find:

a. the general term a_n
b. the sixth term a_6
c. the twentieth term a_{20}
d. the sum of the first 10 terms S_{10}
e. the sum of the first 20 terms S_{20}

11. $a_1 = 4, r = 3$
12. $a_1 = 5, r = 2$
13. $a_1 = -2, r = -5$
14. $a_1 = 6, r = -2$
15. $a_1 = -4, r = \frac{1}{2}$
16. $a_1 = 6, r = \frac{1}{3}$
17. 2, 12, 72, . . .
18. 7, 14, 28, . . .
19. 10, 20, 40, . . .
20. 100, 50, 25, . . .
21. $\sqrt{2}, 2, 2\sqrt{2}, . . .$
22. $5, 5\sqrt{2}, 10, . . .$
23. $a_4 = 40, a_6 = 160$
24. $a_5 = \frac{1}{8}, a_8 = \frac{1}{64}$

25. A savings account earns 6% interest per year. If the account has $10,000 at the beginning of one year, how much will it have at the beginning of the next year? Write a geometric progression that gives the total amount of money in this account every year for 5 years. What is the general term of this sequence?

The purpose of this section is to write and apply the formula for the expansion of expressions of the form $(x + y)^n$ where n is any positive integer. In order to write the formula, we must generalize the information in the following chart:

**12.5
The Binomial
Expansion**

$$(x + y)^1 = x + y$$
$$(x + y)^2 = x^2 + 2xy + y^2$$
$$(x + y)^3 = x^3 + 3x^2y + 3xy^2 + y^3$$
$$(x + y)^4 = x^4 + 4x^3y + 6x^2y^2 + 4xy^3 + y^4$$
$$(x + y)^5 = x^5 + 5x^4y + 10x^3y^2 + 10x^2y^3 + 5xy^4 + y^5$$

Note   The polynomials to the right have been found by expanding the binomials on the left—we just haven't shown the work.

There are a number of similarities to notice among the polynomials on the right. Here is a list of them:

1. In each polynomial, the sequence of exponents on the variable x decreases to zero from the exponent on the binomial at the left. (The exponent 0 is not shown, since $x^0 = 1$.)
2. In each polynomial the exponents on the variable y increase from 0 to the exponent on the binomial on the left. (Since $y^0 = 1$, it is not shown in the first term.)
3. The sum of the exponents on the variables in any single term is equal to the exponent on the binomial to the left.

The pattern in the coefficients of the polynomials on the right can best be seen by writing the right side again without the variables. It looks like this:

$$
\begin{array}{ccccccccccc}
 & & & & 1 & & 1 & & & & \\
 & & & 1 & & 2 & & 1 & & & \\
 & & 1 & & 3 & & 3 & & 1 & & \\
 & 1 & & 4 & & 6 & & 4 & & 1 & \\
1 & & 5 & & 10 & & 10 & & 5 & & 1
\end{array}
$$

This triangular-shaped array of coefficients is called *Pascal's triangle*. Each entry in the triangular array is obtained by adding the two numbers above it. Each row begins and ends with the number 1. If we were to continue Pascal's triangle, the next two rows would be

$$
\begin{array}{ccccccccccccc}
 & 1 & & 6 & & 15 & & 20 & & 15 & & 6 & & 1 \\
1 & & 7 & & 21 & & 35 & & 35 & & 21 & & 7 & & 1
\end{array}
$$

Pascal's triangle can be used to find coefficients for the expansion of $(x + y)^n$. The coefficients for the terms in the expansion of $(x + y)^n$ are given in the nth row of Pascal's triangle. It wouldn't be too much fun, however, to carry out Pascal's triangle much farther. There is an alternate method of finding the coefficients that does not involve Pascal's triangle. The alternate method of finding the coefficients involves notation we have not seen before.

DEFINITION   The expression $n!$ is read "n factorial" and is the product of all the consecutive integers from n down to 1. For example,

$$
\begin{aligned}
1! &= 1 \\
2! &= 2 \cdot 1 = 2 \\
3! &= 3 \cdot 2 \cdot 1 = 6 \\
4! &= 4 \cdot 3 \cdot 2 \cdot 1 = 24 \\
5! &= 5 \cdot 4 \cdot 3 \cdot 2 \cdot 1 = 120
\end{aligned}
$$

The expression 0! is defined to be 1. We use factorial notation to define binomial coefficients as follows.

DEFINITION The expression $\binom{n}{r}$ is called a *binomial coefficient* and is defined by

$$\binom{n}{r} = \frac{n!}{r!(n-r)!}$$

(If you are asking yourself, "What is going on here?"—hang on. We will be back to what we were doing shortly.)

▼ **Example 1** Calculate the following binomial coefficients:

$$\binom{7}{5}, \binom{6}{2}, \binom{3}{0}$$

Solution We simply apply the definition for binomial coefficients:

$$\binom{7}{5} = \frac{7!}{5!(7-5)!}$$

$$= \frac{7!}{5!2!}$$

$$= \frac{7 \cdot 6 \cdot 5 \cdot 4 \cdot 3 \cdot 2 \cdot 1}{(5 \cdot 4 \cdot 3 \cdot 2 \cdot 1)(2 \cdot 1)}$$

$$= \frac{42}{2}$$

$$= 21$$

$$\binom{6}{2} = \frac{6!}{2!(6-2)!}$$

$$= \frac{6!}{2! \cdot 4!}$$

$$= \frac{6 \cdot 5 \cdot 4 \cdot 3 \cdot 2 \cdot 1}{(2 \cdot 1)(4 \cdot 3 \cdot 2 \cdot 1)}$$

$$= \frac{30}{2}$$

$$= 15$$

$$\binom{3}{0} = \frac{3!}{0!(3-0)!}$$

$$= \frac{3!}{0! \cdot 3!}$$

$$= \frac{\cancel{3} \cdot \cancel{2} \cdot \cancel{1}}{(1)(\cancel{3} \cdot \cancel{2} \cdot \cancel{1})}$$

$$= 1 \qquad\qquad \blacktriangle$$

If we were to calculate all the binomial coefficients in the following array, we would find they match exactly with the numbers in Pascal's triangle. That is why they are called binomial coefficients—because they are the coefficients of the expansion of $(a + b)^n$:

$$\binom{1}{0} \qquad \binom{1}{1}$$

$$\binom{2}{0} \qquad \binom{2}{1} \qquad \binom{2}{2}$$

$$\binom{3}{0} \qquad \binom{3}{1} \qquad \binom{3}{2} \qquad \binom{3}{3}$$

$$\binom{4}{0} \qquad \binom{4}{1} \qquad \binom{4}{2} \qquad \binom{4}{3} \qquad \binom{4}{4}$$

$$\binom{5}{0} \qquad \binom{5}{1} \qquad \binom{5}{2} \qquad \binom{5}{3} \qquad \binom{5}{4} \qquad \binom{5}{5}$$

Using the new notation to represent the entries in Pascal's triangle, we can summarize everything we have noticed about the expansion of binomial powers of the form $(x + y)^n$.

The Binomial Expansion

If x and y represent real numbers and n is a positive integer, then the following formula is known as the *binomial expansion* or *binomial formula:*

$$(x + y)^n = \binom{n}{0}x^n y^0 + \binom{n}{1}x^{n-1}y^1 + \binom{n}{2}x^{n-2}y^2 + \cdots + \binom{n}{n}x^0 y^n$$

It does not make any difference, when expanding binomial powers of the form $(x + y)^n$, whether we use Pascal's triangle or the formula

$$\binom{n}{r} = \frac{n!}{r!(n-r)!}$$

to calculate the coefficients. We will show examples of both methods.

▼ **Example 2** Expand $(x - 2)^3$.

Solution The binomial formula for $(x + y)^3$ is

$$(x + y)^3 = \binom{3}{0}x^3y^0 + \binom{3}{1}x^2y^1 + \binom{3}{2}x^1y^2 + \binom{3}{3}x^0y^3$$

Applying this formula to $(x - 2)^3$, we have

$(x - 2)^3$

$$= \binom{3}{0}x^3(-2)^0 + \binom{3}{1}x^2(-2)^1 + \binom{3}{2}x^1(-2)^2 + \binom{3}{3}x^0(-2)^3$$

The coefficients $\binom{3}{0}$, $\binom{3}{1}$, $\binom{3}{2}$, and $\binom{3}{3}$ can be found in the third row of Pascal's triangle. They are 1, 3, 3, and 1:

$$(x - 2)^3 = 1x^3(-2)^0 + 3x^2(-2)^1 + 3x^1(-2)^2 + 1x^0(-2)^3$$
$$= x^3 - 6x^2 + 12x - 8 \qquad \blacktriangle$$

▼ **Example 3** Expand $(3x + 2y)^4$.

Solution The coefficients can be found in the fourth row of Pascal's triangle. They are

$$1, 4, 6, 4, 1$$

Here is the expansion of $(3x + 2y)^4$:

$(3x + 2y)^4$
$$= 1(3x)^4 + 4(3x)^3(2y) + 6(3x)^2(2y)^2 + 4(3x)(2y)^3 + 1(2y)^4$$
$$= 81x^4 + 216x^3y + 216x^2y^2 + 96xy^3 + 16y^4 \qquad \blacktriangle$$

▼ **Example 4** Write the first three terms in the expansion of $(x + 5)^9$.

Solution The coefficients of the first three terms are $\binom{9}{0}$, $\binom{9}{1}$, and $\binom{9}{2}$, which we calculate as follows:

$$\binom{9}{0} = \frac{9!}{0!9!} = \frac{9 \cdot 8 \cdot 7 \cdot 6 \cdot 5 \cdot 4 \cdot 3 \cdot 2 \cdot 1}{(1)(9 \cdot 8 \cdot 7 \cdot 6 \cdot 5 \cdot 4 \cdot 3 \cdot 2 \cdot 1)} = \frac{1}{1} = 1$$

$$\binom{9}{1} = \frac{9!}{1!8!} = \frac{9 \cdot 8 \cdot 7 \cdot 6 \cdot 5 \cdot 4 \cdot 3 \cdot 2 \cdot 1}{(1)(8 \cdot 7 \cdot 6 \cdot 5 \cdot 4 \cdot 3 \cdot 2 \cdot 1)} = \frac{9}{1} = 9$$

$$\binom{9}{2} = \frac{9!}{2!7!} = \frac{9 \cdot 8 \cdot 7 \cdot 6 \cdot 5 \cdot 4 \cdot 3 \cdot 2 \cdot 1}{(2 \cdot 1)(7 \cdot 6 \cdot 5 \cdot 4 \cdot 3 \cdot 2 \cdot 1)} = \frac{72}{2} = 36$$

From the binomial formula, we write the first three terms:

$$(x + 5)^9 = 1 \cdot x^5 + 9 \cdot x^4(5) + 36x^3(5)^2 + \ldots$$
$$= x^5 + 45x^4 + 900x^3 + \ldots \qquad \blacktriangle$$

Problem Set 12.5

Use the binomial formula to expand each of the following:

1. $(x + 2)^4$ 2. $(x - 2)^5$
3. $(x + y)^6$ 4. $(x - 1)^6$
5. $(2x + 1)^5$ 6. $(2x - 1)^4$
7. $(x - 2y)^5$ 8. $(2x + y)^5$
9. $(3x - 2)^4$ 10. $(2x - 3)^4$
11. $(4x - 3y)^3$ 12. $(3x - 4y)^3$
13. $(x^2 + 2)^4$ 14. $(x^2 - 3)^3$
15. $(x^2 + y^2)^3$ 16. $(x^2 - 3y)^4$

17. $\left(\dfrac{x}{2} - 4\right)^3$ 18. $\left(\dfrac{x}{3} + 6\right)^3$

19. $\left(\dfrac{x}{3} + \dfrac{y}{2}\right)^4$ 20. $\left(\dfrac{x}{2} - \dfrac{y}{3}\right)^4$

Write the first four terms in the expansion of the following:

21. $(x + 2)^9$ 22. $(x - 2)^9$
23. $(x - y)^{10}$ 24. $(x + y)^{10}$
25. $(x + 2y)^{10}$ 26. $(x - 2y)^{10}$

Write the first three terms in the expansion of each of the following:

27. $(x + 1)^{15}$ 28. $(x - 1)^{15}$
29. $(x - y)^{12}$ 30. $(x + y)^{12}$
31. $(x + 2)^{20}$ 32. $(x - 2)^{20}$

Write the first two terms in the expansion of each of the following:

33. $(x + 2)^{100}$ 34. $(x - 2)^{50}$
35. $(x + y)^{50}$ 36. $(x - y)^{100}$

37. The third term in the expansion of $(\frac{1}{2} + \frac{1}{2})^7$ will give the probability that in a family with 7 children, 5 will be boys and 2 will be girls. Find the third term.

**Chapter 12
Summary and Review**

Sequences

A *sequence* is a function whose domain is the set of real numbers. The terms of a sequence are denoted by

$$a_1, a_2, a_3, \ldots, a_n, \ldots$$

where a_1 (read "a sub 1") is the first term, a_2 the second term, and a_n the nth or general term.

The notation

$$\sum_{i=1}^{n} a_i = a_1 + a_2 + a_3 + \cdots + a_n$$

is called *summation notation* or *sigma notation*. The letter i as used here is called the *index of summation* or just *index*.

An *arithmetic progression* is a sequence in which each term comes from
the preceding term by adding a constant amount each time. If the first
term of an arithmetic progression is a_1 and the amount we add each
time (called the common difference) is d, then the nth term of the
progression is given by

$$a_n = a_1 + (n - 1)d$$

The sum of the first n terms of an arithmetic progression is

$$S_n = \frac{n}{2}(a_1 + a_n)$$

S_n is called the nth *partial sum*.

A *geometric progression* is a sequence of numbers in which each term
comes from the previous term by multiplying by a constant amount each
time. The constant by which we multiply each term to get the next term
is called the *common ratio*. If the first term of a geometric progression is
a_1 and the common ratio is r, then the formula that gives the general
term, a_n, is

$$a_n = a_1 r^{n-1}$$

The sum of the first n terms of a geometric progression is given by the
formula

$$S_n = \frac{a_1(r^n - 1)}{r - 1}.$$

The notation $n!$ is called n *factorial,* and is defined to be the product of
each consecutive integer from n down to 1. That is,

$$0! = 1 \quad \text{(By definition)}$$
$$1! = 1$$
$$2! = 2 \cdot 1$$

$$3! = 3 \cdot 2 \cdot 1$$
$$4! = 4 \cdot 3 \cdot 2 \cdot 1$$
$$\text{and so on}$$

Binomial Coefficients

The notation $\binom{n}{r}$ is called a *binomial coefficient* and is defined by

$$\binom{n}{r} = \frac{n!}{r!(n-r)!}$$

Binomial coefficients can be found by using the formula above or by Pascal's triangle, which is

```
        1   1
      1   2   1
    1   3   3   1
  1   4   6   4   1
1   5  10  10   5   1
```
and so on

Binomial Expansion

If n is a positive integer, then the formula for expanding $(a + b)^n$ is given by

$$(a + b)^n = \binom{n}{0}a^n b^0 + \binom{n}{1}a^{n-1}b^1 + \binom{n}{2}a^{n-2}b^2 + \cdots + \binom{n}{n}a^0 b^n$$

Chapter 12 Test

Write the first five terms of the sequences with the following general terms:

1. $a_n = 3n - 5$
2. $a_n = 4n - 1$
3. $a_n = n^2 + 1$
4. $a_n = 2n^3$
5. $a_n = \dfrac{n+1}{n^2}$
6. $a_n = (-2)^{n+1}$

Give the general term for each sequence:

7. $6, 10, 14, 18, \ldots$
8. $1, 2, 4, 8, \ldots$
9. $\frac{1}{2}, \frac{1}{4}, \frac{1}{8}, \frac{1}{16}, \ldots$
10. $-3, 9, -27, 81, \ldots$

11. Expand and simplify each of the following:

a. $\displaystyle\sum_{i=1}^{5} (5i + 3)$

b. $\displaystyle\sum_{i=3}^{5} (2^i - 1)$

c. $\displaystyle\sum_{i=2}^{6} (i^2 + 2i)$

12. Find the first term of an arithmetic progression if $a_5 = 11$ and $a_9 = 19$.
13. Find the second term of a geometric progression for which $a_3 = 18$ and $a_5 = 162$.
14. Find the sum of the first 10 terms of the following arithmetic progressions:

    a. 5, 11, 17, . . .
    b. 25, 20, 15, . . .

15. Write a formula for the sum of the first 50 terms of the following geometric progressions:

    a. 3, 6, 12, . . .
    b. 5, 20, 80, . . .

16. Use the binomial formula to expand each of the following:

    a. $(x - 3)^4$
    b. $(2x - 1)^5$
    c. $(3x - 2y)^3$

Appendix A:
Synthetic Division

Synthetic division is a short form of long division with polynomials. We will consider synthetic division only for those cases in which the divisor is of the form $x + k$, where k is a constant.

Let's begin by looking over an example of long division with polynomials as done in Section 3.4:

$$
\begin{array}{r}
3x^2 - 2x + 4 \\
x + 3 \overline{) 3x^3 + 7x^2 - 2x - 4} \\
\underline{3x^3 + 9x^2} \\
-2x^2 - 2x \\
\underline{-2x^2 - 6x} \\
4x - 4 \\
\underline{4x + 12} \\
-16
\end{array}
$$

We can rewrite the problem without showing the variable, since the variable is written in descending powers and similar terms are in alignment. It looks like this:

$$
\begin{array}{r}
3 \quad -2 \quad +4 \\
1 + 3 \overline{) \; 3 \quad\;\; 7 \quad -2 \quad -4} \\
\underline{(3) + 9} \\
-2 \, (-2) \\
\underline{(-2) \, -6} \\
4 \, (-4) \\
\underline{(4) \quad 12} \\
-16
\end{array}
$$

We have used parentheses to enclose the numbers which are repetitions of the numbers above them. We can compress the problem by eliminating all repetitions:

$$
\begin{array}{r}
3 \quad -2 \quad\;\; 4 \\
1 + 3 \overline{) 3 \quad\;\; 7 \quad -2 \quad -4} \\
\underline{9 \quad -6 \quad +12} \\
3 \quad -2 \quad\;\; 4 \quad -16
\end{array}
$$

The top line is the same as the first three terms of the bottom line, so we eliminate the top line. Also, the 1 which was the coefficient of x in the original problem can be eliminated, since we will only consider division problems where the divisor is of the form $x + k$. The following is the most compact form of the original division problem:

$$
+3\overline{)3 \quad 7 \quad -2 \quad -4}
$$
$$
 9 \quad -6 \quad 12
$$
$$
\overline{3 \quad -2 \quad 4 \quad -16}
$$

If we check over the problem, we find that the first term in the bottom row is exactly the same as the first term in the top row—and it always will be in problems of this type. Also, the last three terms in the bottom row come from multiplication by $+3$ and then subtraction. We can get an equivalent result by multiplying by -3 and adding. The problem would then look like this:

$$
\begin{array}{r|rrrr}
-3 & 3 & 7 & -2 & -4 \\
& \downarrow & -9 & 6 & -12 \\
\hline
& 3 & -2 & 4 & \boxed{-16}
\end{array}
$$

We have used the brackets ⌐⌐ to separate the divisor and the remainder. This last expression is synthetic division. It is an easy process to remember. Simply change the sign of the constant term in the divisor, then bring down the first term of the dividend. The process is then just a series of multiplications and additions, as indicated in the following diagram by the arrows:

$$
\begin{array}{r|rrrr}
-3 & 3 & 7 & -2 & -4 \\
& \downarrow & -9 & 6 & -12 \\
\hline
& 3 & -2 & 4 & \boxed{-16}
\end{array}
$$

The last term on the bottom row is always the remainder.

Here are some additional examples of synthetic division with polynomials.

▼ **Example 1** Divide $x^4 - 2x^3 + 4x^2 - 6x + 2$ by $x - 2$.

Solution We change the sign of the constant term in the divisor to get $+2$ and then use the procedure given above:

$$
\begin{array}{r|rrrrr}
+2 & 1 & -2 & 4 & -6 & 2 \\
& \downarrow & 2 & 0 & 8 & 4 \\
\hline
& 1 & 0 & 4 & 2 & \boxed{6}
\end{array}
$$

From the last line we have the answer:

$$
1x^3 + 0x^2 + 4x + 2 + \frac{6}{x - 2}
$$

or

$$
\frac{x^4 - 2x^3 + 4x^2 - 6x + 2}{x - 2} = x^3 + 4x + 2 + \frac{6}{x - 2} \qquad \blacktriangle
$$

▼ **Example 2**   Divide $\dfrac{3x^3 - 4x + 5}{x + 4}$.

Solution   Since we cannot skip any powers of the variable in the polynomial $3x^3 - 4x + 5$, we rewrite it as $3x^3 + 0x^2 - 4x + 5$ and proceed as we did in Example 1:

$$
\begin{array}{r|rrrr}
-4 & 3 & 0 & -4 & +5 \\
   & \downarrow & -12 & 48 & -176 \\
\hline
   & 3 & -12 & 44 & \boxed{-171}
\end{array}
$$

From the synthetic division, we have

$$\frac{3x^3 - 4x + 5}{x + 4} = 3x^2 - 12x + 44 - \frac{171}{x + 4}$$ ▲

▼ **Example 3**   Divide $\dfrac{x^3 - 1}{x - 1}$.

Solution   Writing the numerator as $x^3 + 0x^2 + 0x - 1$ and using synthetic division, we have

$$
\begin{array}{r|rrrr}
+1 & 1 & 0 & 0 & -1 \\
   & \downarrow & 1 & 1 & 1 \\
\hline
   & 1 & 1 & 1 & \boxed{0}
\end{array}
$$

which indicates

$$\frac{x^3 - 1}{x - 1} = x^2 + x + 1$$ ▲

Problem Set A

Use synthetic division to find the following quotients:

1. $\dfrac{x^2 - 5x + 6}{x + 2}$

2. $\dfrac{x^2 + 8x - 12}{x - 3}$

3. $\dfrac{3x^2 - 4x + 1}{x - 1}$

4. $\dfrac{4x^2 - 2x - 6}{x + 1}$

5. $\dfrac{x^3 + 2x^2 + 3x + 4}{x - 2}$

6. $\dfrac{x^3 - 2x^2 - 3x - 4}{x - 2}$

7. $\dfrac{3x^3 - x^2 + 2x + 5}{x - 3}$

8. $\dfrac{2x^3 - 5x^2 + x + 2}{x - 2}$

9. $\dfrac{2x^3 + x - 3}{x - 1}$

10. $\dfrac{3x^3 - 2x + 1}{x - 5}$

11. $\dfrac{x^4 + 2x^2 + 1}{x + 4}$

12. $\dfrac{x^4 - 3x^2 + 1}{x - 4}$

13. $\dfrac{x^5 - 2x^4 + x^3 - 3x^2 + x + 1}{x - 2}$

14. $\dfrac{2x^5 - 3x^4 + x^3 - x^2 + 2x + 1}{x + 2}$

15. $\dfrac{x^2 + x + 1}{x - 1}$

16. $\dfrac{x^2 + x + 1}{x + 1}$

17. $\dfrac{x^4 - 1}{x + 1}$

18. $\dfrac{x^4 + 1}{x - 1}$

19. $\dfrac{x^3 - 1}{x - 1}$

20. $\dfrac{x^3 - 1}{x + 1}$

Appendix C

Powers, Roots, and Prime Factors

n	n^2	$\sqrt{n}$	n^3	$\sqrt[3]{n}$	Prime factors
1	1	1.000	1	1.000	—
2	4	1.414	8	1.260	prime
3	9	1.732	27	1.442	prime
4	16	2.000	64	1.587	$2 \cdot 2$
5	25	2.236	125	1.710	prime
6	36	2.449	216	1.817	$2 \cdot 3$
7	49	2.646	343	1.913	prime
8	64	2.828	512	2.000	$2 \cdot 2 \cdot 2$
9	81	3.000	729	2.080	$3 \cdot 3$
10	100	3.162	1,000	2.154	$2 \cdot 5$
11	121	3.317	1,331	2.224	prime
12	144	3.464	1,728	2.289	$2 \cdot 2 \cdot 3$
13	169	3.606	2,197	2.351	prime
14	196	3.742	2,744	2.410	$2 \cdot 7$
15	225	3.873	3,375	2.466	$3 \cdot 5$
16	256	4.000	4,096	2.520	$2 \cdot 2 \cdot 2 \cdot 2$
17	289	4.123	4,913	2.571	prime
18	324	4.243	5,832	2.621	$2 \cdot 3 \cdot 3$
19	361	4.359	6,859	2.668	prime
20	400	4.472	8,000	2.714	$2 \cdot 2 \cdot 5$
21	441	4.583	9,261	2.759	$3 \cdot 7$
22	484	4.690	10,648	2.802	$2 \cdot 11$
23	529	4.796	12,167	2.844	prime
24	576	4.899	13,824	2.884	$2 \cdot 2 \cdot 2 \cdot 3$
25	625	5.000	15,625	2.924	$5 \cdot 5$
26	676	5.099	17,576	2.962	$2 \cdot 13$
27	729	5.196	19,683	3.000	$3 \cdot 3 \cdot 3$
28	784	5.292	21,952	3.037	$2 \cdot 2 \cdot 7$
29	841	5.385	24,389	3.072	prime
30	900	5.477	27,000	3.107	$2 \cdot 3 \cdot 5$
31	961	5.568	29,791	3.141	prime
32	1,024	5.657	32,768	3.175	$2 \cdot 2 \cdot 2 \cdot 2 \cdot 2$
33	1,089	5.745	35,937	3.208	$3 \cdot 11$
34	1,156	5.831	39,304	3.240	$2 \cdot 17$
35	1,225	5.916	42,875	3.271	$5 \cdot 7$

Powers, Roots, and Prime Factors (Continued)

n	n^2	$\sqrt{n}$	n^3	$\sqrt[3]{n}$	Prime factors
36	1,296	6.000	46,656	3.302	$2 \cdot 2 \cdot 3 \cdot 3$
37	1,369	6.083	50,653	3.332	prime
38	1,444	6.164	54,872	3.362	$2 \cdot 19$
39	1,521	6.245	59,319	3.391	$3 \cdot 13$
40	1,600	6.325	64,000	3.420	$2 \cdot 2 \cdot 2 \cdot 5$
41	1,681	6.403	68,921	3.448	prime
42	1,764	6.481	74,088	3.476	$2 \cdot 3 \cdot 7$
43	1,849	6.557	79,507	3.503	prime
44	1,936	6.633	85,184	3.530	$2 \cdot 2 \cdot 11$
45	2,025	6.708	91,125	3.557	$3 \cdot 3 \cdot 5$
46	2,116	6.782	97,336	3.583	$2 \cdot 23$
47	2,209	6.856	103,823	3.609	prime
48	2,304	6.928	110,592	3.634	$2 \cdot 2 \cdot 2 \cdot 2 \cdot 3$
49	2,401	7.000	117,649	3.659	$7 \cdot 7$
50	2,500	7.071	125,000	3.684	$2 \cdot 5 \cdot 5$
51	2,601	7.141	132,651	3.708	$3 \cdot 17$
52	2,704	7.211	140,608	3.733	$2 \cdot 2 \cdot 13$
53	2,809	7.280	148,877	3.756	prime
54	2,916	7.348	157,464	3.780	$2 \cdot 3 \cdot 3 \cdot 3$
55	3,025	7.416	166,375	3.803	$5 \cdot 11$
56	3,136	7.483	175,616	3.826	$2 \cdot 2 \cdot 2 \cdot 7$
57	3,249	7.550	185,193	3.849	$3 \cdot 19$
58	3,364	7.616	195,112	3.871	$2 \cdot 29$
59	3,481	7.681	205,379	3.893	prime
60	3,600	7.746	216,000	3.915	$2 \cdot 2 \cdot 3 \cdot 5$
61	3,721	7.810	226,981	3.936	prime
62	3,844	7.874	238,328	3.958	$2 \cdot 31$
63	3,969	7.937	250,047	3.979	$3 \cdot 3 \cdot 7$
64	4,096	8.000	262,144	4.000	$2 \cdot 2 \cdot 2 \cdot 2 \cdot 2 \cdot 2$
65	4,225	8.062	274,625	4.021	$5 \cdot 13$
66	4,356	8.124	287,496	4.041	$2 \cdot 3 \cdot 11$
67	4,489	8.185	300,763	4.062	prime
68	4,624	8.246	314,432	4.082	$2 \cdot 2 \cdot 17$
69	4,761	8.307	328,509	4.102	$3 \cdot 23$
70	4,900	8.367	343,000	4.121	$2 \cdot 5 \cdot 7$
71	5,041	8.426	357,911	4.141	prime
72	5,184	8.485	373,248	4.160	$2 \cdot 2 \cdot 2 \cdot 3 \cdot 3$
73	5,329	8.544	389,017	4.179	prime
74	5,476	8.602	405,224	4.198	$2 \cdot 37$
75	5,625	8.660	421,875	4.217	$3 \cdot 5 \cdot 5$

Powers, Roots, and Prime Factors (Continued)

n	n^2	$\sqrt{n}$	n^3	$\sqrt[3]{n}$	Prime factors
76	5,776	8.718	438,976	4.236	$2 \cdot 2 \cdot 19$
77	5,929	8.775	456,533	4.254	$7 \cdot 11$
78	6,084	8.832	474,552	4.273	$2 \cdot 3 \cdot 13$
79	6,241	8.888	493,039	4.291	prime
80	6,400	8.944	512,000	4.309	$2 \cdot 2 \cdot 2 \cdot 2 \cdot 5$
81	6,561	9.000	531,441	4.327	$3 \cdot 3 \cdot 3 \cdot 3$
82	6,724	9.055	551,368	4.344	$2 \cdot 41$
83	6,889	9.110	571,787	4.362	prime
84	7,056	9.165	592,704	4.380	$2 \cdot 2 \cdot 3 \cdot 7$
85	7,225	9.220	614,125	4.397	$5 \cdot 17$
86	7,396	9.274	636,056	4.414	$2 \cdot 43$
87	7,569	9.327	658,503	4.431	$3 \cdot 29$
88	7,744	9.381	681,472	4.448	$2 \cdot 2 \cdot 2 \cdot 11$
89	7,921	9.434	704,969	4.465	prime
90	8,100	9.487	729,000	4.481	$2 \cdot 3 \cdot 3 \cdot 5$
91	8,281	9.539	753,571	4.498	$7 \cdot 13$
92	8,464	9.592	778,688	4.514	$2 \cdot 2 \cdot 23$
93	8,649	9.644	804,357	4.531	$3 \cdot 31$
94	8,836	9.695	830,584	4.547	$2 \cdot 47$
95	9,025	9.747	857,375	4.563	$5 \cdot 19$
96	9,216	9.798	884,736	4.579	$2 \cdot 2 \cdot 2 \cdot 2 \cdot 2 \cdot 3$
97	9,409	9.849	912,673	4.595	prime
98	9,604	9.899	941,192	4.610	$2 \cdot 7 \cdot 7$
99	9,801	9.950	970,299	4.626	$3 \cdot 3 \cdot 11$
100	10,000	10.000	1,000,000	4.642	$2 \cdot 2 \cdot 5 \cdot 5$

Appendix D

Common Logarithms

x	0	1	2	3	4	5	6	7	8	9
1.0	.0000	.0043	.0086	.0128	.0170	.0212	.0253	.0294	.0334	.0374
1.1	.0414	.0453	.0492	.0531	.0569	.0607	.0645	.0682	.0719	.0755
1.2	.0792	.0828	.0864	.0899	.0934	.0969	.1004	.1038	.1072	.1106
1.3	.1139	.1173	.1206	.1239	.1271	.1303	.1335	.1367	.1399	.1430
1.4	.1461	.1492	.1523	.1553	.1584	.1614	.1644	.1673	.1703	.1732
1.5	.1761	.1790	.1818	.1847	.1875	.1903	.1931	.1959	.1987	.2014
1.6	.2041	.2068	.2095	.2122	.2148	.2175	.2201	.2227	.2253	.2279
1.7	.2304	.2330	.2355	.2380	.2405	.2430	.2455	.2480	.2504	.2529
1.8	.2553	.2577	.2601	.2625	.2648	.2672	.2695	.2718	.2742	.2765
1.9	.2788	.2810	.2833	.2856	.2878	.2900	.2923	.2945	.2967	.2989
2.0	.3010	.3032	.3054	.3075	.3096	.3118	.3139	.3160	.3181	.3201
2.1	.3222	.3243	.3263	.3284	.3304	.3324	.3345	.3365	.3385	.3404
2.2	.3424	.3444	.3464	.3483	.3502	.3522	.3541	.3560	.3579	.3598
2.3	.3617	.3636	.3655	.3674	.3692	.3711	.3729	.3747	.3766	.3784
2.4	.3802	.3820	.3838	.3856	.3874	.3892	.3909	.3927	.3945	.3962
2.5	.3979	.3997	.4014	.4031	.4048	.4065	.4082	.4099	.4116	.4133
2.6	.4150	.4166	.4183	.4200	.4216	.4232	.4249	.4265	.4281	.4298
2.7	.4314	.4330	.4346	.4362	.4378	.4393	.4409	.4425	.4440	.4456
2.8	.4472	.4487	.4502	.4518	.4533	.4548	.4564	.4579	.4594	.4609
2.9	.4624	.4639	.4654	.4669	.4683	.4698	.4713	.4728	.4742	.4757
3.0	.4771	.4786	.4800	.4814	.4829	.4843	.4857	.4871	.4886	.4900
3.1	.4914	.4928	.4942	.4955	.4969	.4983	.4997	.5011	.5024	.5038
3.2	.5051	.5065	.5079	.5092	.5105	.5119	.5132	.5145	.5159	.5172
3.3	.5185	.5198	.5211	.5224	.5237	.5250	.5263	.5276	.5289	.5302
3.4	.5315	.5328	.5340	.5353	.5366	.5378	.5391	.5403	.5416	.5428
3.5	.5441	.5453	.5465	.5478	.5490	.5502	.5514	.5527	.5539	.5551
3.6	.5563	.5575	.5587	.5599	.5611	.5623	.5635	.5647	.5658	.5670
3.7	.5682	.5694	.5705	.5717	.5729	.5740	.5752	.5763	.5775	.5786
3.8	.5798	.5809	.5821	.5832	.5843	.5855	.5866	.5877	.5888	.5899
3.9	.5911	.5922	.5933	.5944	.5955	.5966	.5977	.5988	.5999	.6010
4.0	.6021	.6031	.6042	.6053	.6064	.6075	.6085	.6096	.6107	.6117
4.1	.6128	.6138	.6149	.6160	.6170	.6180	.6191	.6201	.6212	.6222
4.2	.6232	.6243	.6253	.6263	.6274	.6284	.6294	.6304	.6314	.6325
x	0	1	2	3	4	5	6	7	8	9

Common Logarithms (Continued)

x	0	1	2	3	4	5	6	7	8	9
4.3	.6335	.6345	.6355	.6365	.6375	.6385	.6395	.6405	.6415	.6425
4.4	.6435	.6444	.6454	.6464	.6474	.6484	.6493	.6503	.6513	.6522
4.5	.6532	.6542	.6551	.6561	.6571	.6580	.6590	.6599	.6609	.6618
4.6	.6628	.6637	.6646	.6656	.6665	.6675	.6684	.6693	.6702	.6712
4.7	.6721	.6730	.6739	.6749	.6758	.6767	.6776	.6785	.6794	.6803
4.8	.6812	.6821	.6830	.6839	.6848	.6857	.6866	.6875	.6884	.6893
4.9	.6902	.6911	.6920	.6928	.6937	.6946	.6955	.6964	.6972	.6981
5.0	.6990	.6998	.7007	.7016	.7024	.7033	.7042	.7050	.7059	.7067
5.1	.7076	.7084	.7093	.7101	.7110	.7118	.7126	.7135	.7143	.7152
5.2	.7160	.7168	.7177	.7185	.7193	.7202	.7210	.7218	.7226	.7235
5.3	.7243	.7251	.7259	.7267	.7275	.7284	.7292	.7300	.7308	.7316
5.4	.7324	.7332	.7340	.7348	.7356	.7364	.7372	.7380	.7388	.7396
5.5	.7404	.7412	.7419	.7427	.7435	.7443	.7451	.7459	.7466	.7474
5.6	.7482	.7490	.7497	.7505	.7513	.7520	.7528	.7536	.7543	.7551
5.7	.7559	.7566	.7574	.7582	.7589	.7597	.7604	.7612	.7619	.7627
5.8	.7634	.7642	.7649	.7657	.7664	.7672	.7679	.7686	.7694	.7701
5.9	.7709	.7716	.7723	.7731	.7738	.7745	.7752	.7760	.7767	.7774
6.0	.7782	.7789	.7796	.7803	.7810	.7818	.7825	.7832	.7839	.7846
6.1	.7853	.7860	.7868	.7875	.7882	.7889	.7896	.7903	.7910	.7917
6.2	.7924	.7931	.7938	.7945	.7952	.7959	.7966	.7973	.7980	.7987
6.3	.7993	.8000	.8007	.8014	.8021	.8028	.8035	.8041	.8048	.8055
6.4	.8062	.8069	.8075	.8082	.8089	.8096	.8102	.8109	.8116	.8122
6.5	.8129	.8136	.8142	.8149	.8156	.8162	.8169	.8176	.8182	.8189
6.6	.8195	.8202	.8209	.8215	.8222	.8228	.8235	.8241	.8248	.8254
6.7	.8261	.8267	.8274	.8280	.8287	.8293	.8299	.8306	.8312	.8319
6.8	.8325	.8331	.8338	.8344	.8351	.8357	.8363	.8370	.8376	.8382
6.9	.8388	.8395	.8401	.8407	.8414	.8420	.8426	.8432	.8439	.8445
7.0	.8451	.8457	.8463	.8470	.8476	.8482	.8488	.8494	.8500	.8506
7.1	.8513	.8519	.8525	.8531	.8537	.8543	.8549	.8555	.8561	.8567
7.2	.8573	.8579	.8585	.8591	.8597	.8603	.8609	.8615	.8621	.8627
7.3	.8633	.8639	.8645	.8651	.8657	.8663	.8669	.8675	.8681	.8686
7.4	.8692	.8698	.8704	.8710	.8716	.8722	.8727	.8733	.8739	.8745
7.5	.8751	.8756	.8762	.8768	.8774	.8779	.8785	.8791	.8797	.8802
7.6	.8808	.8814	.8820	.8825	.8831	.8837	.8842	.8848	.8854	.8859
7.7	.8865	.8871	.8876	.8882	.8887	.8893	.8899	.8904	.8910	.8915
7.8	.8921	.8927	.8932	.8938	.8943	.8949	.8954	.8960	.8965	.8971
7.9	.8976	.8982	.8987	.8993	.8998	.9004	.9009	.9015	.9020	.9025
x	0	1	2	3	4	5	6	7	8	9

Common Logarithms (Continued)

x	0	1	2	3	4	5	6	7	8	9
8.0	.9031	.9036	.9042	.9047	.9053	.9058	.9063	.9069	.9074	.9079
8.1	.9085	.9090	.9096	.9101	.9106	.9112	.9117	.9122	.9128	.9133
8.2	.9138	.9143	.9149	.9154	.9159	.9165	.9170	.9175	.9180	.9186
8.3	.9191	.9196	.9201	.9206	.9212	.9217	.9222	.9227	.9232	.9238
8.4	.9243	.9248	.9253	.9258	.9263	.9269	.9274	.9279	.9284	.9289
8.5	.9294	.9299	.9304	.9309	.9315	.9320	.9325	.9330	.9335	.9340
8.6	.9345	.9350	.9355	.9360	.9365	.9370	.9375	.9380	.9385	.9390
8.7	.9395	.9400	.9405	.9410	.9415	.9420	.9425	.9430	.9435	.9440
8.8	.9445	.9450	.9455	.9460	.9465	.9469	.9474	.9479	.9484	.9489
8.9	.9494	.9499	.9504	.9509	.9513	.9518	.9523	.9528	.9533	.9538
9.0	.9542	.9547	.9552	.9557	.9562	.9566	.9571	.9576	.9581	.9586
9.1	.9590	.9595	.9600	.9605	.9609	.9614	.9619	.9624	.9628	.9633
9.2	.9638	.9643	.9647	.9652	.9657	.9661	.9666	.9671	.9675	.9680
9.3	.9685	.9689	.9694	.9699	.9703	.9708	.9713	.9717	.9722	.9727
9.4	.9731	.9736	.9741	.9745	.9750	.9754	.9759	.9763	.9768	.9773
9.5	.9777	.9782	.9786	.9791	.9795	.9800	.9805	.9809	.9814	.9818
9.6	.9823	.9827	.9832	.9836	.9841	.9845	.9850	.9854	.9859	.9863
9.7	.9868	.9872	.9877	.9881	.9886	.9890	.9894	.9899	.9903	.9908
9.8	.9912	.9917	.9921	.9926	.9930	.9934	.9939	.9943	.9948	.9952
9.9	.9956	.9961	.9965	.9969	.9974	.9978	.9983	.9987	.9991	.9996
x	0	1	2	3	4	5	6	7	8	9

Answers to
Odd-Numbered Exercises
and Chapter Tests

1. $x + 5$ 3. $6 - x$ 5. $2t < y$ 7. $3x/2y > 6$
9. $x + y < x - y$ 11. $3(x - 5) > y$ 13. $s - t \neq s + t$
15. $2(t + 3) \not> t - 6$ 17. $\{0, 1, 2, 3, 4, 5, 6\}$ 19. $\{2, 4\}$
21. $\{-2, -1, 0, 1, 2, 3, 5, 7\}$ 23. $\{1\}$ 25. $\{-2, -1, 0, 1, 2, 4, 6\}$
27. $\{0, 2\}$ 29. $\{0, 1, 2, 3, 4, 5, 6\}$ 31. $\{0, 1, 2, 3, 4, 5, 6\}$
33. $\{0, 2, 4\}$
35. $\{a\}$ $\{b\}$ $\{c\}$ $\{a, b\}$ $\{a, c\}$ $\{b, c\}$ $\{a, b, c\}$ $\emptyset$
37. Any two sets that do not have any elements in common will do.

39.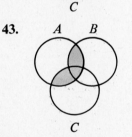

41.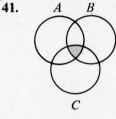

43.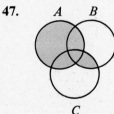

45.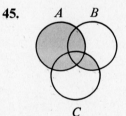

47.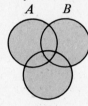

1.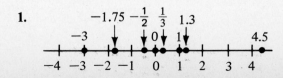

3. $4, -4, \frac{1}{4}$ **5.** $-\frac{1}{2}, \frac{1}{2}, -2$ **7.** $5, -5, \frac{1}{5}$ **9.** $\frac{3}{8}, -\frac{3}{8}, \frac{8}{3}$

11. $-\frac{1}{6}, \frac{1}{6}, -6$ **13.** $3, -3, \frac{1}{3}$ **15.** $\sqrt{3}, -\sqrt{3}, 1/\sqrt{3}$

17. $-1/\sqrt{2}, 1/\sqrt{2}, -\sqrt{2}$ **19.** $x, -x, 1/x$ **21.** $1, -1$

23. 0 **25.** 2 **27.** $\frac{3}{4}$ **29.** π **31.** -4 **33.** -2

35. $-\frac{3}{4}$ **37.** 2 **39.** $-\frac{1}{3}$ **41.** 5 **43.** 8 **45.** 5

47. 6 **49.** 0 **51.** $\{1, 2\}$ **53.** $\{-6, -5.2, 0, 1, 2, 2.3, \frac{9}{2}\}$

55. $\{-\sqrt{7}, -\pi, \sqrt{17}\}$ **57.** $\{0, 1, 2\}$ **59.** False; $1/0$ undefined

61. True   **63.** True   **65.** False; $0/1 = 0$, 0 is rational

67. True   **69.** True

Problem Set 1.3

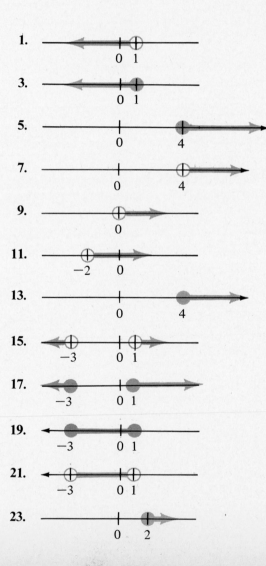

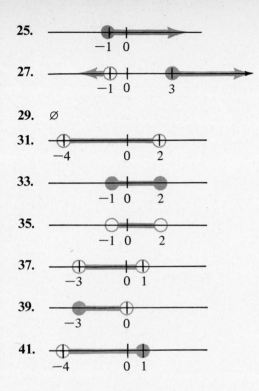

25.
    −1 0

27.
    −1 0   3

29. ∅

31.
    −4   0 2

33.
    −1 0 2

35.
    −1 0 2

37.
    −3   0 1

39.
    −3   0

41.
    −4   0 1

Problem Set 1.4

  1. Reflexive property of equality
  3. Transitive property of inequality
  5. Reflexive property of equality
  7. Symmetric property of equality
  9. Transitive property of equality
11. Commutative property of addition
13. Associative property of multiplication
15. Additive identity property
17. Theorem 1.1
19. Additive inverse property
21. Commutative property of addition
23. Distributive property
25. Closure property for addition
27. Additive inverse property
29. Associative and commutative properties of multiplication
31. Associative and commutative properties of addition
33. $y \cdot 5$ **35.** $a + 3$ **37.** 7   **39.** $(2 + x) + 6$
41. $4(x + y)$ **43.** 0   **45.** $7(a + b + c)$ **47.** $(7 \cdot r)s$ **49.** 11

1. 4   **3.** −4   **5.** −8   **7.** −5   **9.** 5   **11.** −13
13. −14   **15.** 4   **17.** −10   **19.** −4   **21.** 10   **23.** −35
25. 16   **27.** 3   **29.** −5   **31.** 1   **33.** 19   **35.** −14
37. 10   **39.** −1   **41.** 11   **43.** −4   **45.** −3   **47.** 8
49. 16

1. −15   **3.** 15   **5.** −24   **7.** 20   **9.** −12   **11.** −24
13. −24   **15.** 24   **17.** $8(-\frac{1}{4}) = -2$ **19.** $-8(-\frac{1}{4}) = 2$
21. $-7(\frac{1}{3}) = -\frac{7}{3}$ **23.** $5(\frac{1}{20}) = \frac{1}{4}$ **25.** $-9(\frac{1}{18}) = -\frac{1}{2}$
27. $12(-\frac{1}{4}) = -3$ **29.** $28(-\frac{1}{7}) = -4$ **31.** −14   **33.** −7
35. 18   **37.** −44   **39.** −30   **41.** 18   **43.** 4   **45.** $\frac{5}{3}$
47. 11   **49.** $\frac{81}{8}$

1. $2(3x + 4y)$ **2.** $(2a - 3b) < (2a + 3b)$ **3.** $\{1, 3\}$
4. $\{1, 2, 3, 4\}$ **5.** $\varnothing$ **6.** $3, -\frac{1}{3}$ **7.** $-\frac{4}{3}, \frac{3}{4}$
8. $\sqrt{5}, -1/\sqrt{5}$ **9.** 3   **10.** 7   **11.** −2   **12.** $\{-5, 0, 1, 4\}$
13. $\{-5, -4.1, -3.75, -\frac{5}{6}, 0, 1, 1.8, 4\}$ **14.** $\{-\sqrt{2}, \sqrt{3}\}$

15.

16.

17.

18. Commutative property of addition
19. Multiplicative identity property
20. Associative and commutative property of multiplication
21. Associative and commutative property of addition   **22.** −19
23. 14   **24.** −19   **25.** 14   **26.** −26   **27.** $\frac{5}{2}$ **28.** 2
29. 1   **30.** $\frac{2}{5}$ **31.** 1

1. 8   **3.** 5   **5.** 2   **7.** -7 **9.** $-\frac{9}{2}$ **11.** -4

13. $-\frac{4}{3}$ **15.** -4 **17.** $\frac{7}{2}$ **19.** $-\frac{11}{5}$ **21.** -7 **23.** 3

25. -4 **27.** 2   **29.** $\frac{7}{10}$ **31.** 6   **33.** 2   **35.** 0

37. -3 **39.** $\frac{20}{7}$ **41.** $-\frac{5}{2}$ **43.** 17

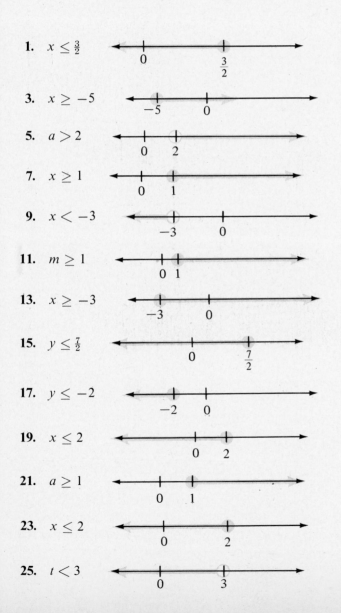

1. $x \le \frac{3}{2}$

3. $x \ge -5$

5. $a > 2$

7. $x \ge 1$

9. $x < -3$

11. $m \ge 1$

13. $x \ge -3$

15. $y \le \frac{7}{2}$

17. $y \le -2$

19. $x \le 2$

21. $a \ge 1$

23. $x \le 2$

25. $t < 3$

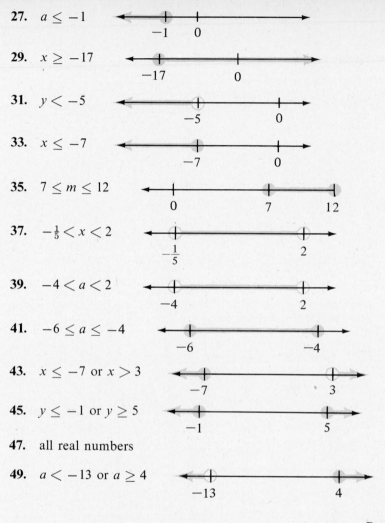

27. $a \leq -1$

29. $x \geq -17$

31. $y < -5$

33. $x \leq -7$

35. $7 \leq m \leq 12$

37. $-\frac{1}{5} < x < 2$

39. $-4 < a < 2$

41. $-6 \leq a \leq -4$

43. $x \leq -7$ or $x > 3$

45. $y \leq -1$ or $y \geq 5$

47. all real numbers

49. $a < -13$ or $a \geq 4$

Problem Set 2.3

1. $4, -4$ 3. $2, -2$ 5. $\varnothing$ 7. $1, -1$ 9. $\varnothing$
11. $6, -6$ 13. $7, -3$ 15. $5, 3$ 17. $2, 4$ 19. $\varnothing$
21. $\frac{4}{3}, -2$ 23. $\frac{2}{3}, -\frac{10}{3}$ 25. $\varnothing$ 27. $4, -3$ 29. $-2, -3$
31. $-5, \frac{3}{5}$ 33. $1, \frac{1}{9}$ 35. $-\frac{1}{2}$ 37. 0 39. $-\frac{1}{2}$
41. $-\frac{1}{6}, -\frac{7}{4}$

Problem Set 2.4

1. $-3 < x < 3$

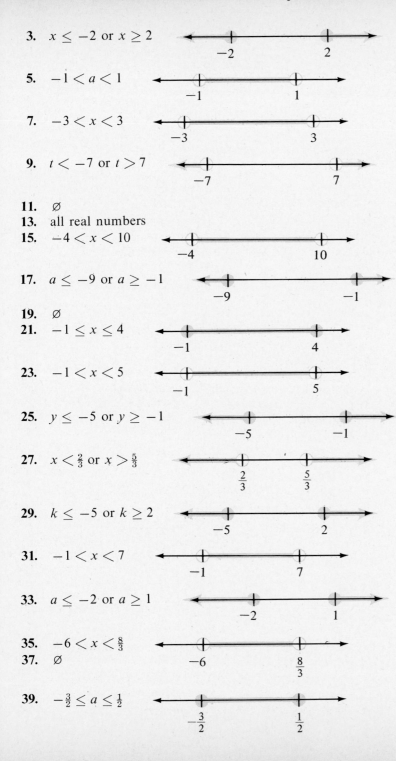

3. $x \leq -2$ or $x \geq 2$

5. $-1 < a < 1$

7. $-3 < x < 3$

9. $t < -7$ or $t > 7$

11. $\varnothing$
13. all real numbers
15. $-4 < x < 10$

17. $a \leq -9$ or $a \geq -1$

19. $\varnothing$
21. $-1 \leq x \leq 4$

23. $-1 < x < 5$

25. $y \leq -5$ or $y \geq -1$

27. $x < \frac{2}{3}$ or $x > \frac{5}{3}$

29. $k \leq -5$ or $k \geq 2$

31. $-1 < x < 7$

33. $a \leq -2$ or $a \geq 1$

35. $-6 < x < \frac{8}{3}$
37. $\varnothing$

39. $-\frac{3}{2} \leq a \leq \frac{1}{2}$

1. 6   **3.** 2   **5.** 4   **7.** 5   **9.** 18.84   **11.** 0   **13.** 8
15. 21   **17.** $l = A/w$ **19.** $t = I/pr$ **21.** $m = E/c^2$
23. $T = PV/nR$ **25.** $b^2 = c^2 - a^2$ **27.** $b = y - mx$
29. $h = \dfrac{2A}{b + B}$ **31.** $c = 2s - a - b$

1. $x + 2 = 2x - 5$ $\{7\}$ **3.** $3(x + 4) = 3$ $\{-3\}$
5. $2(2x + 1) = 3(x - 5)$ $\{-17\}$ **7.** $5x + 2 = 3x + 8$ $\{3\}$
9. $x + (x + 2) = x - (x + 2) + 16$ $\{6, 8\}$
11. $x + (x + 1) = 3x - 1$ $\{2, 3\}$ **13.** $2x + (x + 1) = 7$ $\{2\}$
15. $2x - (x + 2) = 5$ $\{7, 9\}$ **17.** $60 = 4x + 2x$ $\{20 \text{ ft} \times 10 \text{ ft}\}$
19. $28 = 4s$ $\{7 \text{ ft}\}$ **21.** $23 = x + (x + 3) + 2x$ $\{5 \text{ in.}\}$
23. $18 = 2(2x - 3) + 2x$ $\{4 \text{ m}\}$
25. $x + 14 + x + 10 = 36$ $\{6, 10\}$
27. $2x + 5 = 3(x + 5)/2$ $\{5, 10\}$ **29.** $3x + 5 = 2(x + 5) - 2$ $\{3, 9\}$

1. 12   **2.** $-\frac{4}{3}$ **3.** $\frac{17}{3}$ **4.** $-\frac{7}{4}$ **5.** $t \geq -6$ **6.** $x \leq 1$
7. $x < 6$ **8.** $y \geq -52$ **9.** $x = 6$ or $x = 2$
10. $a = -1$ or $a = -6$
11. $x < -1$ or $x > \frac{4}{3}$

12. $-\frac{2}{3} \leq x \leq 4$

13. $w = \dfrac{A - 2l}{2}$ **14.** $B = \dfrac{2A}{h} - b$ **15.** $x + (x + 2) = 18$ $\{8, 10\}$
16. $x + 9 = 2x$ $\{9, 13\}$

CHAPTER 3

1. 16   **3.** -16 **5.** -27 **7.** 32   **9.** $\frac{1}{8}$ **11.** $\frac{25}{36}$
13. -1 **15.** x^9 **17.** a^{10} **19.** x^6 **21.** $-8x^6$
23. $-6x^6$ **25.** $-36x^{11}$ **27.** $324x^{16}$ **29.** $324x^{34}$ **31.** $\frac{1}{9}$
33. $\frac{1}{32}$ **35.** $\frac{16}{9}$ **37.** $\frac{11}{18}$ **39.** $\frac{35}{24}$ **41.** $\frac{1}{9}$ **43.** $2x$

45. $\frac{1}{81}$ **47.** 2   **49.** 1296   **51.** $-\frac{x^5}{3}$ **53.** $\frac{x^{13}}{4}$ **55.** $\frac{4x^{18}}{y^{10}}$

57. $\frac{b^3}{a^4c^3}$ **59.** x **61.** 1   **63.** x^3

Problem Set 3.2

1. trinomial   2   5
3. binomial   1   3
5. trinomial   2   8
7. polynomial   3   4
9. monomial   0 $-\frac{3}{4}$
11. trinomial   3   6
13. $7x + 1$ **15.** 6   **17.** $5a + 7$ **19.** $x^2 - 13x + 3$
21. $3x^2 - 7x - 3$ **23.** $-y^3 - y^2 - 4y + 7$
25. $2x^3 + x^2 - 3x - 17$ **27.** $x^2 + 2xy + 10y^2$
29. $-3a^3 + 6a^2b - 5ab^2$ **31.** $-20x^2 + 10x$
33. $3x^2 - 12xy$ **35.** $17x^5 - 12$ **37.** $14a^2 - 2ab + 8b^2$
39. $2 - x$ **41.** $10x - 5$ **43.** $9x - 35$ **45.** $9y - 4x$
47. $9a + 2$ **49.** -2 **51.** 5
53. $P(-1) = 10;\ P(0) = 3;\ P(2) = 19$
55. $R(-2) = -15;\ R(0) = -1;\ R(3) = 20$
57. $P(-2) = -37;\ P(-1) = -5;\ P(0) = 3$
59. $H(1) = 112,\ H(2) = 192,\ H(3) = 240$

Problem Set 3.3

1. $12x^3 - 10x^2 + 8x$ **3.** $-3a^5 + 18a^4 - 21a^2$
5. $2a^5b - 2a^3b^2 + 2a^2b^4$ **7.** $-28x^4y^3 + 12x^3y^4 - 24x^2y^5$
9. $3r^6s^2 - 6r^5s^3 + 9r^4s^4 + 3r^3s^5$ **11.** $x^2 - 2x - 15$
13. $6x^2 - 19x + 15$ **15.** $x^3 + 9x^2 + 23x + 15$
17. $6a^4 + a^3 - 12a^2 + 5a -$ **19.** $a^3 - b^3$ **21.** $8x^3 + y^3$
23. $2a^3 - a^2b - ab^2 - 3b^3$ **25.** $18x^3 - 15x^2y - 16y^3$
27. $6x^4 - 44x^3 + 70x^2$ **29.** $6x^3 - 11x^2 - 14x + 24$
31. $x^2 + x - 6$ **33.** $x^2 - 5x + 6$ **35.** $6a^2 + 13a + 6$
37. $6x^2 + 2x - 20$ **39.** $5x^2 - 29x + 20$ **41.** $20a^2 + 9a + 1$
43. $20x^2 - 9xy - 18y^2$ **45.** $8x^2 - 22xy + 15y^2$
47. $28a^2 - ab - 2b^2$ **49.** $x^2 + 4x + 4$ **51.** $4a^2 - 12a + 9$
53. $25x^2 + 20xy + 4y^2$ **55.** $x^2 - 16$ **57.** $4a^2 - 9b^2$
59. $9r^2 - 49s^2$ **61.** $25x^2 - 16y^2$ **63.** $x^3 - 6x^2 + 12x - 8$
65. $3x^3 - 18x^2 + 33x - 18$ **67.** $x^{2N} + x^N - 6$ **69.** $x^{4N} - 9$
71. $\frac{3}{8}$

1. $2x^2 - 4x + 3$ **3.** $-2x^2 - 3x + 4$ **5.** $2y^2 + \frac{5}{2} - \dfrac{3}{2y^2}$

7. $-\frac{5}{2}x + 4 + \dfrac{3}{x}$ **9.** $4ab^3 + 6a^2b$ **11.** $-xy + 2y^2 + 3xy^2$

13. $\dfrac{4x^2}{y} - 3x + 5y + \dfrac{2y^2}{x}$ **15.** $a^{6N} - a^{3N}$ **17.** $4x^{5m} - 3x^{2m} + 2$

19. $x - 7 + \dfrac{7}{x + 2}$ **21.** $2x + 5 + \dfrac{2}{3x - 4}$

23. $2x^2 - 5x + 1 + \dfrac{4}{x + 1}$ **25.** $y^2 - 3y - 13 - \dfrac{27}{2y - 3}$

27. $4x + 3y + \dfrac{18y^2}{2x - 3y}$ **29.** $x - 3 + \dfrac{-22x + 1}{2x^2 - 3x + 2}$

31. $3y^2 + 6y + 8 + \dfrac{37}{2y - 4}$ **33.** $x^3 + 7x^2 + 49x + 341 + \dfrac{2392}{x - 7}$

35. $y^3 + 2y^2 + 4y + 8$ **37.** $x^2 + xy + y^2$ **39.** $x^2 - 2x + 1$

1. $5x^2(2x - 3)$ **3.** $9y^3(y^3 + 2)$ **5.** $3ab(3a - 2b)$
7. $7xy^2(3y^2 + x)$ **9.** $2a(4a^2 - 3a + 5)$ **11.** $11x^2(2x^2 - 3 + x^4)$
13. $10x^2y^2(x + 2xy - 3y)$ **15.** $xy(-x + y - xy)$
17. $2xy^2z(2x^2 - 4xz + 3z^2)$ **19.** $5abc(4abc - 6b + 5ac)$
21. $(a - 2b)(5x - 3y)$ **23.** $3(x + y)^2(x^2 - 2y^2)$
25. $11(a - b)^2[(a - b)^2 + 2(a - b) + 3]$ **27.** $(x + 1)(3y + 2a)$
29. $(x + 3)(xy + 1)$ **31.** $(a + b)(1 + 5x)$ **33.** $3(3x - 2y)$
35. $(x - 2)(3y^2 + 4)$ **37.** $(x - 4)(2y^3 + 1)$ **39.** $(x - a)(x - b)$
41. $(b + 5)(a - 1)$ **43.** $(b^2 + 1)(a^4 - 5)$ **45.** $(y - 2)(4x^2 + 5)$

1. $(x + 3)(x + 4)$ **3.** $(x + 3)(x - 4)$ **5.** $(y + 3)(y - 2)$
7. $(x + 2)(x - 8)$ **9.** $(x + 2)(x + 6)$ **11.** $3(a - 2)(a - 5)$
13. $4x(x - 5)(x + 1)$ **15.** $(x + 2y)(x + y)$
17. $(a + 6b)(a - 3b)$ **19.** $3(x + y)(x - 3y)$
21. $2x^3(x^2 + 2xy + 2y^2)$ **23.** $(2x - 3)(x + 5)$
25. $(2x - 5)(x + 3)$ **27.** $(2x + 3)(x + 5)$ **29.** $(2a + 1)(3a + 2)$
31. $(4y + 3)(y - 1)$ **33.** $(3x - 2)(2x + 1)$ **35.** $(2a - 3)(2a - 3)$
37. $(4x + y)(x - 3y)$ **39.** $(2x - 3y)(5x + 6y)$
41. $(3a + 4b)(6a - 7b)$ **43.** $2(2x + 3)(x - 1)$
45. $x^2(3x - 2)(3x + 5)$ **47.** $2a^2(2a - 3)(3a + 1)$
49. $(3x^2 + 1)(x^2 + 3)$ **51.** $(5a^2 + 3)(4a^2 + 5)$
53. $3(4x^2 - 3)(x^2 + 1)$

Problem Set 3.7

1. $(x - 3)^2$ **3.** $(a - 6)^2$ **5.** $(3x + 4)^2$ **7.** $(4a + 5b)^2$
9. $(x + y - 2)^2$ **11.** $[2(a - b - 1)]^2$ **13.** $(x + 3)(x - 3)$
15. $(2a + 1)(2a - 1)$ **17.** $(3x + 4y)(3x - 4y)$
19. $(x - 3)(x + 3)(x^2 + 9)$ **21.** $(2a - 3)(2a + 3)(4a^2 + 9)$
23. $(x - y)(x + y)(x^2 + xy + y^2)(x^2 - xy + y^2)$
25. $(a - 2)(a + 2)(a^2 + 2a + 4)(a^2 - 2a + 4)$ **27.** $5(x - 5)(x + 5)$
29. $3(a^2 + 4)(a - 2)(a + 2)$ **31.** $(x - y)(x^2 + xy + y^2)$
33. $(a + 1)(a^2 - a + 1)$ **35.** $(y - 2)(y^2 + 2y + 4)$
37. $(2x - 3y)(4x^2 + 6xy + 9y^2)$
39. $2x[(x - 4)^2 - (x - 4)(x + 4) + (x + 4)^2] = 2x(x^2 + 48)$
41. $(3b - a)[(a + b)^2 + 2(a^2 - b^2) + 4(a - b)^2] = (3b - a)(7a^2 - 6ab + 3b^2)$

Chapter 3   Test

1. -3 **2.** $\frac{1}{32}$ **3.** $108x^{12}y^{11}$ **4.** $\dfrac{16a^{12}}{5b^{15}}$

5. $3x^3 - 5x^2 - 8x - 4$ **6.** $4x + 75$
7. $P(-3) = 47; P(0) = 2; P(2) = 12$ **8.** $2x^3 + 3x^2 - 26x + 15$
9. $6y^2 + y - 35$ **10.** $16a^2 - 24ab + 9b^2$ **11.** $36y^2 - 1$
12. $3x^4 - \dfrac{3x}{2y} + 2x^2y^2$ **13.** $x^2 - \frac{5}{2}x - \frac{5}{4} + \dfrac{35}{8x - 4}$
14. $2(3x - 1)(2x + 5)$ **15.** $(2a + 3y)(2a - 3y)(4a^2 + 9y^2)$
16. $(x^2 - 2y)(7a - b^2)$ **17.** $(x + 3)(x^2 - 3x + 9)$
18. $4a^3b(a - 8b)(a + 2b)$

Problem Set 4.1   **CHAPTER 4**

1. $\frac{1}{3}$ **3.** $-\frac{1}{3}$ **5.** $3x^2$ **7.** $\dfrac{b^3}{2}$ **9.** $\dfrac{-3y^3}{2x}$ **11.** $\dfrac{18c^2}{7a^2}$

13. $\dfrac{x - 4}{6}$ **15.** $\dfrac{4x - 3y}{x(x + y)}$ **17.** $(a^2 + 9)(a + 3)$ **19.** $y + 3$

21. $\dfrac{a - 6}{a + 6}$ **23.** $\dfrac{2y + 3}{y + 1}$ **25.** $\dfrac{x - 2}{x - 1}$ **27.** $\dfrac{x - 3}{x + 2}$

29. $\dfrac{a^2 - ab + b^2}{a - b}$ **31.** $\dfrac{2x + 3y}{2x + y}$ **33.** $\dfrac{x + 3}{y - 4}$ **35.** $\dfrac{x + b}{x - 2b}$

37. -1 **39.** $-(y + 6)$ **41.** $\dfrac{-(3a + 1)}{3a - 1}$ **43.** -1

Problem Set 4.2

1. $\frac{1}{6}$ **3.** $\frac{9}{4}$ **5.** $\frac{1}{2}$ **7.** $\dfrac{15y}{x^2}$ **9.** $\dfrac{b}{a}$ **11.** $\dfrac{2y^5}{2^3}$ **13.** $\dfrac{x + 3}{x + 2}$

15. $(y + 1)$ **17.** $\dfrac{3(x + 4)}{x - 2}$ **19.** $\dfrac{(a - 2)(a + 2)}{a - 5}$ **21.** $\dfrac{x + 3}{x + 4}$

23. 1 **25.** $\dfrac{(y - 2)(y + 1)}{(y + 2)(y - 1)}$ **27.** $\dfrac{(a + 4)(a - 3)}{(a - 4)(a + 5)}$

29. $\dfrac{(y - 2)(y + 1)}{(y + 2)(y - 1)}$ **31.** $\dfrac{x - 1}{x + 1}$

Problem Set 4.3

1. $\frac{5}{4}$ **3.** $\frac{1}{3}$ **5.** $\frac{41}{24}$ **7.** $\frac{19}{144}$ **9.** $\frac{31}{24}$ **11.** 1 **13.** -1

15. $\dfrac{1}{x + y}$ **17.** $\dfrac{11}{3(x + 2)}$ **19.** $\dfrac{a^2 + 2a + 3}{a^3}$

21. $\dfrac{6 - 5x}{3x^2}$ **23.** $\dfrac{6x + 5}{x^2 - 1}$ **25.** $\dfrac{2(2x - 3)}{(x - 3)(x - 2)}$

27. $\dfrac{a + 9}{(a - 2)(a + 1)(a - 3)}$ **29.** $\dfrac{y^2 - 3y - 34}{(y^2 - 1)(y^2 - y - 12)}$

31. $\dfrac{5a(a + 2)}{(2a + 1)(a - 3)(a + 3)}$ **33.** $\dfrac{x^3 - 10x - 2}{x^2 - 9}$

35. $\dfrac{2(3x - 5)}{(x + 2)(x - 2)(x - 3)}$ **37.** $\dfrac{(x + y)^2}{x^3 - y^3}$ **39.** $p = \frac{51}{10}$

Problem Set 4.4

1. $\frac{9}{8}$ **3.** $\frac{2}{15}$ **5.** $\frac{119}{20}$ **7.** $\dfrac{1}{x + 1}$ **9.** $\dfrac{a + 1}{a - 1}$ **11.** $\dfrac{y - x}{y + x}$

13. $\dfrac{2(x + y)}{x(x - y)}$ **15.** $\dfrac{a - 1}{a + 1}$ **17.** 1 **19.** $\dfrac{-x^2 + x - 1}{x - 1}$

21. $\frac{5}{3}$ **23.** $\dfrac{(2x - 3)(2x + 1)}{(2x - 1)(2x + 3)}$

Problem Set 4.5

1. $-\frac{35}{3}$ **3.** $-\frac{18}{5}$ **5.** $\frac{36}{11}$ **7.** 2 **9.** 5 **11.** 2 **13.** $\varnothing$

15. 5 **17.** $\frac{2}{3}$ **19.** 18 **21.** $\varnothing$ **23.** -6 **25.** -5

27. $\frac{53}{17}$ **29.** $\varnothing$ **31.** $\frac{22}{3}$ **33.** $\frac{15}{8}$ ohms

Problem Set 4.6

1. $\frac{1}{5}, \frac{3}{5}$ **3.** 3 **5.** $\frac{5}{3}$ mi/hr

7. train A—75 mi/hr train B—60 mi/hr **9.** 2 days **11.** 9 hr

13. 16 hr **15.** 15 hr **17.** $\frac{901}{42}$

Chapter 4　Test

1. $x + y$　**2.** $\dfrac{x-1}{x+1}$　**3.** $2(a+4)$　**4.** $4(a+3)$　**5.** $\frac{38}{105}$

6. $\frac{7}{8}$　**7.** $\dfrac{3x-3}{x(x-3)}$　**8.** $\dfrac{x+6}{(x+3)(x-4)(x+4)}$

9. $\dfrac{(3a+8)(a-3)}{(a+3)(3a-8)}$　**10.** $-\dfrac{1}{a}$　**11.** $-\frac{3}{5}$　**12.** $\varnothing$　**13.** $\frac{3}{13}$

14. -7　**15.** $15\ \text{hr}$

Problem Set 5.1　　**CHAPTER 5**

1. 6　**3.** -3　**5.** 2　**7.** -2　**9.** 2　**11.** $\frac{9}{5}$　**13.** $\frac{4}{5}$
15. 9　**17.** 125　**19.** 8　**21.** $\frac{1}{3}$　**23.** $\frac{1}{27}$　**25.** $\frac{6}{5}$　**27.** $\frac{8}{27}$
29. 7　**31.** $\frac{3}{4}$　**33.** 81　**35.** $\frac{1}{12}$　**37.** $x^{\frac{4}{5}}$　**39.** a
41. $1/x^{\frac{2}{3}}$　**43.** $a^{\frac{3}{2}}b^{\frac{1}{3}}$　**45.** $x^{\frac{16}{15}}$　**47.** $x^{\frac{1}{6}}$　**49.** $x^{\frac{9}{25}}y^{\frac{1}{2}}z^{\frac{1}{5}}$
51. $x^{\frac{1}{2}}y^{\frac{1}{3}}$　**53.** $b^{\frac{7}{4}}/a^{\frac{1}{8}}$　**55.** $y^{\frac{3}{10}}$　**57.** $x^{\frac{8}{3}}/y^{\frac{11}{3}}z^{14}$　**59.** $a^{\frac{1}{4}}b^{\frac{1}{18}}c^{\frac{17}{30}}$
61. $25\ \text{mi/hr}$

Problem Set 5.2

1. $2\sqrt{2}$　**3.** $3\sqrt{2}$　**5.** $5\sqrt{3}$　**7.** $12\sqrt{2}$　**9.** $4\sqrt{5}$
11. $4\sqrt{3}$　**13.** $3\sqrt{5}$　**15.** $3\sqrt[3]{2}$　**17.** $4\sqrt[3]{2}$　**19.** $2\sqrt[5]{2}$
21. $3\sqrt[6]{6}$　**23.** $2\sqrt[3]{5}$　**25.** $3\sqrt{11}$　**27.** $3x\sqrt{2x}$　**29.** $4y^3\sqrt{2y}$
31. $2xy^2\sqrt[3]{5xy}$　**33.** $4abc^2\sqrt{3b}$　**35.** $2bc\sqrt[3]{6a^2c}$
37. $2xy^2\sqrt[5]{2x^3y^2}$　**39.** $2x^2yz\sqrt{3xz}$　**41.** $2\sqrt{3}/3$　**43.** $5\sqrt{6}/6$
45. $\sqrt{2}/2$　**47.** $\sqrt{5}/5$　**49.** $2\sqrt[3]{4}$　**51.** $2\sqrt[3]{3}/3$　**53.** $\dfrac{\sqrt{6x}}{2x}$
55. $2\sqrt{2y}/y$　**57.** $\sqrt[3]{36xy^2}/3y$　**59.** $\sqrt[3]{6xy^2}/3y$　**61.** $\sqrt[4]{2x^3}/2x$
63. $3x\sqrt{15xy}/5y$　**65.** $5xy\sqrt{6xz}/2z$　**67.** $2ab\sqrt[3]{6ac^2}/3c$
69. $2xy^2\sqrt[3]{3z^2}/3z$　**71.** $5\sqrt{13}\ \text{ft}$

Problem Set 5.3

1. $7\sqrt{5}$　**3.** $9\sqrt{6}$　**5.** 0　**7.** $4\sqrt{2}$　**9.** $\sqrt{5}$
11. $-32\sqrt{2}$　**13.** $-3x\sqrt{2}$　**15.** $-2\sqrt[3]{2}$　**17.** $8x\sqrt[3]{xy^2}$
19. $3a^2b\sqrt{3ab}$　**21.** $11ab\sqrt[3]{3a^2b}$　**23.** $\sqrt{6}-9$
25. $24\sqrt{3}+6\sqrt{6}$　**27.** $7+2\sqrt{6}$　**29.** $34+20\sqrt{3}$
31. $19+8\sqrt{3}$　**33.** $30+12\sqrt{6}$　**35.** 1　**37.** 15　**39.** 92
41. $2\sqrt{3}-2\sqrt{2}$　**43.** $2\sqrt{3}+2$　**45.** $(2\sqrt{3}+1)/11$
47. $(20+6\sqrt{5})/11$　**49.** $-2-\sqrt{3}$　**51.** $(5-\sqrt{21})/4$
53. $(16+14\sqrt{6})/-23$　**55.** **a.** $5\sqrt{2}/4\ \text{sec}$　**b.** $\frac{5}{2}\ \text{sec}$

Problem Set 5.4

1. 4   **3.** ∅   **5.** 5   **7.** ∅   **9.** $\frac{39}{2}$ **11.** ∅   **13.** 5
15. 3   **17.** $-\frac{32}{3}$ **19.** ∅   **21.** ∅   **23.** ∅   **25.** 8   **27.** 0
29. $-\frac{3}{5}$ **31.** $\frac{392}{121} \approx 3.24$ ft

Problem Set 5.5

1. $6i$ **3.** $-5i$ **5.** $6i\sqrt{2}$ **7.** $-2i\sqrt{3}$ **9.** $9i\sqrt{2}$ **11.** i
13. $-i$ **15.** 1   **17.** -1 **19.** $-i$ **21.** $x = 3, y = -1$
23. $x = -2, y = -\frac{1}{2}$ **25.** $x = -8, y = -5$ **27.** $x = -\frac{6}{5}, y = \frac{3}{4}$
29. $x = 7, y = \frac{1}{2}$ **31.** $x = \frac{3}{7}, y = \frac{2}{5}$ **33.** $5 + 9i$ **35.** $5 - i$
37. $2 - 4i$ **39.** $1 - 6i$ **41.** $9 - 2i$ **43.** $3 - 3i$
45. $2 + 2i$ **47.** $-1 - 7i$ **49.** $6 + 8i$ **51.** $2 - 24i$

Problem Set 5.6

1. $-15 + 12i$ **3.** $7 - 7i$ **5.** $18 + 24i$ **7.** $10 + 11i$
9. $21 + 23i$ **11.** $-26 + 7i$ **13.** $-21 + 20i$ **15.** $-2i$
17. $-7 - 24i$ **19.** 5   **21.** 40   **23.** 13   **25.** 164
27. $-3 - 2i$ **29.** $-2 + 5i$ **31.** $\frac{8}{13} + \frac{12}{13}i$ **33.** $-\frac{18}{13} - \frac{12}{13}i$
35. $-\frac{5}{13} + \frac{12}{13}i$ **37.** $\frac{39}{45} - \frac{18}{45}i$ **39.** $\frac{31}{53} - \frac{24}{53}i$

Chapter 5   Test

1. $\sqrt[5]{8^3}$ **2.** $17\sqrt[3]{17}$ **3.** $3^{\frac{1}{4}}x^{\frac{3}{4}}$ **4.** $(7x^2)^{\frac{1}{3}}$ **5.** $\frac{1}{9}$ **6.** $\frac{7}{5}$
7. $a^{\frac{5}{12}}$ **8.** $\dfrac{x^{\frac{13}{12}}}{y}$ **9.** $5xy^2\sqrt{5xy}$ **10.** $2x^2y^2\sqrt[3]{5xy^2}$
11. $\sqrt{6}/3$ **12.** $2a^2b\sqrt{15bc}/5c$ **13.** $-6\sqrt{3}$ **14.** $-ab\sqrt[3]{3}$
15. $17 - \sqrt{14}$ **16.** $21 - 6\sqrt{6}$ **17.** $\frac{5}{2} + \dfrac{5\sqrt{3}}{2}$ **18.** $\frac{7}{3} - \dfrac{2\sqrt{10}}{3}$
19. 10   **20.** ∅   **21.** -4 **22.** ∅   **23.** $x = \frac{2}{3}, y = \frac{1}{2}$
24. $x = \frac{1}{2}, y = 7$ **25.** $-1 - 9i$ **26.** $6i$ **27.** $17 - 6i$
28. $9 - 40i$ **29.** $2 - 3i$ **30.** $\frac{10}{13} + \frac{15}{13}i$ **31.** $-\frac{5}{13} - \frac{12}{13}i$
32. $i^{38} = (i^4)^9 \cdot i^2$
$\qquad = 1(-1)$
$\qquad = -1$

CHAPTER 6

Problem Set 6.1

1. $6, -1$ **3.** $2, 3$ **5.** $\frac{1}{3}, -4$ **7.** $\frac{2}{3}, \frac{3}{2}$ **9.** $5, -5$
11. $-3, 7$ **13.** $-4, \frac{5}{2}$ **15.** $0, \frac{4}{3}$ **17.** $-\frac{1}{5}, \frac{1}{3}$ **19.** $-\frac{4}{3}, \frac{4}{3}$

21. $-10, 0$ **23.** $-5, 1$ **25.** $1, 2$ **27.** $-2, 3$ **29.** $-2, \frac{1}{4}$
31. $-3, 4$ **33.** $-\frac{4}{3}, 1$ **35.** $-3, 3$ **37.** $-5, 3$ **39.** $-\frac{4}{3}, \frac{1}{2}$
41. 7 **43.** $\frac{8}{5}$ **45.** $-2, \frac{2}{3}$ **47.** $t = 2$ seconds

Problem Set 6.2

1. $2, 8$ **3.** $-3 \pm 3\sqrt{2}$ **5.** $-2, 3$ **7.** $(-3 \pm 3i)/2$
9. $(-2 \pm 2i\sqrt{2})/5$ **11.** $-6, 2$ **13.** $-3, -9$ **15.** $1 \pm 2i$
17. $4 \pm \sqrt{15}$ **19.** $(5 \pm \sqrt{37})/2$ **21.** $1 \pm \sqrt{5}$
23. $(4 \pm \sqrt{13})/3$ **25.** $(3 \pm i\sqrt{71})/8$ **27.** $(-1 \pm \sqrt{13})/3$
29. $-1, \frac{5}{2}$ **31.** $1 \pm \sqrt{2}$ **33.** $-\frac{10}{3}$ **35.** $3 \pm \sqrt{10}$
37. $2 \pm \sqrt{3}$ **39.** $-2 \pm 2i$ **41.** $5 + 2\sqrt{3}$

Problem Set 6.3

1. $-2, -3$ **3.** $1, 2$ **5.** $0, 5$ **7.** $-3 \pm \sqrt{17}$
9. $-\frac{3}{2}, -1$ **11.** 1 **13.** $(1 \pm \sqrt{11})/2$ **15.** $(1 \pm i\sqrt{47})/6$
17. $(3 \pm i\sqrt{7})/2$ **19.** $(2 \pm \sqrt{26})/2$ **21.** $\pm\sqrt{5}, \pm 2i$
23. $\pm\sqrt{5}, \pm 2$ **25.** $\pm\sqrt{6}/3, \pm i\sqrt{2}/2$ **27.** $\pm\sqrt{21}/3, \pm i\sqrt{21}/3$
29. $\pm\sqrt{30}/6, \pm i$ **31.** 4 **33.** 0 **35.** 4 **37.** 6 **39.** 7
41. 25 or 4

Problem Set 6.4

1. $D = 16$, two rational   **3.** $D = 0$, one rational
5. $D = 5$, two irrational   **7.** $D = 17$, two irrational
9. $D = 36$, two rational   **11.** $D = 116$, two irrational   **13.** ± 10
15. ± 12 **17.** 9 **19.** 16 **21.** $\pm 2\sqrt{6}$
23. sum $= -3$
    product $= 2$
25. sum $= \frac{15}{2}$
    product $= 9$
27. sum $= \frac{3}{4}$
    product $= \frac{1}{2}$
29. sum $= 6$
    product $= 9$
31. sum $= -2$
    product $= -\frac{9}{2}$
33. sum $= 0$
    product $= -49$
35. Both are solutions.   **37.** not solutions
39. Both are solutions.   **41.** Both are solutions.

1. 3, 5 or $-5, -3$ **3.** 4, 5 or $-5, -4$ **5.** 5, 13   **7.** 3 or $\frac{1}{3}$
9. 3, 4   **11.** 2 ft   **13.** 1 in.   **15.** 2 cm, 6 cm   **17.** 2 sec
19. $\frac{1}{4}$ sec and 1 sec   **21.** 6 mi/hr   **23.** 15 mi/hr

1. 7, -3 **2.** $\pm\frac{9}{5}$ **3.** $\frac{3}{4}, -2$ **4.** $2 \pm \sqrt{2}$
5. $(4 \pm \sqrt{26})/2$ **6.** $-\frac{5}{3}, -1$ **7.** $\frac{3}{2}, -1$ **8.** $(7 \pm \sqrt{77})/2$
9. $\pm i/2, \pm \sqrt{2}$ **10.** 2   **11.** 0, 4   **12.** 5, $\frac{8}{5}$ **13.** 9
14. $D = 81$, two rational
15. sum $= \sqrt{3}/4$
     product $= \frac{5}{4}$
16. yes   **17.** 3, 5   **18.** 2 mi/hr

CHAPTER 7

1–15 (odd).

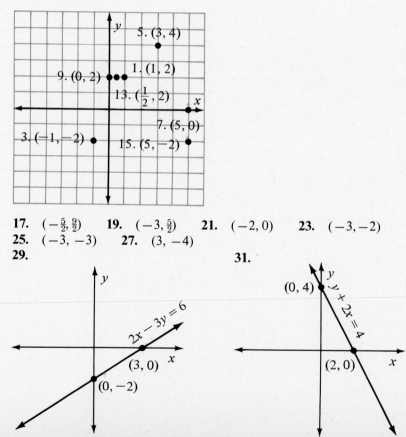

17. $(-\frac{5}{2}, \frac{9}{2})$ **19.** $(-3, \frac{5}{2})$ **21.** $(-2, 0)$ **23.** $(-3, -2)$
25. $(-3, -3)$ **27.** $(3, -4)$
29.   **31.**

33.

35.

37.

39.

41.

43.

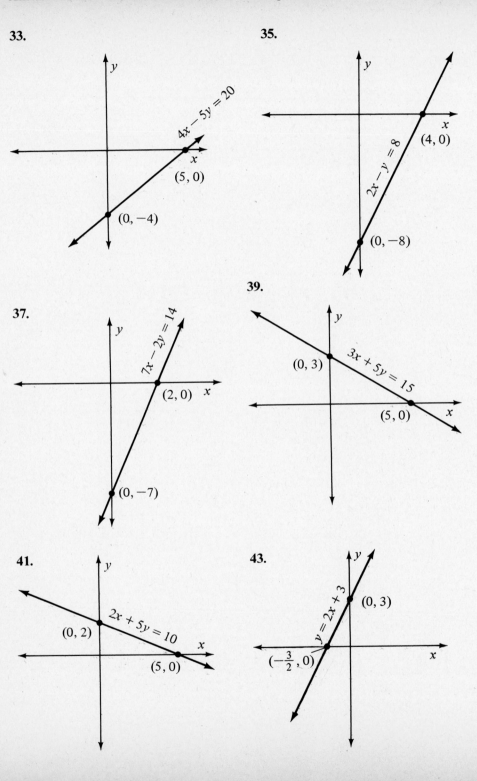

45.

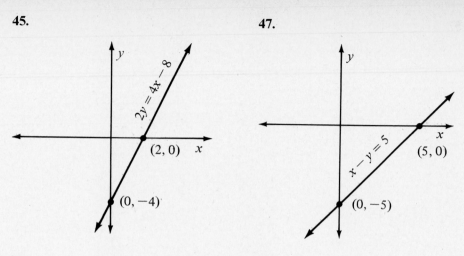

$2y = 4x - 8$
$(2, 0)$
$(0, -4)$

47.

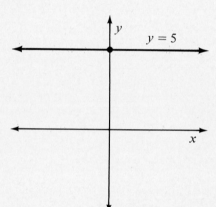

$x - y = 5$
$(5, 0)$
$(0, -5)$

49.

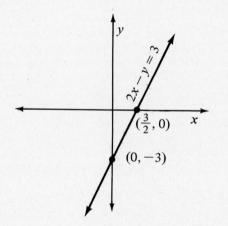

$2x - y = 3$
$\left(\frac{3}{2}, 0\right)$
$(0, -3)$

51.

$y = 5$

53.

$x = -7$

55.

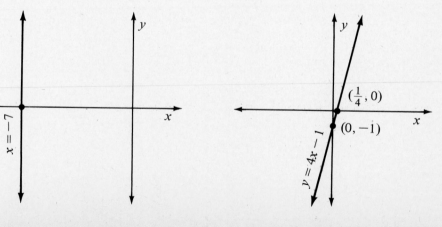

$\left(\frac{1}{4}, 0\right)$
$(0, -1)$
$y = 4x - 1$

57.

59. a.

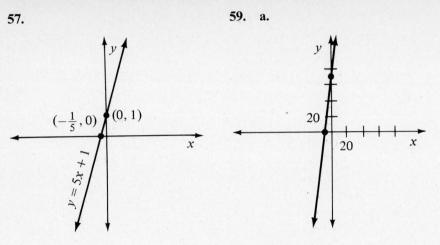

b. $7.65

Problem Set 7.2

1. 1   **3.** no slope   **5.** -1 **7.** -2 **9.** -4 **11.** $-\frac{3}{2}$
13. no slope   **15.** 0   **17.** $-\frac{39}{2}$ **19.** 5   **21.** -1
23. -5 **25.** -8 **27.** $2/3\sqrt{2}$ **29.** 0   **31.** 8
33. $-2, 3$ **35.** 24 feet

Problem Set 7.3

1. $y = 2x + 3$ **3.** $y = x - 5$ **5.** $y = \dfrac{x}{2} + \dfrac{3}{2}$ **7.** $y = 4$
9. $y = -\sqrt{2}x + 3\sqrt{2}$
11. $m = 3$, y-intercept $= -2$, perpendicular slope $= -\frac{1}{3}$

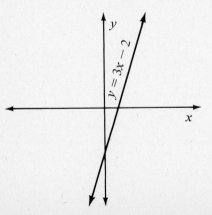

13. $m = 2$, y-intercept $= -4$, perpendicular slope $= -\frac{1}{2}$

15. $m = \frac{2}{3}$, y-intercept $= -4$, perpendicular slope $= -\frac{3}{2}$

17. $m = -\frac{4}{5}$, y-intercept $= 4$, perpendicular slope $= \frac{5}{4}$

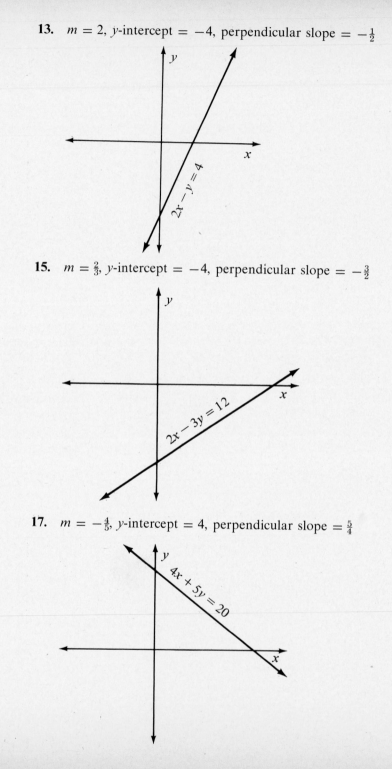

19. $m = \frac{3}{5}$, y-intercept $= -2$, perpendicular slope $= -\frac{5}{3}$

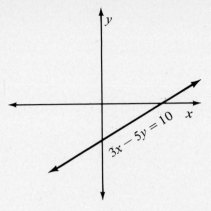

21. $y = 2x + 1$ **23.** $y = -3x + 14$ **25.** $y = -\frac{1}{2}x + 2$
27. $y = \frac{2}{3}x - \frac{11}{6}$ **29.** $y = -2x + 7$ **31.** $y = 2x + 2$
33. $y = -\frac{7}{2}x + \frac{17}{2}$ **35.** $y = \frac{5}{3}x + 5$ **37.** $y = 5$ **39.** $x = 5$
41. slope $= 0$, y-intercept $= -2$ **43.** no slope, x-intercept $= \sqrt{5}$

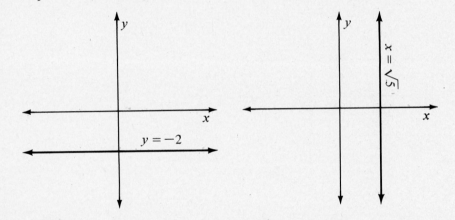

45. $y = 3x + 7$ **47.** $y = -\frac{5}{2}x - 13$ **49.** $y = \frac{1}{4}x + \frac{1}{4}$
51. $y = \frac{2}{3}x - 2$
53. **a.** $d = 1100t$
  b. 4400 ft/sec
  c. 4.8 sec

1.

3.

5.

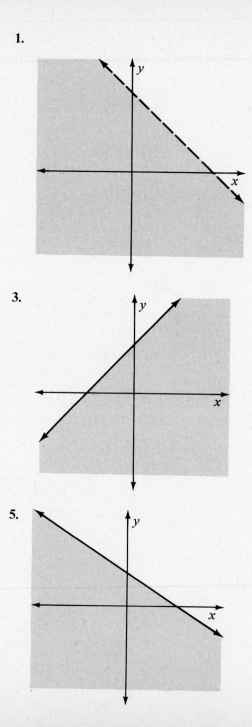

7.

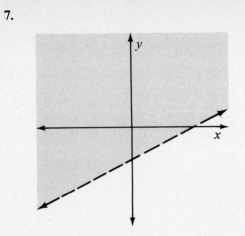

9.

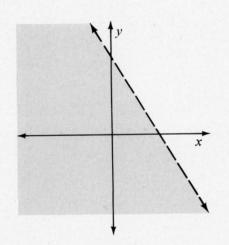

11.

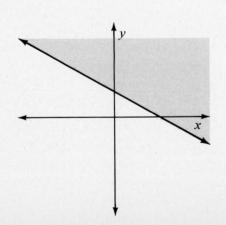

13.

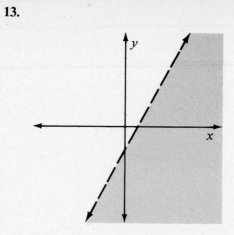

15.

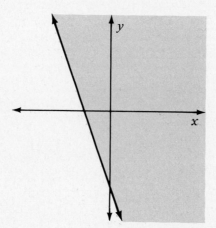

17.

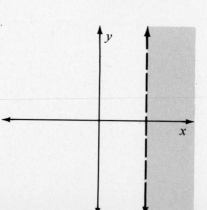

19.

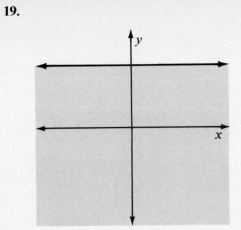

21.

a.

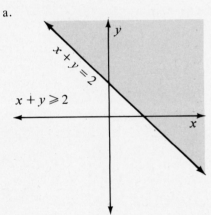

b.

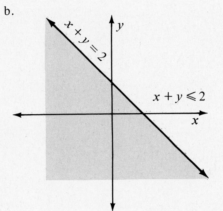

c.

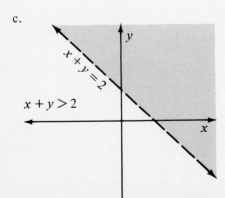

d.

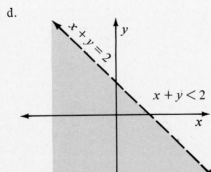

23.

a.

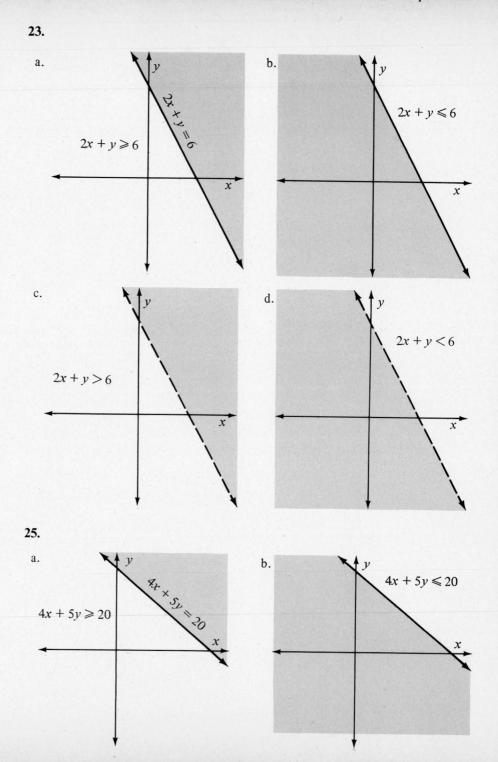

$2x + y = 6$

$2x + y \geqslant 6$

b.

$2x + y \leqslant 6$

c.

$2x + y > 6$

d.

$2x + y < 6$

25.

a.

$4x + 5y = 20$

$4x + 5y \geqslant 20$

b.

$4x + 5y \leqslant 20$

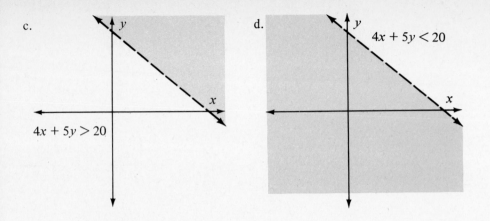

c. $4x + 5y > 20$

d. $4x + 5y < 20$

27.

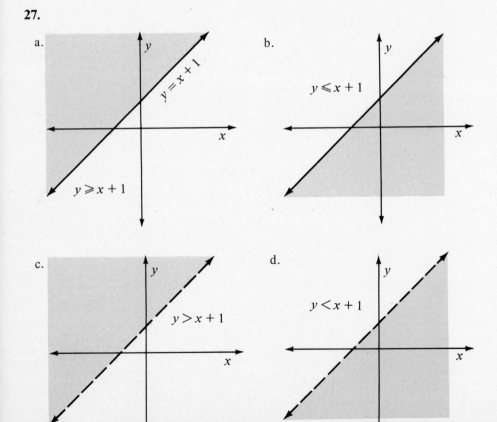

a. $y = x + 1$, $y \geqslant x + 1$

b. $y \leqslant x + 1$

c. $y > x + 1$

d. $y < x + 1$

29.

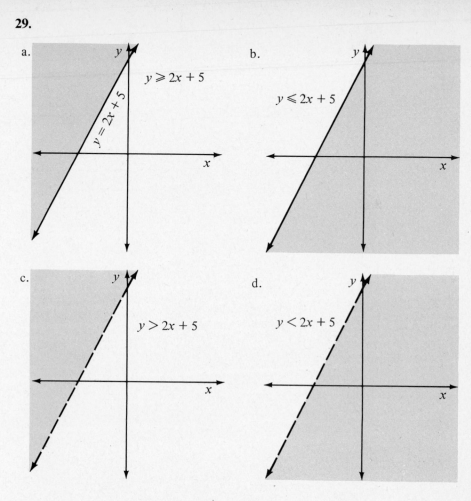

a. $y \geqslant 2x + 5$ $y = 2x + 5$

b. $y \leqslant 2x + 5$

c. $y > 2x + 5$

d. $y < 2x + 5$

31.

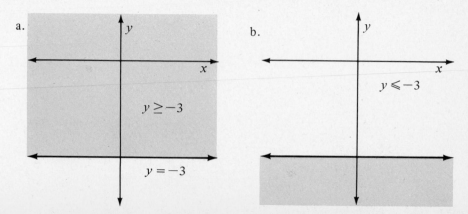

a. $y \geq -3$ $y = -3$

b. $y \leqslant -3$

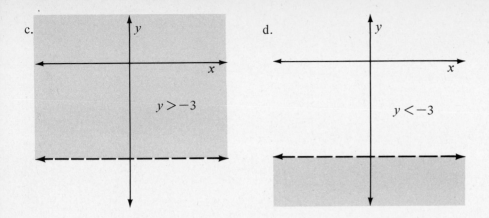

c. $y > -3$

d. $y < -3$

33.

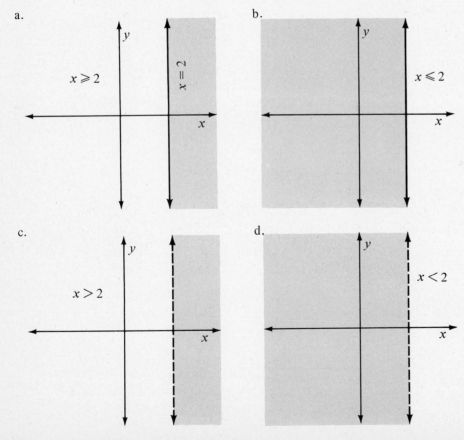

a. $x \geq 2$ $x = 2$

b. $x \leq 2$

c. $x > 2$

d. $x < 2$

35. $y \geq -2x + 4$ **37.** $y < \frac{3}{2}x - 3$

1. 30   **3.** 5   **5.** $\frac{363}{8}$ **7.** -6 **9.** $\frac{1}{2}$ **11.** 12   **13.** 8
15. $4\sqrt{2}$ **17.** $\frac{81}{5}$ **19.** $\pm 3\sqrt{2}$ **21.** 64   **23.** 8   **25.** $\frac{50}{3}$ lb
27. 12 lb   **29.** $\frac{1504}{15}$ sq. in.   **31.** 1.5 ohms

1. x-intercept $= 3$, y-intercept $= 6$,
slope $= -2$

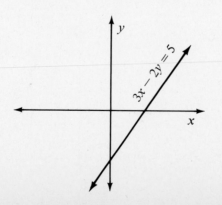

2. x-intercept $= \frac{5}{3}$, y-intercept $= -\frac{5}{2}$,
slope $= \frac{3}{2}$

3. x-intercept $= -\frac{3}{2}$, y-intercept $= -3$,
slope $= -2$

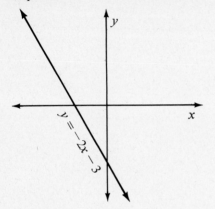

4. x-intercept $= -\frac{8}{3}$, y-intercept $= 4$
slope $= \frac{3}{2}$

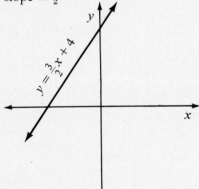

5. x-intercept $= -2$, no y-intercept,
no slope

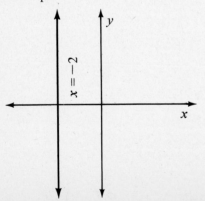

6. no x-intercept, y-intercept $= 3$,
slope $= 0$

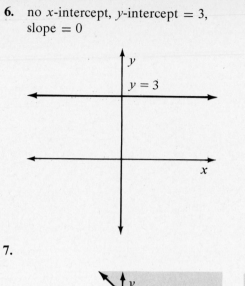

$y = 3$

7.

$x + y \geqslant 4$

8.

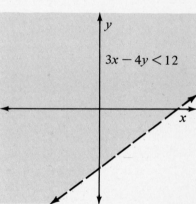

$3x - 4y < 12$

9.

$y \leqslant -x + 2$

10.

$x > 5$

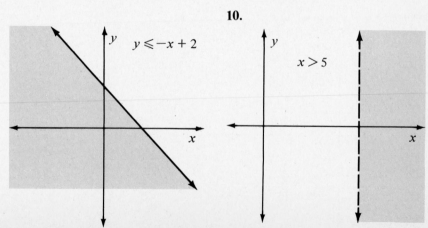

11. $y = 2x + 5$ **12.** $y = -\frac{3}{7}x + \frac{5}{7}$ **13.** $y = \frac{2}{5}x - 5$
14. $y = -\frac{1}{3}x - \frac{7}{3}$ **15.** $x = 4$ **16.** 18   **17.** 5   **18.** $\frac{2000}{3}$ lb

Problem Set 8.1   **CHAPTER 8**

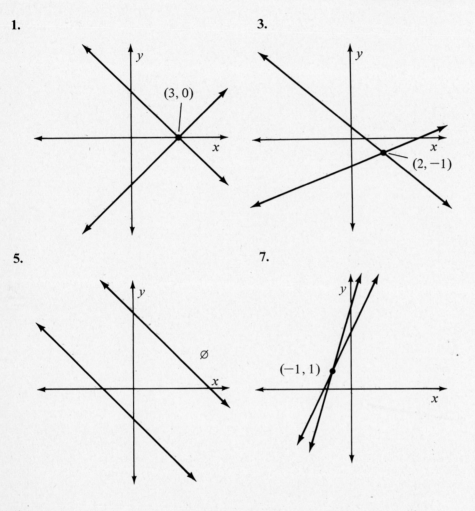

1.   **3.**

5.   **7.**

9. $(2, 3)$ **11.** $(1, 1)$ **13.** lines coincide $\{(x, y) \mid 3x - 2y = 6\}$
15. $(-1, -2)$ **17.** $(0, 0)$ **19.** $(4, -3)$ **21.** $(2, 2)$
23. parallel lines; $\varnothing$ **25.** $(1, 2)$ **27.** $(10, 24)$ **29.** $(4, \frac{10}{3})$
31. $(0, 3)$ **33.** $(6, 2)$ **35.** $(2, 4)$ **37.** $(3, -3)$
39. lines coincide $\{(x, y) \mid 2x - y = 5\}$ **41.** $(1, 1)$ **43.** $(\frac{1}{3}, 1)$
45. graphs coincide $\left\{(x, y) \mid \dfrac{1}{x} + \dfrac{3}{y} = 6\right\}$ **47.** **b.** $y = 16x + 22$

   d. 3 min

Problem Set 8.2

1. $(1, 2, 1)$ **3.** $(2, 1, 3)$ **5.** $(2, 0, 1)$ **7.** no unique solution
9. $(1, 1, 1)$ **11.** no unique solution   **13.** no unique solution
15. $(\frac{1}{2}, 1, 2)$ **17.** $(1, 3, 1)$ **19.** $(-1, 2, -2)$ **21.** 4 amp, 3 amp, 1 amp

Problem Set 8.3

1. 3   **3.** 5   **5.** -1 **7.** 0   **9.** 10   **11.** 2   **13.** -3
15. -2 **17.** $-2, 5$ **19.** 3   **21.** 0   **23.** 3   **25.** 8
27. 6   **29.** -228

Problem Set 8.4

1. $(3, 1)$ **3.** Cramer's Rule does not apply.
5. Cramer's Rule does not apply.   **7.** $(-\frac{15}{43}, -\frac{27}{43})$ **9.** $(\frac{60}{43}, \frac{46}{43})$
11. $(\frac{1}{2}, \frac{5}{2}, 1)$ **13.** $(3, -1, 2)$ **15.** Cramer's Rule does not apply.
17. $(-\frac{10}{91}, -\frac{63}{91}, \frac{107}{91})$ **19.** $(3, -2, 4)$ **21.** $(3, 1, 2)$
23. $x = 50$ items

Problem Set 8.5

1. 5, 13   **3.** 10, 16   **5.** 1, 3, 4   **7.** 225 adults, 700 children
9. 12 nickels, 8 dimes   **11.** $12,000 at 6%, $8,000 at 7%
13. $4000 at 6%, $8000 at 7.5%   **15.** 3 gal of 50%, 6 gal of 20%
17. 5 gal of 20%, 10 gal of 14%
19. 9 mi/hr for boat, 3 mi/hr current
21. 270 mi/hr airplane, 30 mi/hr wind

Chapter 8   Test

1. $(1, 2)$ **2.** $(3, 2)$ **3.** $(-\frac{54}{13}, -\frac{58}{13})$ **4.** $(1, 2)$
5. $(3, -2, 1)$ **6.** -14 **7.** -26 **8.** $(-\frac{14}{3}, -\frac{19}{3})$
9. lines coincide $\{(x, y) \mid 2x + 4y = 3\}$ **10.** $(\frac{5}{11}, -\frac{15}{11}, -\frac{1}{11})$
11. $4000 at 5%, $8000 at 6%   **12.** 3 oz of cereal I, 1 oz of cereal II

1. *x*-intercepts $= -1, 7$;
vertex $= (3, -16)$

3. *x*-intercepts $= -3, 1$;
vertex $= (-1, -4)$

5. *x*-intercepts $= -5, 1$;
vertex $= (-2, 9)$

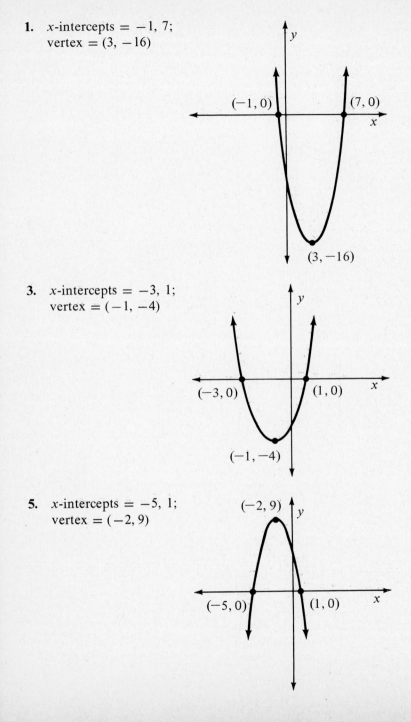

7. *x*-intercepts = 5, −3;
vertex = (1, 16)

9. *x*-intercepts = −1, 1;
vertex = (0, −1)

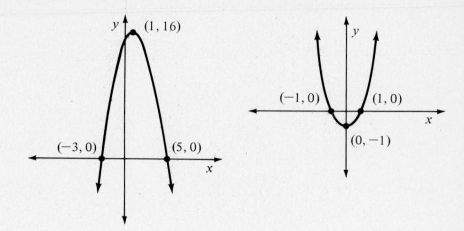

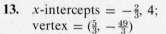

11. *x*-intercepts = 3, −3;
vertex = (0, 9)

13. *x*-intercepts = $-\frac{2}{3}$, 4;
vertex = $(\frac{5}{3}, -\frac{49}{3})$

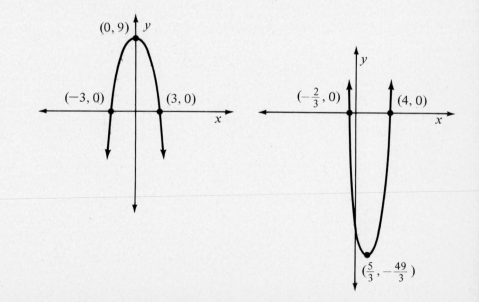

15. x-intercepts $= (1 + 2\sqrt{3}, 0)$, $(1 - 2\sqrt{3}, 0)$;
vertex $= (1, -12)$

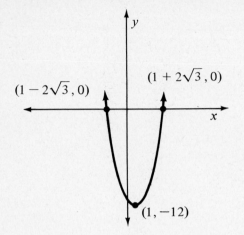

17. x-intercepts $= \left(\dfrac{-3 + 2\sqrt{5}}{2}, 0\right)$, $\left(\dfrac{-3 - 2\sqrt{5}}{2}, 0\right)$;

vertex $= (-\tfrac{3}{2}, -20)$

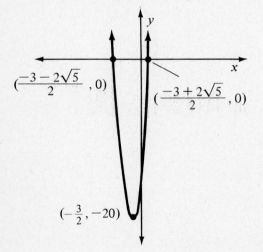

19. x-intercepts $= \left(\dfrac{-1 + 4\sqrt{2}}{2}, 0\right), \left(\dfrac{-1 - 4\sqrt{2}}{2}, 0\right);$

  vertex $= \left(-\tfrac{1}{2}, 32\right)$

21. vertex $= (2, -8)$

23. vertex $= (1, -4)$

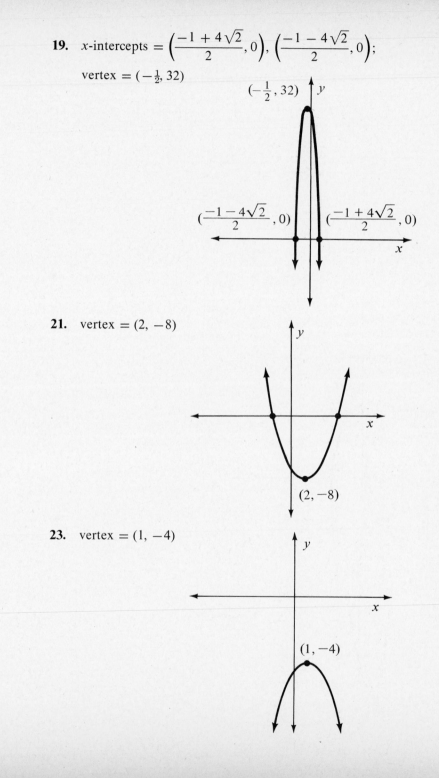

25.   vertex $= (0, 1)$

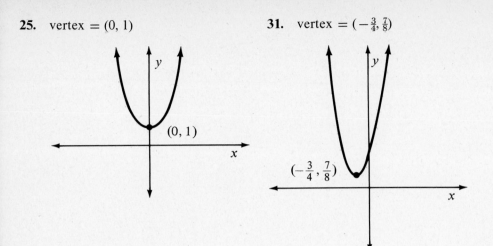

$(0, 1)$

31.   vertex $= (-\frac{3}{4}, \frac{7}{8})$

$(-\frac{3}{4}, \frac{7}{8})$

27.   vertex $= (0, -3)$

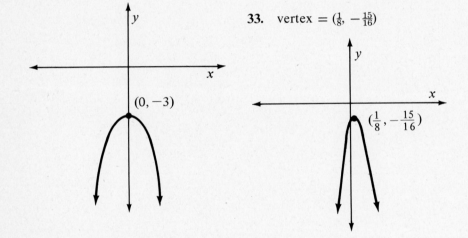

$(0, -3)$

33.   vertex $= (\frac{1}{8}, -\frac{15}{16})$

$(\frac{1}{8}, -\frac{15}{16})$

29.   vertex $= (-\frac{2}{3}, -\frac{1}{3})$

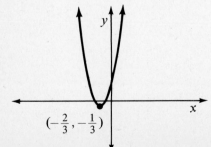

$(-\frac{2}{3}, -\frac{1}{3})$

35.   256 ft

1. 5　　**3.** $\sqrt{106}$　　**5.** $\sqrt{61}$　　**7.** $\sqrt{130}$　　**9.** 3 or -1

11. 3　　**13.** $(x - 2)^2 + (y - 3)^2 = 16$

15. $(x - 3)^2 + (y + 2)^2 = 9$　　**17.** $(x + 5)^2 + (y + 1)^2 = 5$

19. $x^2 + (y + 5)^2 = 1$　　**21.** $x^2 + y^2 = 4$　　**23.** $(x + 1)^2 + y^2 = 12$

25. center = $(0, 0)$, radius = 2

27. center = $(0, 0)$, radius = $\sqrt{5}$

29. center = $(1, 3)$,
　　radius = 5

31. center $= (-2, 4)$,
 radius $= 2\sqrt{2}$

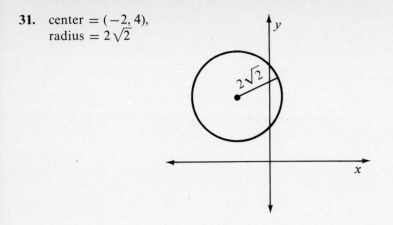

33. center $= (-1, -1)$,
 radius $= 1$

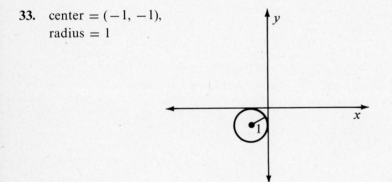

35. center $= (0, 3)$,
 radius $= 4$

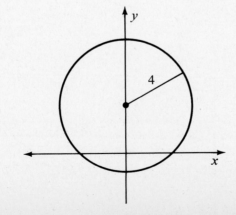

37. center $= (-1, 0)$,
     radius $= \sqrt{2}$

39. center $= (2, 3)$,
     radius $= 1$

41. center $= (-1, -\frac{1}{2})$,
     radius $= \sqrt{17}/2$

43. center $= (\frac{1}{2}, \frac{3}{2})$,
     radius $= 3\sqrt{2}/2$

45.

47.

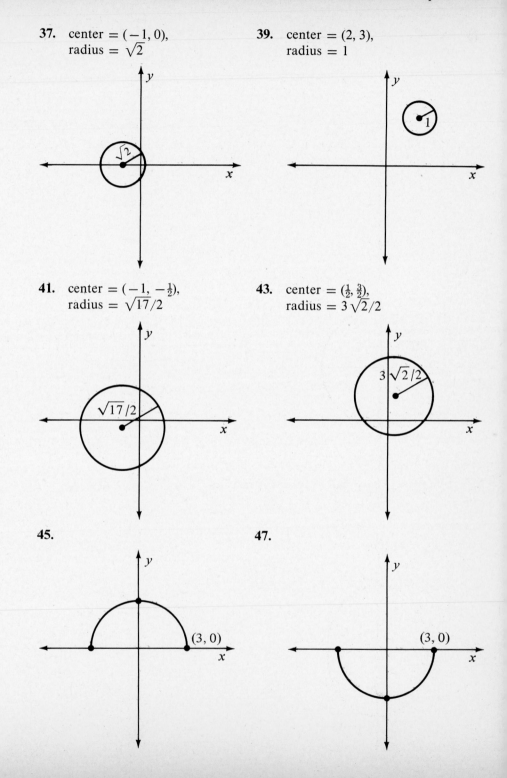

49.

51.

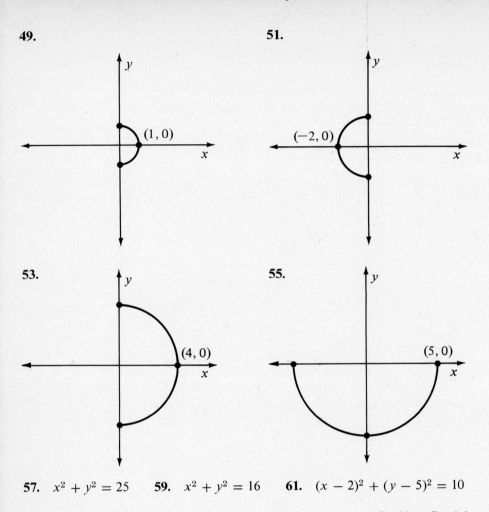

53.

55.

57. $x^2 + y^2 = 25$ **59.** $x^2 + y^2 = 16$ **61.** $(x - 2)^2 + (y - 5)^2 = 10$

Problem Set 9.3

1.

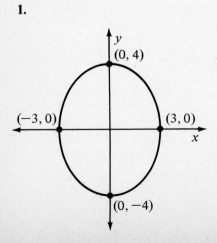

3.

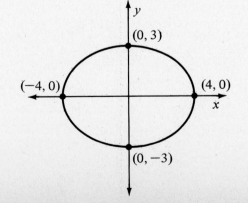

5.

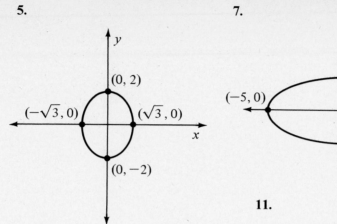

7.

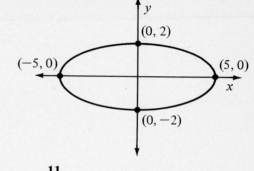

9.

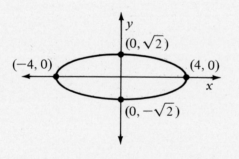

11.

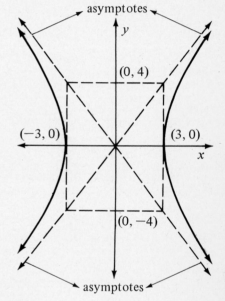

13.

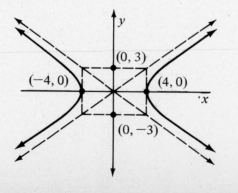

15.

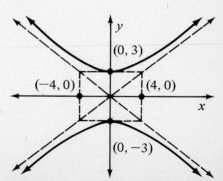

17.

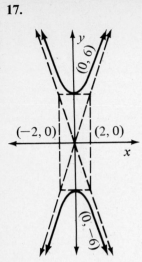

19.

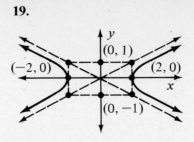

21.

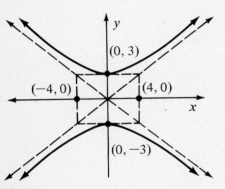

23.

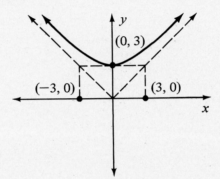

25.

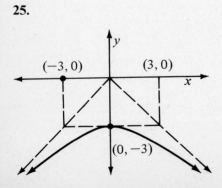

27.

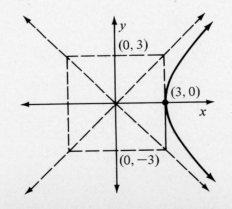

29.

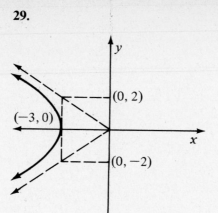

31.

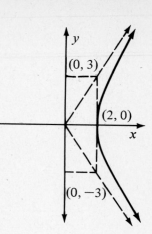

33.

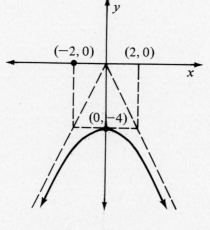

1.

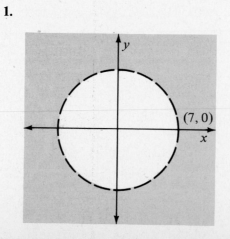

3.

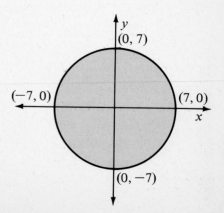

5.

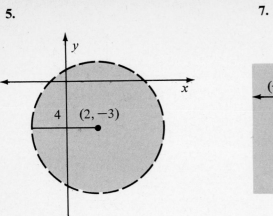

7.

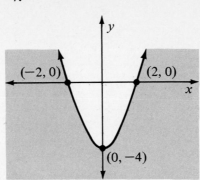

9.

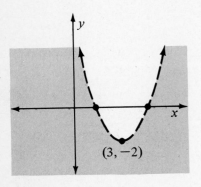

11.

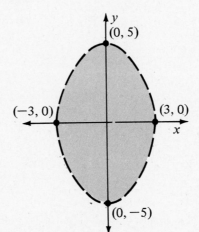

13.

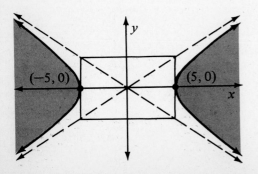

15.

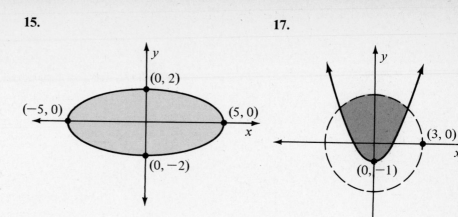

17.

19.

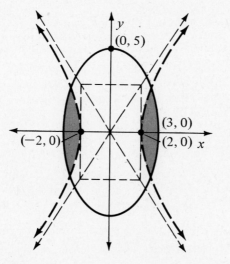

21.

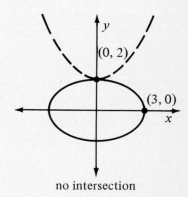

no intersection

23.

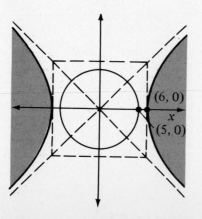

Problem Set 9.5

1. $(0, 3), (\frac{12}{5}, -\frac{9}{5})$ **3.** $(0, 4), (\frac{16}{5}, \frac{12}{5})$ **5.** $(5, 0), (-5, 0)$

7. $(0, -3), (\sqrt{5}, 2), (-\sqrt{5}, 2)$ **9.** $(0, 1)$ **11.** $\left(\dfrac{-3 \pm 3\sqrt{5}}{2}, 3 \pm \sqrt{5}\right)$

13. $(-4, 11), (\frac{5}{2}, \frac{5}{4})$ **15.** $(3, 0), (-3, 0)$

17. $(\sqrt{7}, \sqrt{6}/2), (-\sqrt{7}, \sqrt{6}/2), (\sqrt{7}, -\sqrt{6}/2), (-\sqrt{7}, -\sqrt{6}/2)$

19. $(4, 0), (0, -4)$ **21.** $(8, 5)$ or $(-8, -5)$ **23.** $(6, 3)$ or $(13, -4)$

25. 100 items or 15 items

Chapter 9   Test

1.

(a)

(b)

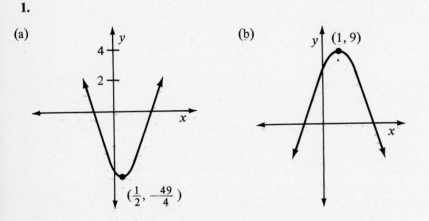

$(\frac{1}{2}, -\frac{49}{4})$

$(1, 9)$

2. $(x + 2)^2 + (y - 4)^2 = 9$ **3.** $x^2 + y^2 = 25$

4. center $= (5, -3)$,
radius $= \sqrt{39}$

$\sqrt{39}$

5.

(a)

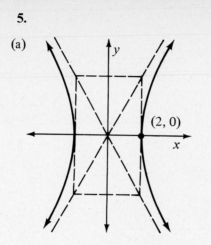

(b)

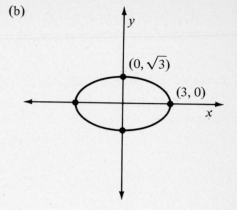

(c) (d)

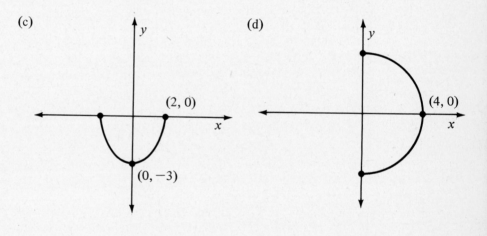

(e)

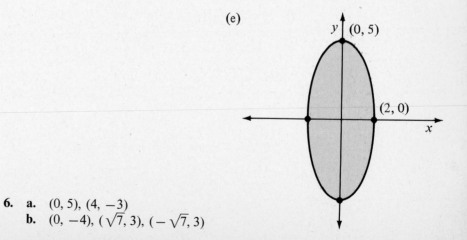

6. a. $(0, 5)$, $(4, -3)$
   b. $(0, -4)$, $(\sqrt{7}, 3)$, $(-\sqrt{7}, 3)$

1. $D = \{1, 2, 4\}$, $R = \{3, 5, 1\}$, yes
3. $D = \{-1, 1, 2\}$, $R = \{3, -5\}$, yes
5. $D = \{7, 3\}$, $R = \{-1, 4\}$, no   **7.** $D = \{4, 3\}$, $R = \{3, 4, 5\}$, no
9. $D = \{5, -3, 2\}$, $R = \{-3, 2\}$, yes   **11.** yes   **13.** no
15. no   **17.** yes   **19.** yes   **21.** $\{x \,|\, x \geq -3\}$
23. $\{x \,|\, x \geq \frac{1}{2}\}$ **25.** $\{x \,|\, x \leq \frac{1}{4}\}$ **27.** $\{x \,|\, x \neq 5\}$
29. $\{x \,|\, x \neq \frac{1}{2}, x \neq -3\}$ **31.** $\{x \,|\, x \neq -2, x \neq 3\}$
33. $\{x \,|\, x \neq -2, x \neq 2\}$

35.

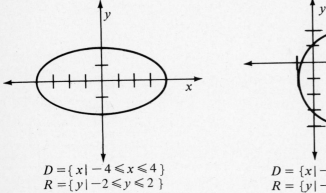

$D = \{x \,|\, -4 \leq x \leq 4\}$
$R = \{y \,|\, -2 \leq y \leq 2\}$

37.

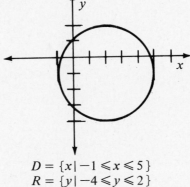

$D = \{x \,|\, -1 \leq x \leq 5\}$
$R = \{y \,|\, -4 \leq y \leq 2\}$

39.

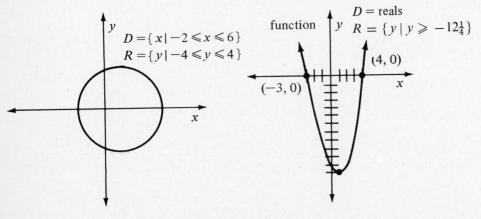

$D = \{x \,|\, -2 \leq x \leq 6\}$
$R = \{y \,|\, -4 \leq y \leq 4\}$

41.

function

$D = \text{reals}$
$R = \{y \,|\, y \geq -12\frac{1}{4}\}$

$(4, 0)$

$(-3, 0)$

43.

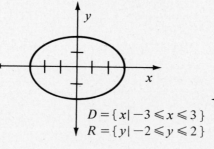

function D = reals
R = reals

45.

$D = \{x \mid x \leqslant -2 \text{ or } x \geqslant 2\}$

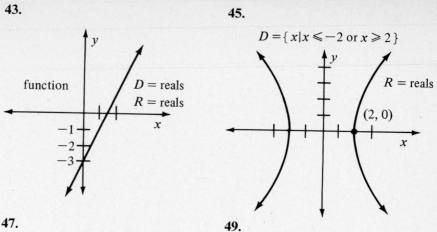

R = reals

$(2, 0)$

47.

$D = \{x \mid -3 \leqslant x \leqslant 3\}$
$R = \{y \mid -2 \leqslant y \leqslant 2\}$

49.

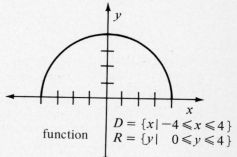

function $D = \{x \mid -4 \leqslant x \leqslant 4\}$
$R = \{y \mid \ 0 \leqslant y \leqslant 4\}$

51.

function D = reals
$R = \{y \mid y \leqslant -3\}$

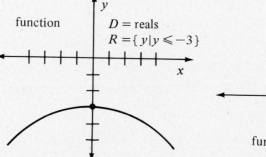

53.

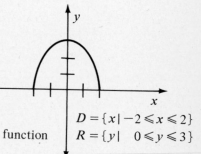

function $D = \{x \mid -2 \leqslant x \leqslant 2\}$
$R = \{y \mid \ 0 \leqslant y \leqslant 3\}$

55.   **a.**   yes
      b. $D = \{t \mid 0 \leq t \leq 6\}$
          $R = \{h \mid 0 \leq h \leq 60\}$
      c.   3
      d.   60
      e.   6

Problem Set 10.2

1. -1 **3.** -11 **5.** -5 **7.** 2 **9.** 4 **11.** 35
13. -13 **15.** 4 **17.** 0 **19.** 2 **21.** 2 **23.** $-\frac{1}{2}$
25. 1 **27.** -9 **29.** $3a^2 - 4a + 1$ **31.** $3a^2 + 14a + 16$
33. 16 **35.** 15 **37.** 1 **39.** $12x^2 - 20x + 8$ **41.** 2
43. $2x + h$ **45.** -4 **47.** $3(2x + h)$ **49.** $4x + 2h + 3$
51. 3 **53.** 4 **55.** $x + a$ **57.** 5 **59.** $x + a$
61. $107, $175, $1000

Problem Set 10.3

1. $6x + 2$ **3.** $-2x + 8$ **5.** $8x^2 + 14x - 15$
7. $(2x + 5)/(4x - 3)$ **9.** $4x - 7$ **11.** $3x^2 - 10x + 8$
13. $-2x + 3$ **15.** $3x^2 - 11x + 10$ **17.** $9x^3 - 48x^2 + 85x - 50$
19. $x - 2$ **21.** $1/(x - 2)$ **23.** $3x^2 - 7x + 3$
25. $6x^2 - 22x + 20$ **27.** 15 **29.** 98 **31.** $\frac{3}{2}$ **33.** 1
35. 40 **37.** 147
39. **a.** $2.49
   b. $1.53
   c. 5 min
   d. $36x + 52$
   e. $2.32
   f. $1.27

Problem Set 10.4

1.

quadratic, $(\sqrt{3}, 0)$, $(0, -3)$

3.

linear

5.

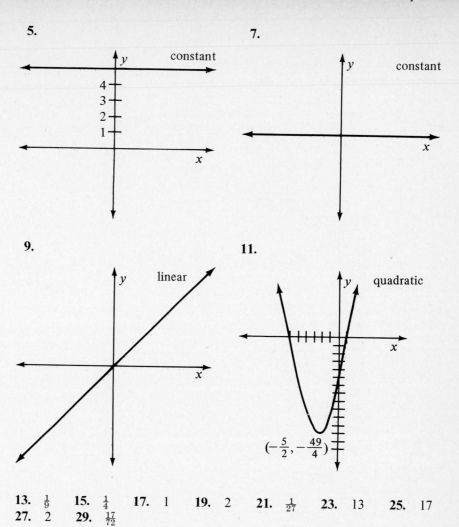

constant

7.

constant

9.

linear

11.

quadratic

$(-\frac{5}{2}, -\frac{49}{4})$

13. $\frac{1}{9}$ **15.** $\frac{1}{4}$ **17.** 1   **19.** 2   **21.** $\frac{1}{27}$ **23.** 13   **25.** 17
27. 2   **29.** $\frac{17}{72}$

31.

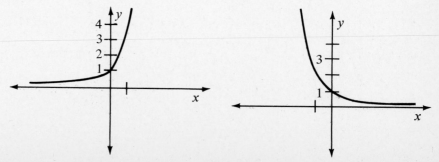

33.

35.   **37.**

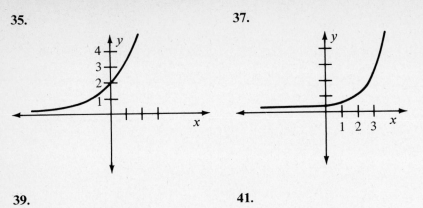

39.   **41.**

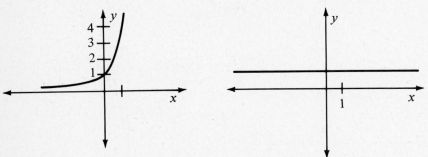

45.   200, 400, 800, 1600, 10 days

Problem Set 10.5

1. $f^{-1}(x) = \dfrac{x+1}{3}$ **3.** $f^{-1}(x) = \dfrac{1-x}{3}$ **5.** $f^{-1}(x) = \pm\sqrt{x-4}$

7. $f^{-1}(x) = 4x + 3$ **9.** $f^{-1}(x) = 2(x+3)$ **11.** $f^{-1}(x) = \pm\sqrt{3-x}$

13.   **15.**

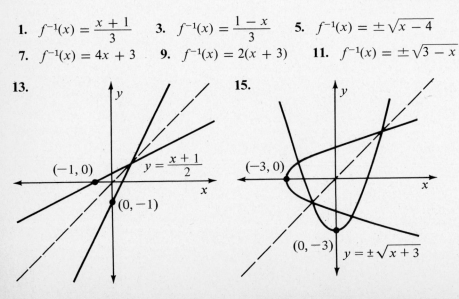

17.

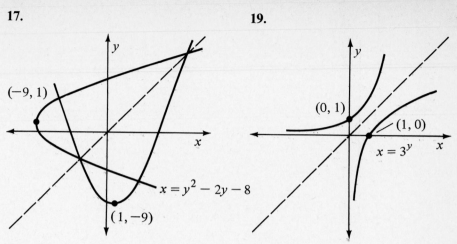

$(-9, 1)$

$x = y^2 - 2y - 8$

$(1, -9)$

19.

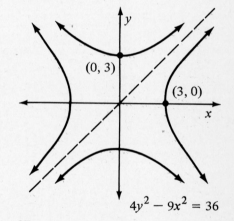

$(0, 1)$

$(1, 0)$

$x = 3^y$

21.

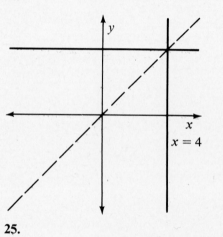

$x = 4$

23.

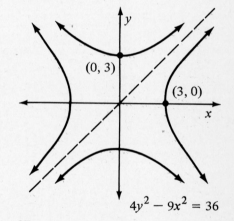

$(0, 3)$

$(3, 0)$

$4y^2 - 9x^2 = 36$

25.

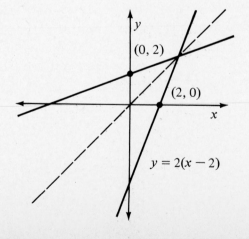

$(0, 2)$

$(2, 0)$

$y = 2(x - 2)$

27.

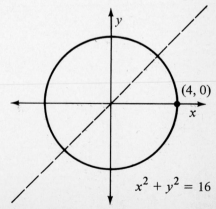

$(4, 0)$

$x^2 + y^2 = 16$

29.

31.

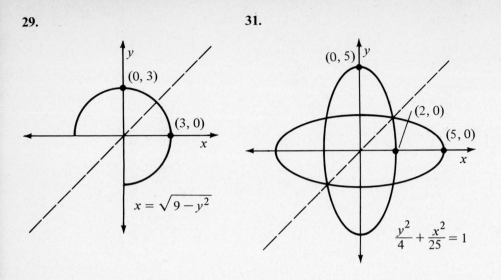

Chapter 10   Test

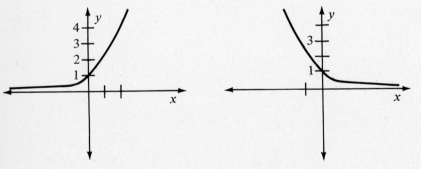

1. $D = \{-3, 2, 3, 0\}$ 2. $D = \{-2, -3\}$
    $R = \{1, 3, 5\}$, yes $R = \{0, 1\}$, no
3. D = all real numbers   4. $D = \{x \mid -3 \leq x \leq 3\}$
    $R = \{y \mid y \geq -9\}$, yes $R = \{y \mid -2 \leq y \leq 2\}$, no
5. $x \geq 4$ 6. $x < 3$ 7. $x \neq -1$ 8. $x \neq 4, x \neq -2$
9. 11   10. -4 11. $4x + 2$ 12. $x - 2$ 13. 15   14. -14
15.   16.

17.   **18.**

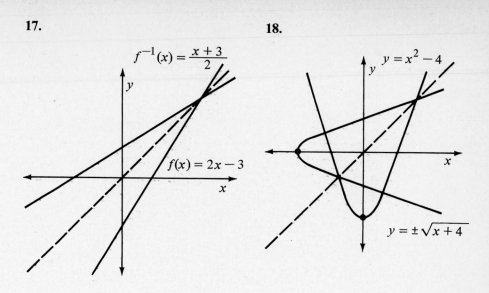

CHAPTER 11   *Problem Set 11.1*

1. $\log_2 16 = 4$ **3.** $\log_5 125 = 3$ **5.** $\log_{10} .01 = -2$

7. $\log_2 \frac{1}{32} = -5$ **9.** $\log_{\frac{1}{2}} 8 = -3$ **11.** $\log_3 27 = 3$

13. $10^2 = 100$ **15.** $2^6 = 64$ **17.** $8^0 = 1$ **19.** $10^{-3} = .001$

21. $6^2 = 36$ **23.** $5^{-2} = \frac{1}{25}$ **25.** 9   **27.** $\frac{1}{125}$ **29.** 4

31. $\frac{1}{3}$ **33.** 2   **35.** $\sqrt[3]{5}$

37.   **39.**

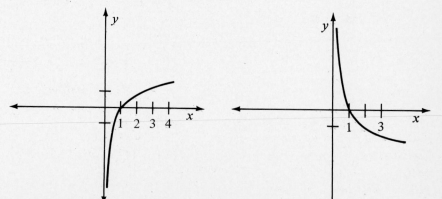

41.   **43.**

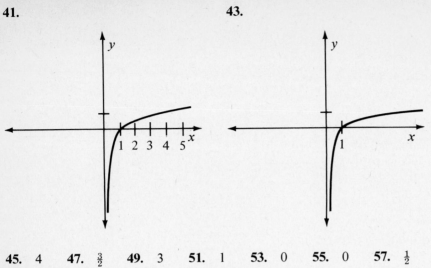

45. 4   **47.** $\frac{3}{2}$ **49.** 3   **51.** 1   **53.** 0   **55.** 0   **57.** $\frac{1}{2}$
59. 0

Problem Set 11.2

1. $\log_3 4 + \log_3 x$ **3.** $\log_6 5 - \log_6 x$ **5.** $5\log_2 y$ **7.** $\frac{1}{3}\log_9 z$
9. $2\log_6 x + 3\log_6 y$ **11.** $\frac{1}{2}\log_5 x + 4\log_5 y$
13. $\log_b x + \log_b y - \log_b z$ **15.** $\log_{10} 4 - \log_{10} x - \log_{10} y$
17. $2\log_{10} x + \log_{10} y - \frac{1}{2}\log_{10} z$
19. $3\log_{10} x + \frac{1}{2}\log_{10} y - 4\log_{10} z$ **21.** $\frac{2}{3}\log_b x + \frac{1}{3}\log_b y - \frac{4}{3}\log_b z$

23. $\log_b xz$ **25.** $\log_3 \dfrac{x^2}{y^3}$ **27.** $\log_{10} \sqrt{x}\sqrt[3]{y}$ **29.** $\log_2 \dfrac{x^3\sqrt{y}}{z}$

31. $\log_2 \dfrac{\sqrt{x}}{y^3 z^4}$ **33.** $\log_{10} \dfrac{x^{\frac{3}{2}}}{y^{\frac{3}{4}}z^{\frac{4}{5}}}$ **35.** $\frac{2}{3}$ **37.** 18   **39.** 3

41. 3   **43.** 4

Problem Set 11.3

1. 2.5775   **3.** 1.5775   **5.** 3.5775   **7.** 8.5775 − 10
9. 4.5775   **11.** 2.7782   **13.** 3.3032   **15.** 7.9872 − 10
17. 8.4969 − 10   **19.** 9.6010 − 10   **21.** 759   **23.** .00759
25. 1430   **27.** .00000447   **29.** .0000000918   **31.** 9260
33. 1.27   **35.** 20   **37.** 10.1   **39.** 386   **41.** 40,200,000
43. 24,800   **45.** .0000075   **47.** 258,000,000   **49.** 31.4

Problem Set 11.4

1. 1.4650   **3.** .6826   **5.** −1.5440   **7.** −.6477   **9.** −.3333
11. 2.0000   **13.** −.1845   **15.** .1845   **17.** 1.6168
19. 2.1131   **21.** 1.3333   **23.** .7500   **25.** 1.3917   **27.** .7186
29. .9650   **31.** 1.0363   **33.** 2.6356   **35.** 4.1632

Problem Set 11.5

1. 2.40   **3.** 5.30   **5.** 4.38   **7.** 1.07   **9.** 3.98×10^{-4}
11. 3.16×10^{-7}
13. **a.** 1.62
        b. .87
        c. .00293
        d. 2.86×10^{-6}
15. 5600   **17.** $8950   **19.** $4120   **21.** 14.2 years
23. 11.9 years

Chapter 11   Test

1. 64   **2.** $\sqrt{5}$
3.   **4.**

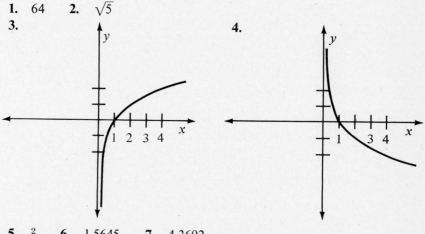

5. $\frac{2}{3}$ **6.** 1.5645   **7.** 4.3692
8. 8.0899 − 10   **9.** 14,200   **10.** 15.6   **11.** 2.19   **12.** 49,200
13. 1.4650   **14.** 1.25   **15.** 15   **16.** 8   **17.** $5100   **18.** 3.52

CHAPTER 12   *Problem Set 12.1*

1. 4, 7, 10, 13, 16   **3.** 3, 7, 11, 15, 19   **5.** 1, 2, 3, 4, 5
7. 4, 7, 12, 19, 28   **9.** $\frac{1}{4}, \frac{2}{5}, \frac{3}{6}, \frac{4}{7}, \frac{5}{8}$ **11.** $\frac{2}{3}, \frac{3}{4}, \frac{4}{5}, \frac{5}{6}, \frac{6}{7}$
13. $1, \frac{1}{4}, \frac{1}{9}, \frac{1}{16}, \frac{1}{25}$ **15.** 2, 4, 8, 16, 32   **17.** $\frac{1}{3}, \frac{1}{9}, \frac{1}{27}, \frac{1}{81}, \frac{1}{243}$

19. $2, \frac{3}{2}, \frac{4}{3}, \frac{5}{4}, \frac{6}{5}$ **21.** $0, \frac{3}{2}, \frac{8}{3}, \frac{15}{4}, \frac{24}{5}$ **23.** $-2, 4, -8, 16, -32$
25. $n + 1$ **27.** $4n$ **29.** $3n + 4$ **31.** n^2 **33.** $3n^2$

35. 2^{n+1} **37.** $(-2)^n$ **39.** $\dfrac{1}{2^{n+1}}$ **41.** $\dfrac{n}{(n+1)^2}$

43. \$50,000; \$50,500; \$51,005; \$51,515; \$52,030; \$52,550; \$53,075;
$a_n = 50{,}000(1.01)^n$; $a_{24} = 50{,}000(1.01)^{24} = 63{,}487$

Problem Set 12.2

1. 36   **3.** 9   **5.** 50   **7.** $\frac{163}{60}$ **9.** $\frac{2897}{315}$ **11.** 60
13. 40   **15.** $5x + 15$
17. $(x+1)^2 + (x+1)^3 + (x+1)^4 + (x+1)^5 + (x+1)^6 + (x+1)^7$
19. $\dfrac{x+1}{x-1} + \dfrac{x+2}{x-1} + \dfrac{x+3}{x-1} + \dfrac{x+4}{x-1} + \dfrac{x+5}{x-1}$
21. $(x+3)^3 + (x+4)^4 + (x+5)^5 + (x+6)^6 + (x+7)^7 + (x+8)^8$
23. $(x+1)^2 + (x+2)^3 + (x+3)^4 + (x+4)^5 + (x+5)^6$

25. $\displaystyle\sum_{i=1}^{4} 2^i$ **27.** $\displaystyle\sum_{i=2}^{6} 2^i$ **29.** $\displaystyle\sum_{i=2}^{6} (i^2 + 1)$ **31.** $\displaystyle\sum_{i=3}^{7} \dfrac{i}{i+1}$

33. $\displaystyle\sum_{i=1}^{4} \dfrac{i}{2i+1}$ **35.** $\displaystyle\sum_{i=3}^{6} (x - i)$ **37.** $\displaystyle\sum_{i=3}^{5} \dfrac{x}{x+i}$

39. $\displaystyle\sum_{i=2}^{4} x^i(x + i)$ **41.** 208 ft, 784 ft

Problem Set 12.3

1. 1   **3.** not an arithmetic progression   **5.** -5
7. not an arithmetic progression   **9.** $\frac{2}{3}$
11. **a.** $a_n = 4n - 1$ **b.** 39 **c.** 95 **d.** 210 **e.** 1176
13. **a.** $a_n = -2n + 8$ **b.** -12 **c.** -40 **d.** -30 **e.** -408
15. **a.** $a_n = 4n + 1$ **b.** 41 **c.** 97 **d.** 230 **e.** 1224
17. **a.** $a_n = -5n + 17$ **b.** -33 **c.** -103 **d.** -105 **e.** -1092
19. **a.** $a_n = n/2$ **b.** 5 **c.** 12 **d.** $\frac{55}{2}$ **e.** 150
21. **a.** $a_n = 6n - 2$ **b.** 58 **c.** 142 **d.** 310 **e.** 1752
23. **a.** $a_n = 2n + 5$ **b.** 25 **c.** 53 **d.** 160 **e.** 720
25. **a.** \$15,000; \$15,850; \$16,700; \$17,550; \$18,400
     b. $a_n = 850n + 14{,}150$
     c. \$22,650

Problem Set 12.4

1. 5   **3.** $\frac{1}{3}$ **5.** not geometric   **7.** -2 **9.** not geometric

11.   **a.** $a_n = 4 \cdot 3^{n-1}$
     b. $4 \cdot 3^5$
     c. $4 \cdot 3^{19}$
     d. $2(3^{10} - 1)$
     e. $2(3^{20} - 1)$

13.   **a.** $a_n = -2(-5)^{n-1}$
     b. $-2(-5)^5$
     c. $-2(-5)^{19}$
     d. $(5^{10} - 1)/3$
     e. $(5^{20} - 1)/3$

15.   **a.** $a_n = -4(\frac{1}{2})^{n-1}$
     b. $-4(\frac{1}{2})^5$
     c. $-4(\frac{1}{2})^{19}$
     d. $8[(\frac{1}{2})^{10} - 1]$
     e. $8[(\frac{1}{2})^{20} - 1]$

17.   **a.** $a_n = 2 \cdot 6^{n-1}$
     b. $2 \cdot 6^5$
     c. $2 \cdot 6^{19}$
     d. $2(6^{10} - 1)/5$
     e. $2(6^{20} - 1)/5$

19.   **a.** $a_n = 10 \cdot 2^{n-1}$
     b. $10 \cdot 2^5 = 320$
     c. $10 \cdot 2^{19}$
     d. $10(2^{10} - 1) = 10230$
     e. $10(2^{20} - 1)$

21.   **a.** $a_n = (\sqrt{2})^n$
     b. 8
     c. 1024
     d. $\sqrt{2}[(\sqrt{2})^{10} - 1]/(\sqrt{2} - 1)$
     e. $\sqrt{2}[(\sqrt{2})^{20} - 1]/(\sqrt{2} - 1)$

23.   **a.** $a_n = 5 \cdot 2^{n-1}$
     b. 160
     c. $5 \cdot 2^{19}$
     d. $5(2^{10} - 1) = 5115$
     e. $5(2^{20} - 1)$

25.   **a.** \$10,600
     b. \$10,000; \$10,600; \$11,236; \$11,910.16; \$12,624.77
     c. $a_n = 10,000(1.06)^{n-1}$

Problem Set 12.5

1. $x^4 + 8x^3 + 24x^2 + 32x + 16$

3. $x^6 + 6x^5y + 15x^4y^2 + 20x^3y^3 + 15x^2y^4 + 6xy^5 + y^6$

5. $32x^5 + 80x^4 + 80x^3 + 40x^2 + 10x + 1$

7. $x^5 - 10x^4y + 40x^3y^2 - 80x^2y^3 + 80xy^4 - 32y^5$

9. $81x^4 - 216x^3 + 216x^2 - 96x + 16$

11. $64x^3 - 144x^2y + 108xy^2 - 27y^3$

13. $x^8 + 8x^6 + 24x^4 + 32x^2 + 16$ **15.** $x^6 + 3x^4y^2 + 3x^2y^4 + y^6$

17. $x^3/8 - 3x^2 + 24x - 64$

19. $x^4/81 + 2x^3y/27 + x^2y^2/6 + xy^3/6 + y^4/16$

21. $x^9 + 18x^8 + 144x^7 + 672x^6$

23. $x^{10} - 10x^9y + 45x^8y^2 - 120x^7y^3$

25. $x^{10} + 20x^9y + 180x^8y^2 + 960x^7y^3$ **27.** $x^{15} + 15x^{14} + 105x^{13}$

29. $x^{12} - 12x^{11}y + 66x^{10}y^2$ **31.** $x^{20} + 40x^{19} + 760x^{18}$

33. $x^{100} + 200x^{99}$ **35.** $x^{50} + 50x^{49}y$ **37.** $\frac{21}{128}$

Chapter 12   Test

1. $-2, 1, 4, 7, 10$ 2. $3, 7, 11, 15, 19$ 3. $2, 5, 10, 17, 26$
4. $2, 16, 54, 128, 250$ 5. $2, \frac{3}{4}, \frac{4}{9}, \frac{5}{16}, \frac{6}{25}$ 6. $4, -8, 16, -32, 64$
7. $a_n = 4n + 2$ 8. $a_n = 2^{n-1}$ 9. $a_n = (\frac{1}{2})^n = 1/2^n$
10. $a_n = (-3)^n$ 11. a. 90 b. 53 c. 130 12. 3 13. 6
14. a. 320 b. 25 15. a. $S_{50} = 3(2^{50} - 1)$ b. $5(4^{50} - 1)/3$
16. a. $x^4 - 12x^3 + 54x^2 - 108x + 81$
   b. $32x^5 - 80x^4 + 80x^3 - 40x^2 + 10x - 1$
   c. $27x^3 - 54x^2y + 36xy^2 - 8y^3$

APPENDIX A

1. $x - 7 + \dfrac{20}{x + 2}$ 3. $3x - 1$ 5. $x^2 + 4x + 11 + \dfrac{26}{x - 2}$

7. $3x^2 + 8x + 26 + \dfrac{83}{x - 3}$ 9. $2x^2 + 2x + 3$

11. $x^3 - 4x^2 + 18x - 72 + \dfrac{289}{x + 4}$ 13. $x^4 + x^2 - x - 1 - \dfrac{1}{x - 2}$

15. $x + 2 + \dfrac{3}{x - 1}$ 17. $x^3 - x^2 + x - 1$ 19. $x^2 + x + 1$

APPENDIX B

1.

3.

5.

7.

9.

11.

13.

15.   all real numbers

17.   no solution, $\varnothing$

19.

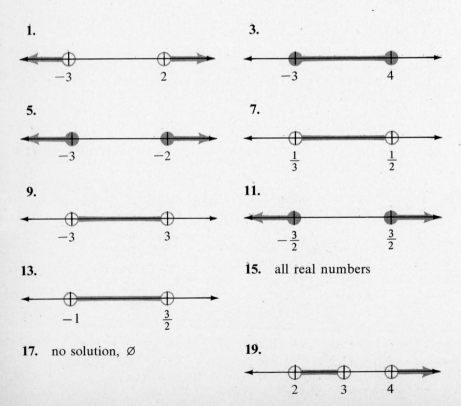

Index